MARINE BIOLOGY

ZOOPLANKTON AND PHYTOPLANKTON: TYPES, CHARACTERISTICS AND ECOLOGY

MARINE BIOLOGY

Additional books in this series can be found on Nova's website under the Series tab.

Additional E-books in this series can be found on Nova's website under the E-books tab.

MARINE BIOLOGY

ZOOPLANKTON AND PHYTOPLANKTON: TYPES, CHARACTERISTICS AND ECOLOGY

GIRI KATTEL
EDITOR

Nova Science Publishers, Inc.
New York

For permission to use material from this book please contact us:
Telephone 631-231-7269; Fax 631-231-8175
Web Site: http://www.novapublishers.com

Library of Congress Cataloging-in-Publication Data

Zooplankton and phytoplankton : types, characteristics, and ecology / editor, Giri Kattel.
p. cm.
Includes index.
ISBN 978-1-61324-508-8 (hardcover)
1. Zooplankton. 2. Zooplankton--Ecology. 3. Phytoplankton. 4. Phytoplankton--Ecology. I. Kattel, Giri.
QL123.Z63 2011
592.177'6--dc22

2011014211

Published by Nova Science Publishers, Inc. † New York

CONTENTS

PREFACE

In this book, the authors present current research in the study of the types, characteristics and ecology of zooplankton and phytoplankton. Topics discussed include the bioaccumulation of cyanobacterial toxins in aquatic organisms and their public health consequences; the use of microalgae for aquaculture; the annual cycle of plankton biomass in the Gulf of Mexico; grazer-periphyton interactions; spatial and temporal distribution patterns of zooplankton in a shallow lowland coastal lake and phytoplankton composition in fish farms. (Imprint: Nova)

Chapter 1 - The occurrence of harmful cyanobacterial blooms in surface waters is often accompanied by a production of a variety of cyanotoxins that are generally classified according to the organs on which they act: hepatotoxins (liver), neurotoxins (nervous system) and dermatotoxins (skin). The presence of such toxins has been reported throughout the world and it appears that liver-toxic microcystins are more commonly found in 50-75 % cyanobacterial blooms. The contamination of surface water by these cyanotoxins can cause water quality problems for fisheries, aquaculture, farming, and sanitary hazard for human and animals. Humans may be exposed to cyanobacterial toxins via several routes, including drinking water, recreational contact, some cyanobacteria-based dietary supplements, and food chain. Information on exposure through aquatic food webs, which is generally scarce, is urgently needed and must not be ignored because aquatic organisms could in a direct or indirect manner contribute to food chain cyanotoxin's transfer, and by the way constitute a potent health risk source.

Chapter 2 - Microalgaes are probably the most commonly used organisms as live feed during the larviculture and nursery of aquacultural organisms, and many researches have been made worldwide in this field. This chapter is a summary of the investigations conducted in the University of Sonora, México, on the use of microalgae for aquaculture of diverse organisms, mainly crustaceans and fishes. The document is divided in five sections related to the most important aspects of the culture such as: production systems; evaluation of particular species and their nutritional value for selected organisms (mollusks and crustaceans), effect of physical and chemical environmental factors on the production and nutritional value, evaluation of commercial laboratories of microalgaes, and evaluation of alternative mediums for their culture. The results of our investigations are discussed when compared to similar studies on the same or different species and in the same or different regions.

Chapter 3 - From April 2007 through May 2008, a total of 12 monthly surveys were carried out at nine oceanographic stations in the National Park Sistema Arrecifal Veracruzano, southwestern Gulf of Mexico. Horizontal tows were taken with a net of 120 μm mesh size for zooplankton, and with a net of 30 μm mesh size for phytoplankton. In total, 216 samples were analyzed. ANOVA and a principal component analysis showed temporal variation in biomass and parameters influencing it but no spatial pattern. Phytoplankton species richness was highest in June and October 2007 and lowest in April, November and December 2007. Zooplankton taxa richness was highest in April and June 2007 and lowest in March 2008. A list of 36 major zooplankton taxa and 275 phytoplankton species and infraspecific taxa is given. Plankton biomass ranged from 23 to 3637 mg wet weight m^{-3} (mean 552.6 mg m^{-3}), equaling 1.3-202.8 mg C m^{-3} (mean 30.7 mg C m^{-3}). The biomass within the park was unimodal in annual dynamics, with peak production in September-October at the end of the rainy season when the currents changed direction. Taxa richness contributed to the seasonal variation, and the calanoid copepods (mostly copepodite stages) and centric diatoms were the most conspicuous plankton components. Annual plankton biomass values showed a rather homogeneous spatial distribution.

Chapter 4 - Periphyton ecology is affected by abiotic factors like light limitation, waterflow, nutrients and physical and chemical properties of substrates. However, the significant influence of biotic factors is contributed through grazing, for algal dominant periphytic communities, the herbivory. Grazing usually causes a reduction to periphytic algal biomass, but individual algal populations can exhibit a positive response to increased grazing intensity. Such positive responses can overweight negative responses in certain nutrient environments. Differential grazing on overstory periphyton canopy stimulate competitive interaction among algae for enhanced understory nutrient. Grazing resistant algae, in addition, contributes to periphytic standing crop with higher nutrient retaining algae. Further, selective grazing on competitively dominant algal species clears the substrates for colonization by slow-growing and deterrent species increasing periphytic micro-algal diversity. Thereby, a heterogeneic periphytic community could be initiated on the grazed surface. These producer-consumer interactions vary specifically on spatial gradients. Stoichiometrically, periphytic C:P and N:P ratios are decreased under grazing pressure, whereas, C:N and C:Chl ratios remain unaffected. Low levels of grazing, in turn, stimulate increase in nutrient cycling and hence algal productivity. Extensive grazing has negative impact since such grazing rates could exceed reproductive rates of periphytic algae. Grazer mouth morphology and feeding preferences may further determine the direction of ecological growth of periphyton.

Chapter 5 - Physical and chemical characteristics of the head waters of the Olifant's, Lephalala and Mokolo rivers and their tributaries were assessed over a period of one year to compare effects of seasonal, spatial and human environmental factors on epilithic filamentous green algae communities as indicators of different sources of pollution. The epilithic filamentous green algae assemblage differed among the three rivers and their tributaries as a function of drainage basin characteristics, but exhibited common seasonal changes related to temperature and electrical conductivity (EC). Human impacts were assumed to cause higher epilithic algal densities in stream basins with increase in TP due to progressive eutrophication. In the study frequent floods in the raining season truncate filamentous green algal successional sequence and thereby favouring either tightly attached, resistant forms or those recolonization and growth rates high enough to accumulate biomass between flood periods. It was evident from the study that certain filamentous green algal species was indicators of land use activities and can therefore be used as indicators of pollution during medium and high flow regimes.

Chapter 6 - Communities in shallow coastal lakes are inclined to change with environmental variations quite frequently due to their proximity to marine environments. The pelagic zooplankton community of shallow lowland coastal lake, Lake Waihola in South Island, New Zealand was examined by monthly day and night sampling by hand-operated bilge pump. Over 15 sampling trips, each consisted of diel (D/N) measurement at 50 m apart with five replicate samples at four positions, shallow-inshore (Sh/In), shallow-offshore (Sh/Of), deep-inshore (Dp/In) and deep-offshore(Dp/Of) corresponded a total of 600 samples being collected. Zooplankton individuals collected in D/N samples in four positions were pooled and averaged for each season. Distribution of zooplankton community over temporal and spatial scales were visualized in ordination using non-metric multi-dimensional scaling (MDS) followed by permutational multivariate ANOVA in PRIMER. Four-factor interaction (Date*D/N*Sh/Dp*In*Of) reveals that the effect of sampling dates was significant on distribution of zooplankton, but the distribution across the diel timeframe over spatial scales (D/N*In/Of*Sh*Dp) was insignificant. Some degree of distribution was observed along vertical (Sh/Dp) gradient, but no significant patterns were evident across the horizontal (In/Of) gradient. Amongst zooplankton, cladocerans such as *Daphnia*, *Ceridaphnia* and *Bosmina* showed a poor and patchy distribution pattern where *Bosmina* being dominant in January 1998. Except the Sh/Of day sample of February 1998, when calanoid copepods were absent, the distribution of calanoid copepods, cyclopoid copepods and amphipods was relatively common throughout the period of study. The overall distribution patterns of zooplankton in Lake Waihola indicate that there may be significant differences amongst zooplankton community (e.g., cladocerans vs non-cladocerans) to respond to local environmental and seasonal changes, consequently our understanding of shallow lentic coastal ecosystems in South Island, New Zealand is becoming increasingly complex.

Chapter 7 - Fast development of different aquaculture in the world in 90s caused different studies of effects of aquaculture on the environment. Aquaculture activity has a serious impact on environment, e.g. the enrichment of the water column in dissolved organic and inorganic material. This may subsequently affect populations of phytoplankton differentially. In this study we report on short-term changes in the water column in relation to fish feeding and differences along transect from the centre of fish farm towards open waters. The authors took samples at different sites around the fish cage. They also undertake the enriched enclosures experiment. With HPLC (High Performance Liquid Chromatography) pigment analyses we determine the phytoplankton community structure. Comparing the pigment fingerprints at fish farm area before and after feeding the authors observed only minor differences. The main phytoplankton group were diatoms. They noticed differences on the profile from the centre of the fish cage outward.

Chapter 8 - The relative importance of local and regional processes in shaping natural communities within a metacommunity context has been a focus of intense debate in recent years. We examine the variation of the zooplankton communities structure attributable to local (water temperature, dissolved oxygen, pH, suspended matter, nutrients, chlorophyll a, phytoplankton and fish abundances) and regional factors (dispersal and connectivity) in four freshwater reservoirs with (Beni Mtir) and without (Lakhmes, Nabhana, and Sidi Sâad) artificial connections to others reservoirs in Mediterranean area. Fourteen zooplankton species (four copepods: *Copidodiaptomus numidicus*, *Acanthocyclops robustus*, *Acanthocyclops viridis*, *Cyclops strenuous*; four cladocerans: *Bosmina longirostris*, *Ceriodaphnia quadrangula*, *Diaphanosoma brachyurum*, *Daphnia longispina* and six rotifers: *Asplanchna* sp., *Filinia longiseta, Hexarthra mira, Keratella quadrata, Keratella cochlearis, Brachionus urceolaris*) were identified in the studied reservoirs. Seasonal changes in the density ($P < 0.01$) and biomass of zooplanktonic organisms ($P < 0.001$) revealed clear differences between the four reservoirs. The results indicated that the zooplankton communities were significantly related to local factors in Lakhmes, Nabhana, and Sidi Saâd reservoirs. However, in Beni Mtir reservoir, connectivity and dispersal with different local conditions determine the zooplankton composition and dynamics. This study demonstrated that artificial connections to other reservoirs alter the metacommunity structure of zooplankton communities.

Chapter 9 - Parasites hijack hosts and alter the latters' behaviour so as to enhance their dispersion and transmission to new hosts and new environments. *Daphnia,* as one of the major zooplankton members in freshwater environments, are targets of parasitic fungi, bacteria, and even waterborne human pathogens. Although the effects of infestation on *Daphnia* have been studied for over 120 years, scarce information is available regarding behavioural consequences. Conspicuous swimming behaviour of *Daphnia* can increase predation rates by visual predators such as fish. Therefore, if the swimming behaviour of *Daphnia*are modified by infestation, their predation risk could increase. Here we observed the effect of *Vibrio cholerae* infestation on the swimming performances of *Daphnia pulicaria* in dark and light conditions to simulate the risk of detection by visual predators. Our results show that *D. pulicaria*both with and without infestation by *V. cholerae* display similar swimming patterns in the dark. However, in light conditions, *D. pulicaria* infected by *V. cholerae* swim faster and travel more convoluted trajectories than those without infestation. Our study is the first to indicate that a microbial infestation can modify the swimming behaviour of *D. pulicaria* and that, consequently, infected *D. pulicaria* could experience more extensive predation by fish.

Chapter 10 - Body size is an important parameter in ecological processes, at both individual and community levels suggest that density of the biomass in any ecosystem is dependent on the community structure that is reflected in body sizes of the constituent organisms. Model behaviour and dynamics are mainly regulated by physiological rate parameters of the living state variables. The physiological processes of living organisms show an allometric relationship to body sizes and the calculation follow algebraic rules that are homologus to those used with logarithms. The role of body size in planktonic system has significant contribution in the system dynamics. In this article an alternative approach has been developed in two generic models of planktonic systems and in this approach the rate processes vary according to the body sizes of planktons following allometric principle and these are incorporated in the dynamic model. In the first model different body sizes of phytoplankton and zooplankton are used and their rate parameters particularly the phtotosynthetic rate of phytoplankton and grazing rate of zooplankton vary accordingly. Optimization tool has been used to determine the perfect combination of phytoplankton and zooplankton to obtain the maximum fish yield in different trophic conditions such as oligotrophic, mesotrophic and eutrophic conditions. Model result shows in different trophic conditions the combinations of phytoplankton and zooplankton are different. In any ecosystem the equilibrium condition may gradually turn into a chaotic situation for different reasons. In the later part of the article a three species (phytoplankton, zooplankton and fish) model is proposed. Rate parameters are changed according to the change of the size of the organisms. The model is run in different conditions with different sizes of zooplankton by increasing the grazing rate and consequently decreasing the half-saturation constant of this organisms following allometric principles. The system exhibits different states (equilibrium point – stable limit cycle – doubling and ultimately chaos) by gradual increment of zooplankton grazing rate and decrease of half-saturation constant.

In: Zooplankton and Phytoplankton
Editor: Giri Kattel, pp. 1-33
ISBN 978-1-61324-508-8

Chapter 1

BIOACCUMULATION OF CYANOBACTERIAL TOXINS IN AQUATIC ORGANISMS AND ITS CONSEQUENCES FOR PUBLIC HEALTH

***A. Ettoumi*[1], *F. El Khalloufi*[1], *I. El Ghazali*[1], *B. Oudra*[1], *A. Amrani*[2], *H. Nasri*[2], *and N. Bouaïcha*[*,3]**

[1]Laboratory of Biology and Biotechnology of Microorganisms (LBBM), Faculty of Sciences Semlalia, University Cadi Ayyad, Marrakesh, Morocco
[2]Laboratoire de Recherche: Biodiversité et Pollution des Écosystèmes, Centre Universitaire El Taref, Institut de Biologie, 36 000, El Tarf, Algérie
[3]Laboratoire Ecologie, Systématique et Evolution, UMR 8079-Univ. Paris-Sud/CNRS/AgroParisTech, Bâtiment 362, 91405 Orsay Cedex, France

ABSTRACT

The occurrence of harmful cyanobacterial blooms in surface waters is often accompanied by a production of a variety of cyanotoxins that are generally classified according to the organs on which they act: hepatotoxins (liver), neurotoxins (nervous system) and dermatotoxins (skin). The presence of such toxins has been reported throughout the world and it appears that liver-toxic microcystins are more commonly found in 50-75 % cyanobacterial blooms. The contamination of surface water by these cyanotoxins can cause water quality problems for fisheries, aquaculture, farming, and sanitary hazard for human and animals. Humans may be exposed to cyanobacterial toxins via several routes, including drinking water, recreational contact, some cyanobacteria-based dietary supplements, and food chain. Information on exposure through aquatic food webs, which is generally scarce, is urgently needed and must not be ignored because aquatic organisms could in a direct or indirect manner contribute to food chain cyanotoxin's transfer, and by the way constitute a potent health risk source.

[*]Laboratoire Ecologie, Systématique et Evolution, UMR8079-Univ. Paris-Sud/CNRS/AgroParisTech, Bâtiment 362, 91405 Orsay Cedex, France, email: noureddine.bouaicha@u-psud.fr

1. INTRODUCTION

The occurrence of harmful cyanobacterial blooms in surface waters is often accompanied by a production of a variety of cyanotoxins that are generally classified according to the organs on which they act: hepatotoxins (liver), neurotoxins (nervous system) and dermatotoxins (skin). The presence of such toxins has been reported throughout the world and it appears that liver-toxic microcystins are more commonly found in 50-75 % cyanobacterial blooms. The contamination of surface water by these cyanotoxins can cause water quality problems for fisheries, aquaculture, farming, and sanitary hazard for human and animals. Since the first report of livestock poisoning associated with cyanobacterial blooms (Francis, 1878), numerous cases of animal deaths have been reported worldwide due to contact with cyanobacterial scums, indicating that this is a widespread phenomenon in Mediterranean, continental, and temperate climates in both hemispheres (Codd *et al.,* 2005; Falconer, 2005). Cyanotoxins are known to bioaccumulate in common aquatic vertebrates and invertebrates, including fish, mussels and zooplankton (Eriksson *et al.,* 1989; Kotak *et al.,* 1996; Prepas *et al.,* 1997; Watanabe *et al.,* 1997; Williams *et al.,* 1997; Thostrup and Christoffersen, 1999; Magalhães *et al.,* 2001; Sipiä *et al.,* 2001; Mohamed *et al.,* 2003; Chen and Xie, 2005; Chen *et al.,* 2005, 2007), which pose a risk to both animal and human heath if such aquatic animals are consumed (Carmichael and Falconer, 1993; Ibelings and Chorus, 2007; Smith *et al.,* 2008). Although no case of poisoning by these products has been reported in the world, this eventuality must not be ignored. Indeed, a recent epidemiological study showed that the excessive incidence of amyotrophic lateral sclerosis in the population of the islands of Guam in the Pacific was linked to a consumption of the seeds of cycas contaminated by a neurotoxin, β-methylamino-L-alanine (BMAA), produced by a species of cyanobacteria of the genus *Nostoc* living in symbiosis in the roots of this plant (Cox *et al.,* 2003). The source of BMAA for human exposure is not limited to cycad seed exposure or animals which have fed on cycad seeds but it was also found in freshwater cyanobacterial samples (Metcalf *et al.,* 2008; Esterhuizen and Downing, 2008; Karamyan and Speth, 2008; Jonasson *et al.,* 2008). Given the possibility of widespread human exposure and the possibility that BMAA may be linked to neurodegenerative disease this hypothesis deserves further consideration in freshwater food webs. Therefore, the accumulation of cyanotoxins in the food chain is at present remains more worrying and the proposed quality limits are rare, indeed, many aspects concerning these toxins remain unknown, notably those relative to their bioconcentration, bioaccumulation, bioamplification, and the role of detoxication and covalent binding of microcystins on transfer of toxins in the foodweb. This chapter reviews recent advances in the effects of cyanotoxins and their accumulation on aquatic organisms and discusses the current events of human exposure to cyanobacterial toxins collected from freshwater and coastal areas.

2. CYANOTOXINS AND THEIR PRODUCERS

Microorganisms incriminated in the poisonings in freshwater belong essentially to the order of cyanobacteria that forms the most wide-spread group of photobacteria with 150 genera and more than 2000 species (Bourrelly, 1985; Holte *et al.*, 1998). Cyanobacteria are photosynthetic prokaryotes with no structured nucleus. Their photosynthetic membranes contain chlorophyll-*a* and the pigment phycocyanin, which provides the characteristic blue-green color of many species (Whitton and Potts, 2000). Cyanobacteria appear under unicellular shape, cells are then solitary or formed colonies in a mucilaginous gel matrix, or under pluricellular shape in filamentous species (Kaebernick and Neilan, 2001). They have a cosmopolitan distribution and colonize a great variety of ecosystems on the surface of the planet, including Antarctica lakes, thermal springs, dry deserts, wet tropical grounds and acid peat bogs (Kaebernick and Neilan, 2001). Cyanobacteria are generally living, however, in moist or aquatic environments in the water column of lakes, but they can be benthic, fixed or very close to different substrates (corals, algae, animals) or developing within sediments (Couté and Bernard, 2001). However, cyanobacteria are armed to flourish in lakes and rivers where they can produce blooms (Figure 1) usually in summer to late autumn, in both subtropical and temperate latitudes, when water temperature and irradiance as well as meteorological conditions and nutrient supply are favorable (Park *et al.*, 1993; Kaebernick and Neilan, 2001; Haider *et al.*, 2003). Some species of these microorganisms, belonging essentially to the genera *Microcystis, Anabaena, Aphanizomenon, Planktothrix, Oscillatoria* and less often *Gomphosphaeria, Coelosphaerium, Gloeotrichia, Nodularia* and *Nostoc,* produce a diverse range of toxins called cyanotoxins that are recognized responsible for human and animal health hazards (Carmichael and Falconer, 1993; Falconer, 1996; 1999; Kuiper-Goodman *et al.*, 1999; Hitzfeld *et al.*, 2000). The cyanotoxins, with currently identified being predominantly alkaloids and peptides, are generally classified according to the organs on which they act: neurotoxins (nervous system), hepatotoxins (liver), and dermatotoxins (skin).

Four types of neurotoxic alkaloid (anatoxins, anatoxin-a(s), saxitoxins and β-methylamino-alanine) have been isolated from cyanobacteria, as illustrated in Figure 2 (for review see Aráoz *et al.,* 2010). The anatoxins, initially isolated from *Anabaena*, comprise anatoxin-a and homoanatoxin-a, which are a neuromuscular junction blocking agent (Aráoz *et al.,* 2010) and anatoxin-a(s), which resembles an organophosphate insecticides, with effects exerted through irreversible inhibition of acetylcholinesterase at the nerve synapse (Mahmood and Carmichael, 1986). The third family of cyanobacterial neurotoxins is saxitoxins that are a group of structurally related molecules that block voltage-gated sodium channels (Aráoz *et al.,* 2010). Saxitoxins were identified for the first time in freshwater in a strain of *Aphanizomenon flos-aquae* (Ikawa *et al.*, 1982; Haney *et al.*, 1995). They are also synthesized by the other species of the genera *Anabaena, Aphanizomenon, Lyngbya, Cylindrospermopsis* and *Planktothrix* (Mahmood and Carmichael, 1986; Humpage *et al.*, 1994; Pomati *et al.*, 2001). Anatoxins are specific of cyanoabcteria, however, saxitoxins are also synthesized by marine dinoflagellates (*Alexandrium spp.*, *Gymnodinium catenatum*, *Pyrodinium hahamense* var. *Copressum*) and associated with the human disease paralytic shellfish poisoning or PSP (Carmichael, 1994; Falconer, 1996). Additionally, the fourth cyanobacterial neurotoxin type is the unusual non protein neurotoxic amino acid L-beta-N-methylamino-L-alanine (BMAA) which initially isolated from a symbiotic cyanobacteria resident in specialized roots within the cycad tree (Cox *et al.,* 2003), and recently it has been associated to the neurological disorder amyaotrophic lateral sclerosis/Parkinsonium dementia complex (ALS/PDC) among the indigenous Chamorro people of Guam and other Marianas islands (Monson *et al.,* 2003; Banack and Cox, 2003; Murch *et al.,* 2004). Its neurotoxicity may be mediated via glutamate regulation (Papapetropoulos, 2007).

Figure 1. Cyanobacterial blooms. Upper : *Microcystis* sp. blooms associated with fish mortality. Lower; *Microcystis* sp. Blooms in aquaculture ponds.

Contrary to the other neurotoxins which their production depends on the phylogeny of the species, the BMAA can be produced by almost all groups of cyanobacteria from freshwater, brackish, and marine environments (Cox *et al.,* 2005; Banack *et al.,* 2007).

Cyanobacterial hepatotoxins (Figure 3) can be divided into two groups: cyclic peptides of weak molecular weight (from 800 to 1200 Da), microcystins (MCs) and nodularins (NODs) named from *Microcystis* and *Nodularia*, the first two genera of toxic cyanobacteria from which they were isolated, and the alkaloid cylindrospermopsins initially isolated from *Cylindrospermopsis raciborskii* (Ohtani *et al.*, 1992).

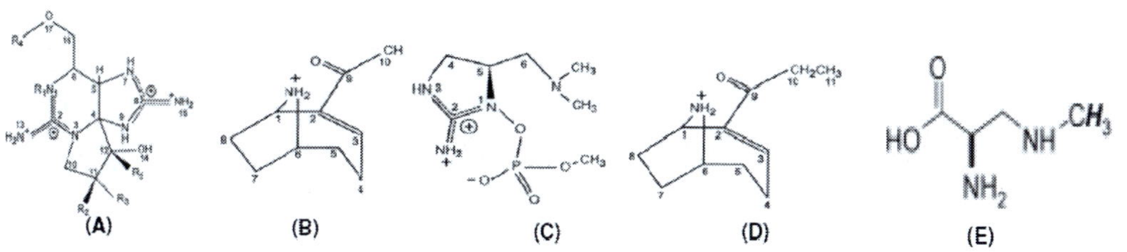

Figure 2. Chemical structures of cyanobacterial neurotoxins, (A) Saxitoxins, (B) Anatoxine-a, (C) Anatoxine-a(S), (D) Homoanatoxine-a, (E) β-N-methylamine-L-alanine.

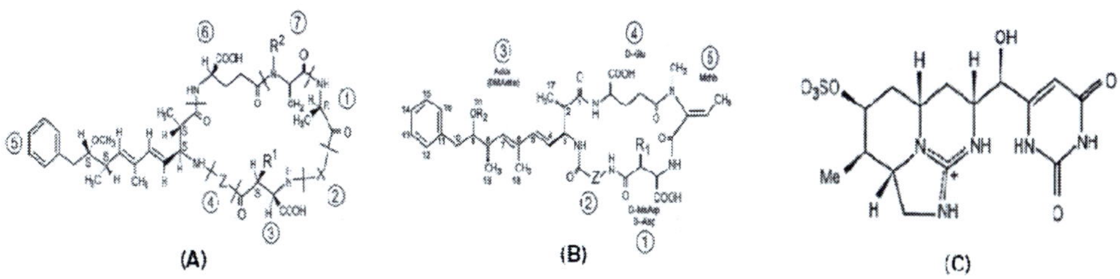

Figure 3. Chemical structures of cyanobacterial hepatotoxins, (A) Microcystins, (B) Nodularins, (C) Cylindrospermopsin.

MCs and NODs are respectively, cyclic heptapeptides with the general structure (-D-Ala-X-D-MeAsp-Z-Adda-D-Glu-Mdha-) and pentapeptides with the general structure (-D-MeAsp-Z-Adda-D-Glu-Mdhb-), where X and Z are variable L-amino acids, D-MeAsp is D-erythro-β-methyl aspartic acid, Mdha is *N*-methyldehydroalanine, Mdhb is *N*-methyldehydrobutyrine, and Adda is 3-amino-9-methoxy-2,6,8-trimethyl-10-phenyldeca-4,6-dienoic acid (Sivonen and Jones, 1999). Multiple combinations of the variable amino acids (X and Z for MCs, and only Z for NODs) make the difference between more than 80 MC variants already reported (Sivonen and Jones, 1999; Codd *et al.,* 2005; Del Campo and Ouahid, 2010) and only 9 NODs (Codd *et al.,* 2005). Each toxin variant is clearly identified by the initials of variable amino acids X and Z, for example, the most toxic variant MC-LR has leucine in X position and arginine in Z position. In the same way, NOD having arginine as variable amino acid Z should be identified as NOD-R. Both MCs and NODs are water-soluble molecules and their cyclic structure provides them a high chemical stability. Indeed, these hepatotoxins are known to resist to boiling, chemical hydrolysis and oxidation at pH close to neutrality (Sivonen and Jones, 1999). Their toxicity resulted on a potent and specific inhibition of serine/threonine protein phosphatases (Mackintosh *et al.,* 1990). They also have known to induce oxidative stress (Bouaïcha and Maatouk, 2004). Therefore, these toxins modulate the expression of oncogenes, early-response genes, and tumour necrosis factor α, and affect cell division, and apoptosis (Yoshizawa *et al.,* 1990; Sueoka *et al.,* 1997; Humpage and Falconer, 1999; Chen *et al.,* 2005), and they are considered as power tumour promoters in experimental animals (Nishiwaki-Matsushima *et al.,* 1992; Falconer and Humpage, 1996). However, the significance of this for humans, who may be subjected to chronic exposure via drinking water or food chain, remains unclear. Overall, the evidence for carcinogenecity of these hepatotoxins is today considered inadequate in humans and limited in animals (Chorus and Bartram, 1999). Recently, MC-LR has been classified as “possibly carcinogenic to humans" (group 2B), and NOD-R as "not classifiable as to their carcinogenicity" (group 3) (Grosse, 2006). Cylindrospermopsins (CYLs) are cytotoxic alkaloids consisting of a tricyclic guanidine moiety combined with hydroxymethyluracil that inhibits the synthesis of protein, resulting in wide spread necrosis of the tissues of many organs (Ohtani *et al.,* 1992; Ohtani *et al.,* 1992; Terao *et al.,* 1994; Falconer *et al.,* 1999; Seawright *et al.,* 1999; Runnegar *et al.,* 2002). Today, 2 variants of the cylindrospermopsin (CYL) toxin have been identified. The first one, called 7-epicylindrospermopsin, differs from the CYL only by the orientation of the hydroxyl group close to the uracil moiety (Banker *et al.,* 2000). The second one, called deoxycylindrospermopsin, seems to be non-toxic and is characterized by a missing oxygen atom related to the initial hydroxyl group close to uracil moiety (Norris *et al.,* 1999; Li *et al.,* 2001).

Freshwater cyanobacterial dermatotoxins such as lipopolysaccharides (LPS) are frequently cited in the cyanobacteria literature as toxins or endotoxins responsible for a variety of health effects in humans, from skin rashes to gastrointestinal, respiratory and allergic reactions (for review see Stewart *et al.,* 2006). LPS endotoxins are external components of cell membranes of most cyanobacteria as well as Gram-negative bacteria. They consist of three regions: an internal acylated glycolipid (termed lipid A), and a central area of liposaccharides, linking the internal subunit with the external specific carbohydrate polymer (O-specific chain) (Jann and Jann, 1984; Mayer and Weckesser, 1984; Kaya, 1996). Cyanobacterial LPS are credibly less toxic than those of Gram-negative enteric bacteria (*Salmonella* sp.), and they are nevertheless the cause of allergies, fever and gastroenteritis (Kuiper-Goodman *and al.*, 1999).

3. Effects of Cyanotoxins on Aquatic Organisms

3.1. Effects of Cyanotoxins on Algae Growth and Development

Algae play an important role in all aquatic ecosystems. They are considered as form of food and base of energy for all organisms living in different kind of aquatic ecosystems. Their ability to absorb sunlight for photosynthesis process confers to them a vital role in trophic chain. However, the fluctuations of algal biomass are related to diverse factors such as: temperature, pH, sunlight, nutrients (the amounts of phosphorus and nitrogen) and competition from other algae and aquatic organisms. The stability of algal biomass in water bodies depends on the balance between all these last factors that govern growth and development of algal species. However, several biotic and abiotic factors may interact in causing disturbances in the algal populations. Recently, several studies have focused in interactions between species of algae or more precisely the interactions between algae and substances they produce (toxins). The effects of these biotoxins on different species of algae have received great importance in order to evaluate the allelopathic action exerted by cyanotoxins on algae.

The majority of investigations into the toxicity of the microcystins have focused on animals and higher plants. Very few studies have highlighted the effects induced by cyanotoxins on algae, which are important in aquatic ecosystems. The abundance of cyanobacteria in aquatic ecosystems and their toxins may affect not only the structures of aquatic communities but also their composition. Sedmak and Kosi (1998) have first reported that microcystin-RR influenced phytoplankton proliferation dependent on light conditions and toxin concentrations. Recently, they have also shown that exposure of microcystin-producing and nonproducing *Microcystis aeruginosa* strains to the most common microcystins MC-LR, -RR and -YR produced several morphological and physiological changes. The main modifications recorded were, an important cells aggregation, increase in cells volume in addition to an overproduction of photosynthetic pigment. The same effects were also noted on the green alga *Scenedesmus quadricauda* (Sedmak and Elersek, 2005). The exposure of *Synechococcus elongatus* to MC-RR induced a significant inhibition in growth. The toxin-treated alga was reduced by 53.6% after 6 days of exposure to 100 μg L^{-1} of MC- RR (Hu *et al.,* 2005). However, in another study the growth of filamentous alga, *Chladophora fracta* was not significantly affected when exposed to microcystin-LR concentrations up to 10 μg mL^{-1} (Mitrovic *et al.,* 2005). In addition, Valdor and Aboal (2007) have reported that the growth of many species *Nostoc* sp., *Pseudocapsa* sp., *Scytonema* sp., are also inhibited after exposure to cyanophytes extracts and pure microcystins except *Klebsormidium* sp. which seemed to be resistant in all cases. After exposure of *Nostoc* sp. and *Pseudocapsa* sp. a colony forming species, cells were disaggregate and isolated with many deformations. Many others cells modifications were also detected in exposed *Scytonema* sp. (cells elongation and death; trichomes fragmentation, increased cytoplasmic inclusions, etc.). Other morphological modifications induced by microcystins, have recently been noted by exposing a filamentous cyanobacterium *Trichormus variabilis* to extract containing 2 or 20 nM of microcystins within 10 days of exposure (Bártová *et al.*, 2010). These changes are manifested by a significant decrease of heterocyst and akinete. Mohamed (2008) has shown that the exposure of two representatives of green algae, *Chlorella vulgaris* and *Scenedesmus quadricauda,* to microcystins resulting to decrease in both cell density and pigment contents (chlorophyll-*a*, chlorophyll-*b* and carotenoids) of the two algae species. On the other hand, the effects of microcystins on the physiology of algae were revealed. Singh *et al.* (2001) showed that photosynthetic process of *Nostoc muscorum* and *Anabaena* BT1 was inhibited by high dose of microcystin-LR (50 μg mL^{-1}), besides to the loss of nitrogenase activity. Pflugmacher (2002) reported that MC-LR at a concentration of 0.5 μg L^{-1} significantly inhibits the photosynthesis of *Cladophora* sp. A 50% inhibition of photosynthesis activity was also observed after exposure of the strain HUB 041 of *Senedesmus armatus* Chodat to the crude extract of cyanotoxins but no significant effect was shown after exposure to the purified toxins (Pietsch *et al.*, 2001).

3.2. Effects of Cyanotoxins on Aquatic Plants Physiology

The first work investigating the effects of blue-green algal blooms on an emergent reed plant, *Phragmites australis* (Cav.) Trin. ex Steud., was carried out in a eutrophic lake in central Japan. The emergent plants in the area of cyanobacterial bloom showed reduction on shoot length and dry weight in comparison with *P. australis* without bloom manifestation (Yamakasi, 1993). A clear dose dependent inhibition of the growth of the submerged macrophyte *Ceratophyllum demersum* was also observed at MC-LR concentrations of 0.1 to 5.0 μg L^{-1} (Pflugmacher, 2002). Exposure to MC-LR at high concentrations (5 mg L^{-1}) for 24 h induced the plant death. Whereas, the exposure to lower concentration of MC-LR (5 μg L^{-1})

caused faster reduction of the macrophyte growth (Pflugmacher, 2002). The photosynthetic oxygen production in the species *Ceratophyllum demersum,* was significantly inhibited after exposure to crude extract of microcystins at 0.25 µg L^{-1}, while the pure toxin MC-LR caused a slight but insignificant inhibition (Pietsch *et al.*, 2001). Similarly, the inhibition of the photosynthetic oxygen production in *C. demersum* was also confirmed when this species was exposed to MC-LR at concentration of 0.5 µg L^{-1} and the effect was more pronounced with a concentration ten time higher (Pflugmacher, 2002). The photosynthesis of other aquatic plants, the submergent macrophyte *Myriophyllum spicatum,* and the emergent macrophyte *Phragmites australis*, was also significantly inhibited by MC-LR at 0.5 µg L^{-1}. The inhibitory effect was more pronounced in *Elodea canadensis* and *C. demersum* exposed to the same MC-LR concentration (Pflugmacher, 2002). Further, the changes in pigment pattern of *C. demersum* after exposure to different concentrations of MC-LR raised the evident damage of photosynthesis process. The same effects on photosynthesis process were reported on the water moss *Vesicularia dubyana.* The exposure of this species to both microcystin-LR (inhibitor of protein phosphatases 1 and 2A) and microcin SF608 (inhibitor of serine proteases) affected strongly the production of photosynthetic oxygen. Besides, the pigment pattern illustrated the typical stress reaction of plants. The chlorophyll ratio *a/b* increased in both toxins concentrations tested (Wiegand *et al.*, 2002).

The Growth of aquatic plant *Lemna minor* (weight and frond number) and root length were significantly reduced after 5 days of exposure to 10 and 20 µg $mL^{-1}$of MCs. Similarly, *Wolffia arrhiza* growth (frond number) decreased significantly when exposed to 15 µg mL^{-1} (Mitrovic *et al.*, 2005). The potential allelopathic effects of *Microcystis aeruginosa* on the duckweed plant, *Lemna gibba* L. were examined by LeBlanc *et al.* (2005). No significant dose-dependent effects were observed with MC-LR or crud extract of a toxic species of *M. aeruginosa.* In addition, photosynthesis oxygen evolution was not affected after direct exposure of chloroplasts to MC-LR. Otherwise Saqrane et al. (2007) reported that chronic exposure to MCs caused an important negative effect on *Lemna gibba* growth, with a net reduction of frond number. The inhibitory effect was time and concentration dependent, and the chlorophyll ($a + b$) content of the cells was also affected by MC exposure at 0.3 µg mL^{-1} with a maximum decrease of 65%. In addition, the exposure of *L. gibba* to different microcystin concentrations showed a concentration-dependent effect on the measured chlorophyll fluorescence *in vivo* with significant differences except for the lower concentration at 0.01 µg equivalent MC-LR mL^{-1}(Saqrane *et al.*, 2009). In another study, direct exposure of *Lemna japonica* Landolt to toxic monoclonal strains of *Microcystis aeruginosa* Küzing (NIES strains 103 and 107) induced reciprocal allelopathic effects on both aquatic plant and cyanobacteria strains (Jang *et al.*, 2007). The growth of *L. japonica* and *M. aeruginosa* was inhibited and significant increase in intra- and extra- cellular MC production was registered, which could be related to the possible release of allelochemicals from the aquatic plant.

Not only *M. aeruginosa* toxic strains seem to have allelopathic effects on aquatic plants. Recently, Nam *et al.* (2008) indicate that the aquatic vascular plant, *Myriophyllum spicatum*, inhibited the growth of the cyanobacterium *Microcystis aeruginosa* resulting on a decrease of the chlorophyll-*a* and metabolic changes in the plant *M. spicatum*. Li *et al.* (2009) reported that the growth and the photosynthetic activity of the submerged aquatic plant *Ceratophyllum oryzetorum* Kom. were changed in presence of different cell densities of cyanobacterial bloom. The plant length and fresh mass were promoted by low cyanobacterial cell densities.

In the contrary to medium and high cyanobacterial cell densities which acted as growth inhibitor, and the photosynthetic activity of *C. oryzetorum* was strongly inhibited by high cyanobacterial cell densities.

3.3. Effects of Cyanotoxins on Various Aquatic Animals

Aquatic organism mortalities due to cyanobacterial toxins can occur by two routes: through consumption of cyanobacterial cells from the water, or indirectly through consumption of other animals that have themselves fed on cyanobacteria and accumulated cyanotoxins. Consequently, there is considerable potential for toxic effects to be magnified in aquatic food chains (Malbrouck and Kestemont, 2006). Cyanotoxins have caused detrimental effects in aquatic vertebrates and invertebrates under laboratory and field conditions. Hepatotoxins, MCs, and CYLs are regarded as the main cyanotoxins involved in ecotoxicological events.

Much of what we know about the effects of cyanobacterial toxins on various aquatic animal organisms derives from laboratory studies. Aqueous exposure of fish to MCs and *Microcystis* cells has been reviewed in the literature; however, the methods used, in most cases, were not comparable to an environmental exposure scenario. For example, the majority of studies conducted with aqueous microcystins was in acute conditions during early developmental stages (Best *et al.*, 2002, 2003; Liu *et al.*, 2002), and fewer studies of chronic exposures have been conducted with adult fish (Adamovsky *et al.*, 2007; Li *et al.*, 2007; Mares *et al.*, 2009; Qiu *et al.*, 2009). Another type of acute exposures found in the literature is the microinjection of microcystins (Huynh-Delerme *et al.*, 2005; Jacquet *et al.*, 2004; Wang *et al.*, 2005). The purpose of this last procedure is to mimic uptake of toxins from the surrounding water by the embryo or their transfer from females to eggs; however, the occurrence of maternal transfer has not been established for MCs, and this type of exposure is somewhat presumptive (Malbrouck and Kestemont, 2006). Finally, in studies addressing effects of MCs via ingestion, toxins have been administered through oral gavage (Fischer and Dietrich, 2000; Fischer *et al.*, 2000) and intraperitoneal injection (Fournie and Courtney, 2002; Malbrouck *et al.*, 2003); however some studies have incorporated *Microcystis* into fish feed (Dong *et al.*, 2009; El Ghazali *et al.*, 2010), which is more comparable to a natural feeding situation.

Numerous studies reported that the exposure of carp (*Cyprinus carpio*) and southern catfish (*Silurus meridionalis*) embryos to crude extracts of *Microcystis* resulted in high mortality rates, delayed hatching, malformations, and lesions in the liver detected by histopathology (Palikova *et al.*, 2003; Palikova *et al.*, 2004; Zhang *et al.*, 2008). Jacquet *et al.* (2004) showed that microinjection in medaka fish embryos (*Oryzias latipes*) of MC-LR at 1 or 10 µg mL^{-1} for 10 days, reduced survival rate by 90 %. This result was confirmed by Deng (2010) deducing a growth inhibition of the Japanese fish medaka (*Oryzias latipes*) after exposure to MC-LR for 8 weeks, and a decrease of the embryos survival. The exposure of zebrafish embryos (*Danio rerio*) to 3 µg mL^{-1} eq. MC-LR for 24h affects the embryos survival (El Ghazali *et al.*, 2009). The proteomic analysis of exposed zebrafish (*D. rerio*) to MC-LR (2 or 20 µg L^{-1}) for 30 days revealed an induction of oxidative stress, a dysfunction of cytoskeleton assembly and macromolecule metabolism, with a concomitant interference with signal transduction and other functions (e.g. protein degradation, transport, apoptosis and translation (Wang *et al.*, 2010). Li *et al.* (2008) reported that the initial significant increases in reactive oxygen species (ROS) contents in the liver of bighead carp (*Aristichthys nobilis*) at 3 h postinjection with MCs suggests that these toxins induced oxidative stress. When the tilapia fish was exposed to orally single dose of cyanobacterial cells containing 120 µg MC-LR g^{-1} for 12 and 72 h an oxidative damage and a reduction of LPO and protein oxidation were observed (Prieto *et al.*, 2007). Li *et al.* (2009) showed that the injection of crude extracts of MCs to the crucian carp (*Carassius auratus*) at sublethal and lethal doses (150 and 600 µg MC kg^{-1} bw) resulted in prominent hypotension, increase in heart rate, and decreases in heart rate. In a previous study, they reported that direct MCs exposure to the phytoplanktivorous silver carp generated deep physiological and patho-cytological effects (Li *et al.,* 2007). On the other hand, in laboratory conditions, the culture of *Moina micrura* (cladoceran) for bioassays using raw water from the Funil reservoir contaminated by MCs, showed adverse effects including death, paralysis, and reduced population growth rate (Ferrao-Filho, 2009). Buryskovà *et al.* (2006) indicate that the exposure of *Xenopu*s *laevis* embryos with various cyanobacterial fractions showed high mortalities of embryos and minor effects on viability of tadpoles and malformations in surviving embryos accompanied with inhibition of the growth. In aquatic invertebrates, microcystin exposure disrupted osmoregulation of an estuarine crab (*Chasmagnathus granulata*) through the inhibition of Na^+/K^+-ATPase activity in the anterior gills (Vinagre *et al.,* 2002) and caused oxidative stress and histopathological damage in the crab's hepatopancreas (Pinho *et al.,* 2003).

The purified alkaloid hepatotoxin, cylindrospermopsin (CYL) caused mortality of the brine shrimp, *A. salina* (Metcalf *et al.,* 2002). Whole cells of the cyanobacterial species *C. raciborskii*, containing 200–232 µg CYL L^{-1}, were toxic to the tadpole *Bufo marinus* (White *et al.,* 2007) and snail embryos, *Melanoides tuberculata* (Kinnear *et al.,* 2007). The purified neurotoxin saxitoxin concentrations as low as 10 µg L^{-1} delayed hatching of zebrafish (*Danio rerio*), while the highest concentration, 500 µg L^{-1}, caused malformations and mortality (Oberemm *et al.,* 1999). Whole cells of *Anabaena* sp. (10^7 cells ml^{-1} with 970 µg anatoxin-a g^{-1} dw) were also toxic to fish as all juvenile carp (*Cyprinus carpio*) died after 26–29 h of exposure (Osswald *et al.,* 2007). Nogueira *et al.* (2004) reported that the cyanoabcterial species *A. issatschenkoi*, containing PSP toxins, decreased *Daphnia magna* growth and survival.

The microbial food web may also suffer in the presence of toxic cyanobacterial blooms, as exposure decreased the survival of the protozoan, *Tetrahymena pyriformis* (Ward and Codd, 1999) and reduced the growth rate of nanoflagellates (Christoffersen, 1996). In the presence of hepatotoxin-producing cyanobacteria, the zooplankton community in extensive aquaculture ponds can shift to less desirable species as a result of reduced fecundity, decreased feeding rates, and increased mortality of large-bodied species (Thostrup and Christoffersen, 1999; Engström *et al.,* 2000; Liu *et al.,* 2006). cyanobacterial hepatotoxins also have proven harmful to crustaceans, decreasing grazing rates in the amphipod, *Gammarus zaddachi*, and mysid shrimp, *Mysis mixta* (Engström *et al.,* 2001; Korpinen *et al.,* 2006) and causing mortality in fairy shrimp, *Thamnocephalus platyurus* (Keil *et al.,* 2002), brine shrimp, *Artemia salina* (Metcalf *et al.,* 2002), and the microcrustacean, *Kalliapseudes schubartii* (Montagnolli *et al.,* 2004).

In natural environment and since the first report of livestock poisoning associated with cyanobacterial blooms (Francis, 1878), numerous cases of animal deaths have been reported worldwide due to contact with cyanobacterial scums, indicating that this is a widespread phenomenon in Mediterranean, continental, and temperate climates in both hemispheres (Codd *et al.,* 2005; Falconer, 2005). In natural systems, toxic cyanobacterial blooms were associated with mass fish mortality (Rodger *et al.,* 1994; Lindholm *et al.*, 1999); however, this mass mortality is more likely to be the result of multiple stress factors that co-occur during cyanoabcteria blooms, and therefore, the exact cause could not be determined. Nevertheless, direct evidence has been presented, however, for hepatotoxin-induced mortality in intensive aquaculture systems specifically, pond farming of catfish and white shrimp (Zimba *et al.,* 2001b) and net-pen farming of Atlantic salmon, chinook salmon, and rainbow trout (Andersen *et al.,* 1993; Kent *et al.,* 1996). Lugomela et *al.* (2006) have reported the mortality of 15 and 50 individuals per day, during the study period, of the Lesser Flamingos in Lake Manyara and Lake Big Momela, Tanzania. The analysis of gut content indicated that *Arthrospira fusiformis* was the main food item in moribund flamingos. The injection of extract dominated by *A. fusiformis* in mice at different dose, suggested that the all of them become lethargic, with loss of balance, uncoordinated movements, intermittent tremors and respiratory arrest. This result confirmed the toxicity of *A. fusiformis* at high concentrations to the Lesser Flamingo. Nasri *et al.* (2008) reported for the first time the death of two species of turtles from Lake Oubeira (Algeria) during a toxic *Microcystis bloom*. Recently, Miller *et al.* (2010) reported the death of 21 southern sea otters. Microcystin-poisoned sea otters were commonly recovered near river mouths and harbors and contaminated marine bivalves were implicated as the most likely source of this potent hepatotoxin for wild otters. Therefore, this is the first report of deaths of marine mammals due to cyanotoxins and confirms the existence of a novel class of marine harmful algal bloom in the pacific environment, that of hepatotoxin shellfish poisoning (HSP), suggesting that animals and humans are at risk from microcystin poisoning when consuming shellfish at the land-sea interface.

4. Bioaccumulation of Cyanotoxins on Aquatic Organisms

Many studies have suggested that cyanotoxins bioaccumulate in aquatic biota and that this may enhance the risk of exposure of biota higher up in the food web (Li *et al.,* 2004; Sipia *et al.,* 2001; Negri and Jones 1995). Biota may take up and accumulate cyanotoxins via two main routes: dissolved toxins via a transdermal route or ingestion via the intake of food. Filter feeding organisms like *Daphnia* or mussels as well as phytoplanktivorous fish may directly graze on toxic cyanobacteria (Kurmayer, 2001; Dioniso Pires *et al.,* 2004; Rohrlack *et al.,* 2005), but most seafood is found at higher trophic levels, and toxins may reach these organisms through vectorial transport, a concept known from the marine environment (Tester *et al.,* 2000), via the aquatic foodweb.

4.1. Bioaccumulation of Cyanotoxins in Algae Cells and Aquatic Plants

Very few studies have reported the possible accumulation of cyanotoxins in algae cells when exposed to those molecules. A filamentous alga *Chladophora fracta* was found to accumulate MC-LR concentrations of 0.042 ± 0.015 ng mg^{-1} wet wt over the 5 days of the experiment, with an accumulation rate of 0.008 ng mg^{-1} day^{-1} (Mitrovic *et al.*, 2005). Another study showed that few amounts of MCs were detected in *Chlorella vulgaris* and *Scenedesmus quadricauda* cells only during the first 3 days of exposure, but not during the remaining period of the experiment which indicates absorption and biotransformation of MCs in these algae (Mohamed, 2008). The inhibitory effect of cyanotoxins on the growth of some microalgae was more persistent in pure microcystins than in the crude extracts of cyanobacteria, which lost their properties eight days after exposure; this fact could be related to toxins accumulation and depuration process of algae (Valdor and Aboal, 2007).

The possible uptake of cyanobacterial toxins by aquatic plants, especially on *Phragmites australis*, was highlighted by Pflugmacher *et al.* (2001). The emergent reed plant *P. australis* showed an apparent distribution of MC-LR in the different parts of the plant, after exposure to this toxin at 0.5 µg L^{-1}. Highest uptake was detected in the stem and then the rhizome (Pflugmacher *et al.*, 2001). In addition, *Lemna minor* has also been shown to accumulate MC-LR up to a concentration of 0.2887±0.009 ng mg^{-1} wet wt plant material, after 5 days of exposure to this toxin at 20 µg L^{-1} with an accumulation rate equivalent to 0.058 ng mg^{-1} day^{-1} (Mitrovic *et al.*, 2005). While, it was demonstrated that *Lemna gibba* could take up and bio-transforms microcystins (Saqrane *et al.*, 2007). The chronic exposure of plant led to dose-dependent MC accumulation which reached 2.24 µg g^{-1} dry weight after being exposed to 0.3 µg mL^{-1} of MCs (Saqrane *et al.*, 2007). It has been shown that collected water chestnut (*Trapa natans*) from Lake Tai accumulated MCs at highest level of 7.02 ng g^{-1} (Xiao *et al.*, 2009).

4.2. Bioaccumulation of Cyanotoxins in Aquatic Animals

Microcystins can be transferred along food chain (Ibelings *et al.*, 2005; Smith and Haney, 2006), suggesting a potential risk to high trophic level species and human consumption of contaminated aquatic products. Chen *et al.* (2009) reported that MCs were identified for the first time in the serum of a chronically exposed human populations (fishermen at Lake Chaohu, China) together with indication of hepatocellular damage. Extensive laboratory experiments and field investigations have been conducted to document bioaccumulation and distribution of MCs in a variety of aquatic organisms (Xie *et al.*, 2004; Chen and Xie, 2005; Zhang *et al.*, 2007). Crayfish (*Procambarus clarkii*) collected in Massaciuccoli Lake (Italy), accumulated cyanotoxins in all the organs analyzed. The highest concentration of MCs was found gradually in intestine > hepatopancreas=stomach > abdominal muscle (Tricarico *et al.*, 2008). In another study, the MC-RR was detected in the liver of *Carassius auratus* and *Cyprinus carpio* at dose of 0.82 and 2.06 µg.g^{-1} dw, respectively (Xie *et al.*, 2007). The two species of fish (*Carassius auratus*, *Linnaeus auratus*) accumulated MCs in the liver from 461.8 to 3628.6 ng.g^{-1} dw, and in the muscle, ranged from no detectable concentration to 377.8 ng.g^{-1} dw (Song *et al.*, 2007). Besides, the mollusks (prosobranchs, pulmonates and bivalves) collected at three French sites (Frémur, Pleine-Fougères, Combourg) showed an ability to accumulate microcystins. The Pulmonates accumulated 34.95±11.4 ng MC-LR g^{-1} fresh weight, in opposition to prosobranchs and bivalves witch accumulated respectively 8.23 ± 4.42 and 7.96 ± 3.46 ng MC-LR g^{-1} fresh weight (Gérard *et al.*, 2009). The recent study reported by Lance *et al.* (2010) concerns accumulation and elimination of both free and bound microcystins (MCs) in tissues of a gastropod exposed to MCs producing-cyanobacteria or dissolved MC-LR., taking into consideration the MCs cell-bounding fraction will constitute a new challenge for cyanotoxins monitoring and sanitary risk. Another species of snail, *Sinotaia histrica* accumulated the highest concentration of MCs in the digestive tract content (9.03 μg g^{-1} dw), followed by the gonad (6.90 μg g^{-1} dw) and hepatopancreas (5.38 μg g^{-1} dw), whereas the less concentration in the foot (2.48 μg g^{-1} dw) (Xie *et al.,* 2007). On the other hand, the snail, *Bellamya aeruginosa* accumulated MCs in the foot ranged from below detectable levels to 380.8 ng.g^{-1} dw, but the high toxin levels were detected in the visceral sac with the values of 2315.4 ng.g^{-1} dw (Song *et al.,* 2007). The shrimp, *Macrobrachium nipponensis* accumulated MCs in two organs: the abdomen and cephalothorax with maximum values of 388.9 and 754.3 ng.g^{-1} dw, respectively (Song *et al.,* 2007).

Chen *et al.* (2009) reported that the water bird, *Nycticorax nycticorax* and the duck, *Anas platyrhynchos* accumulate high MC levels in the gonad, egg yolk and egg white. The highest MC concentrations were identified for the first time in the spleens of *N. nycticorax* and *A. platyrhynchos* (6.850 and 9.462 ng g^{-1} dw, respectively). The native species of clam, *Cristaria plicata* accumulated MCs in the foot to a concentration of 730.3 ng.g^{-1} dw (Song *et al.,* 2007). The turtle *Pelodiscus sinensis* accumulated highest concentration of MCs in liver (20.8 ng.g^{-1} dw) and intestine (19.7 ng.g^{-1} dw), and decreased in the order of Kidney > lung > muscle> gallbladder> heart >gonad (Chen *et al.,* 2009). In laboratory experiments, exposure of zebrafish (*D. rerio*) to MC-LR at 2 and 20 μg L^{-1} for 30 days, showed an accumulation of this toxin to a concentration of 0.030 and 0.053 μg mg^{-1} dw in the brain (Wang *et al.,* 2010).

The freshwater mussel *Anodonta cygnea* accumulated the cylindrospermopsin (CYL) to concentrations up to 2.52 μg g^{-1} dw tissue (Saker *et al.,* 2005). Similarly, the snail, *Pomacea patula catemacensis* accumulates this hepatotoxin to a concentration of 3.35 (± 1.90) ng/g and neurotoxins STX/PST (paralytic shellfish toxins) to 1.04±0.42 ng/g (Berry *et al.,* 2010). The crayfish, *Cherax quadricarinatus* accumulates more CYL in their hepatopancreas (4.3 μg g^{-1} dw) and abdominal muscle tissue (0.9 μg g^{-1} dw) when exposed to a natural population of *C. raciborskii* in an aquaculture pond than when exposed to a laboratory culture for 14 days. No histological abnormalities were identified in exposed crayfish (Saker and Eaglesham, 1999). Rainbow fish, also collected from the aquaculture pond, accumulated 1.2 μg g^{-1} dw in viscera tissue. After 16 days of exposure, the freshwater mussel *Anodonta cygnea* accumulated up to 2.52 μg CYL g^{-1} dw, 50% of which was retained in tissues after two weeks of depuration (Saker *et al.,* 2004). The alkaloid neurotoxins were rapidly accumulated in fish (0.768 μg ANA g^{-1} dw) and mussels (6.2 μg PSP toxins g^{-1} fw), thereby opening the possibility for transfer of these cyanotoxins up the food chain (Negri and Jones, 1995).

5. Extraction and Analysis of Cyanotoxins from Aquatic Organisms

The global occurrence of cyanobacterial toxins and associated toxicosis has resulted in the development of vast array of methods for their detection. These methods fall into three main categories: (i) Bioassays with broad specificity (Falconer, 1993) (ii) Class-specific screening methods based on immunoassays and enzymatic assays (An and Carmichael, 1994; McDermott, 1995; Fontal *et al.,* 1999; Bouaïcha *et al.,* 2002), (iii) Trace analyses of individual compounds, mainly based on high-performance liquid chromatography (HPLC) and gas chromatography (GC) combined with conventional detection or mass spectrometric (MS) detection (Lawton *et al.,* 1994; McElhiney and Lawton, 2005; Msagati *et al.,* 2006; Osswald *et al.,* 2007; Aráoz *et al.,* 2010; Li *et al.,* 2010). These last techniques allow separation, highly selective identification and sensitive quantification of the different toxins present into a sample. Moreover, limits of detection are generally one order of magnitude lower than those obtained with the mouse bioassay. However, it is well known that they require expensive equipment and trained personnel, and that they are laborious and time-consuming. The lack of standards is also a serious difficulty, since non-certified or unknown toxins cannot be evaluated. Additionally, all of these methods are laboratory based, many of them requiring sample pre-treatment/concentration to achieve adequate sensitivity. Therefore, much effort has been put forth over the last two decades towards developing efficient extraction techniques, cleanup steps, and quantification methods to evaluate accumulation of cyanotoxins in aquatic organisms. For water-soluble neurotoxins (anatoxins and saxitoxins) and the alkaloid hepatotoxin, cylindrospermopsin those not bound covalently to macromolecules in animal tissues, extraction solvents have included various percentages of aqueous or acidified methanol, acidified water, or various ratios of butanol, water, and methanol (Henriksen *et al.,* 1997; Osswald *et al.,* 2007; Turner *et al.,* 2011). Extraction has been facilitated through the use of extended incubation with solvents, bath sonication at various wattages, or tissue homogenization. Solid phase extraction (SPE), protein precipitation, immunoaffinity chromatography, hexane wash, centrifugation, and filtration have been used to prepare samples for analysis (Harada *et al.,* 1993; Rapala *et al.,* 1994; James *et al.,* 1998; Namikoshi *et al.,* 2003; Rellán *et al.,* 2007). The recovery of free anatoxins and saxitoxins from animal tissues, however, can vary from 70 to 100%, depending upon the extraction and preparatory procedures utilized, and the tissue type extracted (Hormazabal *et al.,* 2000). Currently, routine analysis of these toxins is most commonly carried out using high-performance liquid chromatography with UV or fluorescence detection or coupled to mass spectrometry (Poon *et al.,* 1993; Draisci *et al.,* 2001; Osswald *et al.,* 2007; Turner *et al.,* 2011).

Since microcystins and BMAA were found in the free fraction of both plant and animal tissues but were also concentrated in the protein associated fraction within these tissues (Williams *et al.,* 1997; Cox *et al.,* 2003; Murch *et al.,* 2004; Banack *et al.,* 2006; Esterhuizen and Downing, 2008; Nasri *et al.,* 2008; Brand *et al.,* 2010), conventional methods such as organic extraction only extract the unbounded portions of the toxin, which will lead to a false determination of total toxin concentration. BMAA was generally extracted from complex matrices by sonication with 0.1M trichloroacetic acid. Therefore, free BMAA was obtained in the supernatant after centrifugation to precipitate proteins and protein-associated BMAA was released by liquid acid hydrolysis in 6M HCl at 110 °C for 24h (Cox *et al.,* 2005; Esterhuizen and Downing, 2008; Brand *et al.,* 2010). Several methods have been reported to detect BMAA in cyanobacterial, plant, and animal tissue samples including high performance liquid chromatography (HPLC) with UV or fluorescence detection (Banack and Cox, 2003; Cox *et al.,* 2005; Jonhson *et al.,* 2008), gas chromatography-mass spectrometry (GC–MS) (Guo *et al.,* 2007; Esterhuizen and Downing, 2008), and liquid chromatography-mass spectrometry (LC-MS) (Banack *et al.,* 2007; Kubo *et al.,* 2008; Rosen and Hellenas, 2008).

Free fraction of microcystins was generally extractable from dried samples using organic solvents (Harada *et al.,* 1997; Fastner *et al.,* 1998; Krienitz *et al.,* 2003; Cazenave *et al.,* 2005;

Smith and Boyer, 2009). Soares *et al.* (2004) have reported the extraction of microcystins from fish liver and muscles using of 100% methanol. The methanol extract was then mixed with an equal volume of hexane before discarding the hexane layer. The extract was then loaded onto C18 cartridges for pre-concentration and further clean-up step using solid phase extraction (SPE). Xie and coworkers have reported the extraction and analysis of hepatotoxins in the tissues and organs of aquatic organisms (shrimps, snails, etc.) using a mixture of solvents comprised of butanol/methanol/water (1:4:15) for 24 h (Xie *et al.,* 2004; Chen and Xie, 2005). The same extracting solvent and ratios was used by Lehtonen *et al.* (2003) in extracting nodularins in the clam *Macoma balthica*. Fastner *et al.* (1998) showed that, for lyophilized field samples of cyanobacteria, extraction with a mixture of methanol and water (75:25, w/w) was most effective. Extraction from lyophilized tissue samples of aquatic organisms has been facilitated through the use of ultrasonication and tissue homogenization (Nasri *et al.,* 2008). Solid phase extraction (SPE) has been then used to prepare samples for analysis. Once covalently bound to either protein phosphatases or other cysteine containing peptides (e.g., glutathione), MCs can be oxidized to release a portion of the Adda moiety as 2-methyl-3-methoxy-4- phenylbutyric acid (MMPB), which can then be quantified by liquid chromatography or gas chromatography coupled to mass spectrometry (Williams *et al.,* 1997; Ott and Carmichael, 2006; Nasri *et al.,* 2008; Wu *et al.,* 2009; Neffling *et al.,* 2010). Therefore, the oxidation method reveals the total level of MCs (free and bound) accumulated in tissues.

6. Exposure to Cyanotoxins through Aquatic Food Webs and its Consequences for Public Health

Exposure to phycotoxins produced by harmful algal blooms through the consumption of food is very well known from the marine environment. Among the toxins involved, some are comparable to toxins produced by cyanobacteria (like saxitoxin derivatives that cause paralytic shellfish poisoning, PSP). Human poisonings through phycotoxins in marine shellfish highlight that biomagnifications of toxins produced in the phytoplankton in the aquatic foodweb may cause illness in humans when biota from higher trophic levels are consumed, and that this risk needs to be assessed in freshwater ecosystems for cyanotoxins as well. Freshwater aquacultures are the fastest growing animal food sector in the world, and cultured fish supply now provides over 13% of the animal protein intake for the human population (WHO, 2007). In China, they are providing about 20% of the total animal protein intake for Chinese residents (Wang and Tian, 2009). Therefore, aquaculture farming has grown to offset the increased demand. This increased demand has pushed aquaculture from extensive aquaculture, towards intensively operated production systems such as high-density ponds in which growth of cultured species is stimulated with fertilizer additions. Therefore, this intensive aquaculture is commonly resulting in eutrophic conditions and toxic cyanobacterial blooms (Magalhães *et al.,* 2001). Several studies have been mounting evidence to suggest that a number of widespread cyanobacterial toxins may also accumulate throughout various trophic levels of freshwater food webs, and thereby present an additional route of possible exposure and a consequent threat to human health (see review by Ibelings and Chorus, 2007). Although no case of poisoning by these products has been reported in the literature, this eventuality must not be ignored. Indeed, a recent epidemiological study showed that the excessive incidence of amyotrophic lateral sclerosis in the population of the islands of Guam in the Pacific was linked to a consumption of the seeds of cycas contaminated by a neurotoxin, ß-methylamino-L-alanine (BMAA), produced by a species of cyanobacteria of the genus *Nostoc* living in symbiosis in the roots of this plant. This last cited fact is gaining importance since plants could in a direct or indirect manner contribute to food chain cyanotoxin's transfer, and by the way constitute a potent health risk source. Recently, several studies reported that the source of BMAA for human exposure is not limited to cycad seed exposure or animals which have fed on cycad seeds (Karamyan and Speth, 2008; Jonasson *et al.,* 2008). For example, BMAA was also found in cyanobacterial blooms, scums and mats, collected in British water bodies that are used for drinking water, recreation, fisheries and livestock (Metcalf *et al.,* 2008). In addition, both free and bound BMAA were found in freshwater cyanobacterial samples collected in South Africa (Esterhuizen and Downing, 2008). Other studies reported that all cyanobacteria species produce BMAA in the environment (Cox *et al.,* 2005; Metcalf *et al.,* 2008; Esterhuizen and Downing, 2008). Brand *et al.* (2010) reported that a wide range of BMAA concentrations were found in animals in South Florida waters, ranging from below assay detection limits to approximately 7000 µg g^{-1}, a concentration associated with a potential long-term human health hazard. These data indicate that the situation in Guam is not unique, and further suggest that BMAA could be found in high concentrations in aquatic animals in many areas of the world where cyanobacteria blooms occur. However, exposure guidelines have not been estimated for this neurotoxin due to insufficient data.

For cyanobacterial neurotoxins type saxitoxin (Paralytic shellfish poisoning toxins, PSTs) several cases of mortality animal were indicated further to the consumption of freshwater contaminated by this toxin (Humpage *et al.*, 1994; Negri and Jones, 1995) but no case of human poisoning was reported. However, the poisonings associated to these neurotoxins are not limited to the direct poisoning by ingestion of contaminated water but can also appear after consumption of animals of freshwater which accumulated this toxin. Recently, Berry and Lind (2010) reported the first evidence in Lago Catemaco (Veracruz, Mexico) the bioaccumulation of saxitoxins in a freshwater snail that known locally as tegogolo and it's widely consumed and also represents a major fishing product in this region. Numerous studies reported that the levels of PSTs bioaccumulation in the freshwater mussels were comparable with those often reported in field sample of marine bivalves exposed to naturel dinoflagellate blooms (Sasner *et al.,* 1984; Negri and Jones, 1995; Pereira *et al.,* 2000; Pereira *et al.,* 2004; Deeds *et al.,* 2008). Although these freshwater bivalves are little consumed, they are a part of the traditional food of certain populations in some region in the World and could be a risk factor for the human health (Negri and Jones, 1995; Christoffersen, 1996). The recommendation for cyanobacterial neurotoxin type PSTs stays the same that recommended for seafoods which is of 80 µg equivalent saxitoxin by 100 g of bivalve's flesh.

For the most common cyanobacterial hepatotoxins type microcystin, numerous studies have proved that high concentration of these toxins could be accumulated in different organs of aquatic organisms, including the edible parts, such as fish muscle, abdomen of shrimp, whole body of bivalves, etc. (Vasconcelos, 1999; Magalhães *et al.,* 2001; Zimba *et al.,* 2001a; Ibelings *et al.,* 2005; Chen and Xie, 2005; Song *et al.,* 2007; Ibelings and Chorus, 2007; Smith *et al.,*2008; Galvão *et al.,* 2009 ; Peng *et al.,* 2010). The lifetime tolerable daily intake (TDI) value for humans, as determined by the World Health Organization, is 0.04 µg of MC-

LR/kg bw/d (Sivonen and Jones, 1999). Recently, Garcia *et al.* (2010) reported that the highest concentration of microcystins occurring in crab tissue was 105 μg MCs/kg (wet weight), representing an EDI (estimated daily intake) based on human body weight and amount of crab tissue consumed, of 0.46 μg MCs/kg body weight/day, a value more than 10 times the WHO-TDI guideline. Several other studies reported hepatotoxin accumulation in edible tissues of various organs of aquatic animals that exceeded the WHO-TDI guideline, suggesting that human intoxication is possible through both intensive and extensive aquaculture (Smith *et al.,* 2008). These studies highlighted that gastropods, bivalves and crayfish provide a greater risk than fish to human consumers as they accumulate higher concentrations of cyanobacterial toxins, and are frequently consumed whole. In a recent study, Peng *et al.* (2010) indicate that most of the aquatic products from three large Chinese lakes seem to be unsafe for human consumption due to MC accumulations, with the estimated daily intake (EDI) values 2-148 times higher than the lifetime WHO-TDI guideline. A previous study showed that the MC accumulations in the muscle of crucian carp and common carp, harvested from Lake Taihu in China, varied from below detection limit to 82% of TDI values, and 0.93-3.55 times of TDI values, respectively (Zhang *et al.,* 2009). However, another study conducted in the same year on the same lake found MC accumulations in the edible organs of aquatic products seemed to be unsafe for consumption due to the high concentration of toxin accumulations. Chen *et al.* (2009) estimated that a 60 kg adult who ingests on average 100 g of fish or turtle or water bird or duck muscle per day from Lake Taihu (China), daily uptake of MC-LR equivalent would be only 0.01– 0.11 μg, much lower than the lifetime TDI proposed by WHO, and that consumption of the muscle of these animals was still safe to human health. What emerges from this study is that as both abundance and toxicity of cyanobacteria may show temporal variation, and also as intracellular MC concentrations in the lake water in this study was only 0.15 μg L^{-1}, while in other study the maximum MC concentrations in Meiliang Bay of Lake Taihu reached as high as 10.4 μg L^{-1} (Shen *et al.,* 2003). Therefore, it is recommended that regular monitoring of MC contamination in aquatic vertebrates is needed from the viewpoint of human health protection.

Since microcystins are known to bind covalently to protein phosphatases or other cysteine containing peptides (e.g., glutathione), conventional methods such as organic extraction only extract the unbounded portions of the toxin from the aquatic tissues, which will lead to a false determination of total toxin concentrations. With the exception of a few studies (Williams *et al.,* 1997; Dionisio Pires *et al.,* 2004; Ibelings *et al.,* 2005; Ott and Carmichael, 2006; Nasri *et al.,* 2008), MCs are extracted from animal tissue using organic solvents (e.g., methanol). Extraction in this condition only releases free MCs (i.e., MCs that are dissolved or in reversible interactions with protein phosphatases within the organism); however, the covalently bound portion of MCs can be significant, representing up to 99% of the total MCs contained within in the organism's tissues (Williams *et al.,* 1997a, 1997b, 1997c; Ibelings *et al.,* 2005; Nasri *et al.,* 2008). Williams *et al.* (1997a) found that 60% of the total radio-labeled microcystin-LR in salmon liver was not extractable with methanol 5h following intraperitoneal injection, likely due to covalent bonding to protein phosphatases. Extraction with only methanol may explain the lower recovery rates in tissue samples, thereby significantly underestimating those microcystin concentrations. In a recent study, Lance *et al.* (2010) follow the accumulation of total (free and bound) MCs in the aquatic snail, *Lymnaea stagnalis* exposed to i) dissolved MC-LR (33 and 100 μg L/) and ii) *Planktothrix agardhii* suspensions producing 5 and 33 μg MC-LR equivalents/L over a 5-week period, and after a 3-week depuration period. Results of this study indicate that snails exposed to dissolved MC-LR accumulated up to 0.26 μg total MCs g^{-1} dw, with no detection of bound MCs. However, snails exposed to *P. agardhii*-producing MCs accumulated up to 69.9 μg total MCs g^{-1} dw, of which from 17.7 to 66.7% were bound. After depuration, up to 15.3 μg g^{-1} dw of bound MCs were detected in snails previously exposed to toxic cyanobacteria, representing a potential source of MCs transfer through the food web. Nasri *et al.* (2008) showed that 99% of the total level of microcystins in dead terrapin tissues from Lake Oubeira (Algeria) was not detectable by the methanol extraction-PPase methodology.

Nevertheless, free MCs have been generally accepted to represent the total risk to organisms and humans due to the untested assumption that the covalently bound portion was not bioavailable to food web transfer. MCs can be transferred to higher trophic levels, from zooplankton and mollusks to herbivorous and carnivorous fish species in natural and aquaculture systems (Vasconcelos, 1999; Magalhães *et al.,* 2001; Zimba *et al.,* 2001a, 2001b; Mohamed *et al.,* 2003; Smith *et al.,* 2006; 2008) however, the bioavailability of the covalently bound portion is unknown. In addition, complexes of microcystin covalently bound to protein phosphatase enzymes are thought to be less toxic when compared to unbound microcystins, but questions remain largely unanswered (Ibelings and Havens, 2008). To test if digestion products could be biologically active to the consumer, four predicted MC-peptides resulting from the digestion of a PP1-MC complex by typical digestion enzymes and GSH-MC were synthesized and assayed for activity against protein phosphatase type 1 (PP1). This *in vitro* study demonstrated that free MCs are resistant to digestion and that the predicted MC-peptides and GSH-MC retain a significant fraction (58%) of the parent toxin's inhibitory activity (Smith *et al.,* 2010). Therefore, if covalently bound MCs are released during gut passage in an active form, they may represent a significant risk to consumers and should be considered when evaluating the impact of microcystins on the food web.

CONCLUSIONS

The studies reviewed in this chapter reveal that cyanotoxins are known to bioaccumulate in common aquatic vertebrates and invertebrates, including fish, mussels and zooplankton, which poses a risk to both animal and human heath if such aquatic animals are consumed. Evidence suggests that gastropods, bivalves and crayfish provide a greater risk than fish to human consumers as they accumulate higher concentrations of cyanobacterial toxins, are relatively more tolerant of the toxins, and are frequently consumed whole. Although there is much basic information on the concentrations of these toxins found in freshwater foodwebs, there are very significant gaps in our knowledge of their bioconcentration, bioaccumulation, bioamplification, and the role of detoxication and covalent binding of microcystins on transfer in the foodweb. Several studies demonstrated that a large part of microcystins in biota is covalently bound. This means that almost all of the concentrations given in these studies seriously underestimate the total amount of MCs present in biota. Nevertheless, information on this particular aspect is urgently needed for risk assessment purposes.

REFERENCES

Adamovsky , O., Kopp, R., Hilscherova, K., Babica, P., Palikova , M., Paskova, V., et al., 2007. Microcystin kinetics (bioaccumulation and elimination) and biochemical responses in common carp (Cyprinus carpio) and silver carp (Hypophthalmichthys molitrix) exposed to toxic cyanobacterial blooms. Environ. Toxicol. Chem., 26(12): 2687-2693.

An, J., Carmichael, W.W., 1994. Use of a colorimetric protein phosphatase inhibition assay and enzyme linked immunosorbent assay for the study of microcystins and nodularins. Toxicon, 32 (12), 1495-1507.

Andersen, R.J., Luu, H.A., Chen, D.Z.X., Holmes, C.F.B., Kent, M. L., Le Blanc, M., Taylor, F.J.R.M., Williams, D.E., 1993. Chemical and biological evidence links microcystins to salmon 'netpen liver disease'. Toxicon, 31(10), 1315-1323.

Aráoz, R., Molgó, J., Tandeau de Marsac, N., 2010. Neurotoxic cyanobacterial toxins. Toxicon, 56(5), 813-828.

Banack, S.A., Cox, P.A., 2003. Distribution of the neurotoxic nonprotein amino acid BMAA in Cycas micronesica. Bot. J. Linn. Soc. 143, 165–168.

Banack, S.A., Murch, S.J., Cox, P.A., 2006. Neurotoxic flying foxes as dietary items for the Chamorro people, Marianas Islands.Journal of Ethnopharmacology, 106 (1), 97-104.

Banack, S.A., Johnson, H.E., Cheng, R., Cox, P.A., 2007. Production of the Neurotoxin BMAA by a marine cyanobacterium. Mar. Drugs 5, 180–196.

Banker, R., Teltsch, B., Sukenik, A., and Carmeli, S., 2000. 7-Epicylindrospermopsin, a toxic minor metabolite of the cyanobacterium *Aphanizomenon ovalisporum* from Lake Kinneret, Israel. *J. Natural Products,* 63, 387–389.

Bártová, K., Hilscherová, K., Babica, P., Maršálek, B., 2010. Extract of Microcystis water bloom affects cellular differentiation in filamentous cyanobacterium *Trichormus variabilis* (Nostocales, Cyanobacteria). *Journal of Applied Phycology*. DOI: 10.1007/s10811-010-9624-5.

Berry J. P., Lind O., 2010. First evidence. *Toxicon*, 55, 930-938.

Best, J.H., Pflugmacher, S., Wiegand, C., Eddy, F.B., Metcalf, J.S., Codd, G.A., 2002. Effects of enteric bacterial and cyanobacterial lipopolysaccharides, and of microcystin-LR, on glutathione S-transferase activities in zebra fish (*Danio rerio*). *Aquat. Toxicol* ., 60(3-4), 223-231.

Best, J.H., Eddy, F.B., Codd, G.A., 2003. Effects of Microcystis cells, cell extracts and lipopolysaccharide on drinking and liver function in rainbow trout *Oncorhynchus mykiss* Walbaum. *Aquat. Toxicol.,* 64(4), 419-426.

Bouaïcha, N., Maatouk, I., Vincent, G., Levi Y., 2002. A colorimetric and fluorometric microplate assay for the detection of microcystin-LR in drinking water without preconcentration. *Food and Chemical Toxicology*, 40, (11), 1677-1683

Bouaïcha N., Maatouk I. 2004. Microcystin-LR and Nodularin induce intracellular glutathione alteration, reactive oxygen species production and lipid peroxidation in primo-culture of rat hepatocytes. *Toxicology Letters*, 148, 53-63.

Bourrelly, P., 1985. Les algues d'eau douce. Tome III. – N. Boubée and Cie., Paris. 606p.

Brand L.E., Pablo J., Compton A., Hammerschlag N., Mash D.C., 2010. Cyanobacterial blooms and the occurrence of the neurotoxin, beta-N-methylamino-L-alanine. *Harmful Algae*, 9, 620-635.

Buryskova, B., Hilscherova, K., Babica, P., Vrskova, D., Marsalek, B., Blaha, L., 2006. Toxicity of complex cyanobacterial samples and their fractions in *Xenopus laevis* embryos and the role of microcystins. *Aquatic Toxicology,* 80, 346–354.

Carmichael, W.W., Falconer, I.R., 1993. Diseases related to freshwater blue-green algal toxins, and control measures. In: Falconer, I.R. (Ed.), Algal Toxins in Seafood and Drinking Water. *Academic Press*, London, pp. 187–209.

Carmichael, W.W., 1994. The toxins of cyanobacteria, *Scientific American*, 270(1), 78-86.

Cazenave, J., Wunderlin, D.A., Bistoni, M.A., Amé, M.V., Krause, E., Pflugmacher, S., Wiegand, C., 2005. Uptake, tissue distribution and accumulation of microcystin-RR in *Corydoras paleatus*, *Jenynsia multidentata* and *Odontesthes bonariensis*: A field and laboratory study.*Aquatic Toxicology*, 75 (2), 178-190.

Chen, J., Xie, P., 2005. Seasonal dynamics of the hepatotoxic microcystins in various organs of four freshwater bivalves from the large eutrophic lake Taihu of subtropical China and the risk for human consumption. *Environmental Toxicology*, 20, 572-584.

Chen, J., Xie, P., Guo, L.G., Zheng, L., Ni, L.Y., 2005. Tissue distributions and seasonal dynamics of the hepatotoxic microcystins-LR and -RR in a freshwater snail (*Bellamya aeruginosa*) from a large shallow, eutrophic lake of the subtropical China. *Environmental Pollution* 134, 423-430.

Chen, J., Xie, P., Zhang, D., Lei, H., 2007. In situ studies on the distribution patterns and dynamics of microcystins in a biomanipulation fish-bighead carp (*Aristichthys nobilis*). *Environ. Pollut.* 147, 150–157.

Chen, J., Zhang, D., Xie, P., Wang, Q., Ma, Z., 2009. Simultaneous determination of microcystin contaminations in various vertebrates (fish, turtle, duck and water bird) from a large eutrophic Chinese lake, Lake Taihu, with toxic *Microcystis* blooms. *Science of the Total Environment*, 407, 3317–3322.

Chen, J., Xie, P., Li, L., Xu, J., 2009. First identification of the hepatotoxic microcystins in the serum of a chronically exposed human population together with indication of hepatocellular damage. *Toxicological Science,* 108, 81–89.

Chorus, I., Bartram, J., 1999. Toxic cyanobacteria in water. Aguideline to their public health consequences, monitoring and management. Published on Behalf of WHO. London : E. & FN. Spon. P. 416.

Christoffersen, K., 1996. Ecological implications of cyanobacterial toxins in aquatic ecosystems. *Phycologia,* 36, 42-50.

Codd, G.A., Morrison, L.F., Metcalf, J.S., 2005. Cyanobacterial toxins: risk. *Toxicology and Applied Pharmacology*, 203(3), 264-272.

Couté, A., Bernard, C., 2001. Les cyanobactéries toxiques. In : Frémy J.-M.and Lassus P. (Eds), Toxines d'algues dans l'alimentation.. Plouzané, Editions Ifremer. p. 21-37.

Cox, P.A., Banack, S.A., Murch, S.J., 2003. Biomagnification of cyanobacterial neurotoxins and neurodegenerative disease among the Chamorro people of Guam. *Proceedings of the National Academy of Sciences USA* 100, 3380-3383.

Cox, P. A., Banack, S. A., Murch, S. J., Rasmussen, U., Tien, G., Bidigare, R. R., Metclaf, J. S., Morrison, L. F., Codd, G. A., Bergman, B., 2005. Diverse taxa of cyanobacteria produce ß-N-methylamino-L-alanine, a neurotoxic amino acid. *Proceedings of the National Academy of Sciences - PNAS*, 102, 5074–5078.

Deeds, J.R., Landsberg, J.H., Etheridge, S.M., Pitcher, G.C., and Watt Longan, S., 2008. Non-traditional vectors for paralytic shellfish poisoning. *Marine Drugs*, 6, 308-348.

Del Campo, F.F., Ouahid, Y., 2010. Identification of microcystins from three collection strains of *Microcystis aeruginosa. Environmental Pollution,* 158, 2906-2914.

Deng , D , F, Zheng K, Teh F C, Lehman W P , Teha J S. 2010. Toxic threshold of dietary microcystin (-LR) for quart medaka. *Toxicon*, 55, 787–794.

Dionisio Pires, L.M., Karlsson, K.M., Meriluoto, J.A.O., Kardinaal, E., Visser, P.M., Siewertsen, K., Van Donk, E., Ibelings, B.W., 2004. Assimilation and depuration of microcystin–LR by the zebra mussel, *Dreissena polymorpha. Aquatic Toxicology*, 69 (4), 385-396.

Dong, G.F., Zhu, X.M., Han, D., Yang, Y.X., Song, L.R., Xie, S.Q., 2009. Effects of dietary cyanobacteria of two different sources on growth and recovery of hybrid tilapia (*Oreochromis niloticus* x *O. aureus*). *Toxicon,* 54(3), 208-216.

Draisci, R., Ferretti, E., Palleschi, L., Marchiafava, C., 2001. Identification of anatoxins in blue-green algae food supplements using liquid chromatography-tandem mass spectrometry. *Food Addit. Contam*., 18(6), 525–31.

El Ghazali, I., Saqrane, S., Carvalho, A, P., Ouahid, Y., Oudra, B., Del Campo, F, F., Vasconcelos, V., 2009. Compensatory Growth Induced in Zebrafish Larvae after Pre-Exposure to a *Microcystis aeruginosa* Natural Bloom Extract Containing Microcystins. *Int. J. Mol. Sci,* 10, 133-146.

El Ghazali, I., Saqrane, S., Carvalho, A.P., Ouahid, Y., Del Campo, F.F., Oudra, B., Vasconcelos, V., 2010. Effect of different microcystin profiles on toxin bioaccumulation in common carp (*Cyprinus carpio*) larvae via *Artemia* nauplii. *Ecotoxicology and Environmental Safety*, 73(5), 762-770.

Engstrom, J., Koski, M., Viitasalo, M., Reinikainen, M., Repka, S. *et al.*, 2000. Feeding interactions of the copepods *Eurytemora affinis* and *Acartia bifilosa* with the cyanobacteria *Nodularia* sp. *J. Plankton. Res.,* 22(7), 1403–1409.

Engström, J., Lehtiniemi, M., Green, S., Kozlowsky-Suzuki, B., Viitasalo, M., 2001. Does cyanobacterial toxin *Journal of Experimental Marine Biology and Ecology*, 276, 95-107.

Eriksson, J.E., Meriluoto, J.A.O., Lindholm, T., 1989. Accumulation of a peptide toxin from the cyanobacterium *Oscillatoria agardhii* in the freshwater mussel *Anadonta cygnea. Hydrobiologia*, 183, 211 –216.

Esterhuizen M., Downing T.G., 2008. *β-N*-methylamino-L-alanine. *Ecotoxicology and Environmental Safety*, 71, 309-313.

Falconer, I.R., 1993. Measurement of toxins from blue-green algae in water and foodstuffs. In: *Falconer, I.R. (Ed.)*, Algal Toxins in Seafood and Drinking Water. Academic Press, London, pp. 165– 175.

Falconer, I.R., 1996. *I. Phycologia* 1996, 35 ŽSuppl 6, 6-11.

Falconer, I.R., Humpage, A.R., 1996. Tumour promotion by cyanobacterial toxins. *Phycologia*, 35, 74–79.

Falconer, I.R., Hardy, S.J., Humpage, A.R., Froscio, S.M., Tozer, G.J., Hawkins, P.R., 1999. Hepatic and renal toxicity of the blue-green alga (Cyanobacterium) *Cylindrospermopsis raciborskii* in male Swiss Albino mice. *Environ. Toxicol.* 14, 143–150.

Falconer, I.R., Humpage, A.R., 2005. Health risk assessment of cyanobacterial (blue-green algal) toxins in drinking water. *Int. J. Environ. Res. Public Health,* 2, 43-50.

Fastner, J., Flieger, I., Neumann, U., 1998. Optimised extraction of microcystins from field samples—a comparison of different solvents and procedures. *Water Res.* 32, 3177–3181.

Filho, F. A., Maria Soares, S. C., Magalhaes, V. de F., Sandra, M.F.O. Azevedoc, 2009. Biomonitoring of cyanotoxins in two tropical reservoirs by cladoceran toxicity bioassays. *Ecotoxicology and Environmental Safety,* 72, 479–489.

Fischer, W.J., Dietrich, D.R., 2000. Pathological and biochemical characterization of icrocystininduced hepatopancreas and kidney damage in carp (*Cyprinus carpio*). *Toxicol. Appl. Pharmacol.,* 164(1), 73-81.

Fischer, W.J., Hitzfeld, B.C., Tencalla, F., Eriksson, J.E., Mikhailov, A., Dietrich, D.R., 2000. Microcystin-LR toxicodynamics, induced pathology, and immunohistochemical localization in livers of blue-green algae exposed rainbow trout (*Oncorhynchus mykiss*). *Toxicol. Sci.* 54(2), 365-373.

Fontal, O.I., Vieytes,M.R.,de Sousa,J.,Louzao, M.C.,Botana, L.M.,1999. A fluorescent microplate assay for microcystin-LR. *Analytical Biochemistry* 269,289–296.

Fournie, J.W., Courtney, L.A., 2002. Histopathological evidence of regeneration following hepatotoxic effects of the cyanotoxin microcystin-LR in the hardhead catfish and gulf killifish. *J. Aquat. Anim. Health*, 14(4), 273-280.

Francis, G., 1878. Poisonous Australian lake. *Nature,* 18,11–12.

Galvão J.A., Oetterer M., Bittencourt-Oliveira M.C., Gouvêa-Barros S., Hiller S., Erler K., Luckas B., Pinto E., Kujbida P., 2009. Saxitoxins accumulation by freshwater. *Toxicon*, 54, 891-894.

Garcia A. C., Bargu S., Dash P., Rabalais N. N., Sutor M., Morrison W., Walker N. D., 2010. Evaluating the potential risk. *Harmful Algae*, 9, 134-143.

Gérard, C., Virginie, P., Lance, E., Acou, A., Brient, L. , Carpentier, A., 2009. Influence of toxic cyanobacteria on community structure and microcystin accumulation of freshwater molluscs. *Environmental Pollution*, 157, 609–617.

Grosse, Y., Baan, R., Straif, K., Secretan, B., El Ghissassi, F., Cogliano, V. and on behalf of the WHO International Agency for Research on Cancer Monograph Working Group, 2006. Carcinogenicity of nitrate, nitrite, and cyanobacterial peptide toxins. *The Lancet Oncology*, 7, 628-629.

Guo, T., Geis, S., Hedman, C., Arndt, M., Krick, W., Sonzogni, W., 2007. Characterization of ethyl chloroformate derivative of b-methylamino- L-alanine. *J. Am. Soc. Mass Spectrom.* 18, 817–825.

Haider, S. V., Naithani, P.N., Viswanathan, P., Kakkar, 2003. Cyanobacterial toxins: a growing environmental concern. *Chemosphere* 52: 1-21.

Haney, J. F., Sasner, J. J. and Ikawa, M., 1995. Effects of products released by Aphanizomenon flos-aquae and purified saxitoxin on the movements of *Daphnia carinata* feeding appendages. *Limnology and Oceanography*, 40, 263-272.

Harada, K.I., Nagai, H., Kimura, Y., Suzuki, M., Park, H., Watanabe, M.F., *et al.,* 1993. Liquid chromatography/mass spectrometric detection of anatoxin-a, a neurotoxin from cyanobacteria. *Tetrahedron,* 49(41), 9251–60.

Harada, K.-I., Oshikata, M., Shimada, T., Nagata, A., Ishikawa, N., Suzuki, M., Kondo, F., Shimizu, M., Yamada, S., 1997. High performance liquid separation of microcystins derivatized with a highly fluorescent dienophile. *Nat. Toxins*, 5, 201–207.

Henriksen, P., Carmichael, W.W., An, J., Moestrup, Ø., 1997. Detection of an anatoxin-a(s)-like anticholinesterase in natural blooms and cultures of cyanobacteria/blue-green algae from danish lakes and in the stomach contents of poisoned birds. *Toxicon*, 35 (6), 901-913.

Hitzfeld, B.C., Lampert, C.S., Spaeth, N., Mountfort, D., Kaspar, H., Dietrich, D.R., 2000. Toxin production in cyanobacterial mats from ponds on the McMurdo Ice Shelf, Antarctica. *Toxicon,* 38, 1731-1748.

Holte, H.R., Eriksen, S., Skulberg, O., Aas, P., 1998. The effect of water soluble cyanotoxin(s) produced by two species of *Anabaena* on the release of acetylcholine from the peripheral cholinergic nervous system of the rat airway. *Environmental Toxicology and Pharmacology,* 5, 51-59.

Hormazabal, V., Ostensvik, O., Underdal, B., Skulberg, O.M., 2000. Simultaneous determination of the cyanotoxins anatoxin-a, microcystin desmethyl-3, LR, RR, and YR in fish muscle using liquid chromatography-mass spectrometry. *J. Liq. Chromatogr. Rel. Technol.,* 23(2), 185–96.

Hu, Z.-Q., Liu, Y.-D., Li, D.-H., Dauta, A., 2005. Growth and antioxidant system of the cyanobacterium *Synechococcus elongatus* in response to microcystin-RR. *Hydrobiologia,* 534, 23–29.

Humpage, A. R., Rositano, J., Bretag, A. H., Brown, R., Baker, P. D., Nicholson, B. C. and Steffensen, D. A., 1994. Paralytic shellfish poisons from Australian cyanobacterial blooms. *Aust. J. Mar. Freshwater Res.*, 45, 761-771.

Humpage, A.R., Falconer, I.R., 1999. Microcystin-LR and liver tumor promotion: effects on cytokinesis, ploidy, and apoptosis in cultured hepatocytes. Environ Toxicol, 14:-1–75.

Huynh-Delerme, C., Edery, M., Huet, H., Puiseux-Dao, S., Bernard, C., Fontaine, J.J., *et al.,* 2005. Microcystin-LR and embryo-larval development of medaka fish, *Oryzias latipes*. I. Effects on the digestive tract and associated systems. *Toxicon,* 46(1), 16-23.

Ibelings, B.W., Bruning, K., de Jonge, J., Wolfstein, K., Pires, L.M.D., Postma, J., Burger, T., 2005. Distribution of microcystins in a lake foodweb: No evidence for biomagnification. *Microbial Ecology*, 49, 487-500.

Ibelings, B.W., Chorus, I. 2007. Accumulation of cyanobacterial toxins in freshwater ''seafood'' and its consequences for public health: A review. *Environmental Pollution,* 150, 177-192.

Ibelings, B.W., Havens, K.H., 2007. Cyanobacterial toxins: a qualitative meta-analysis of concentrations, dosage and effects in freshwater, estuarine and marine biota. In: Hudnell, K.H. (Ed.), Proceedings of the Interagency, International Symposium on Cyanobacterial Harmful Algal Blooms. *Advances in Experimental Medicine and Biology*, 675-732.

Ibelings, B. W., Havens, K. E., 2008. Cyanobacterial toxins: a qualitative meta-analysis of concentrations, dosage and effects in freshwater, estuarine and marine biota. *Advances in Experimental Medicine and Biology*, 619, 675-732.

Ikawa, M., Wegener, K., Foxall, T.L., Sasner, J.J., 1982. Comparison of the toxins of the blue-green alga *Aphanizomenon flos-aquae* with the Gonyaulax toxins. *Toxicon,* 20, 747-752.

Jacquet, C., Thermes, V., Luze, A. , Daoa, P. S., Bernarda, C., Jolyb, J. S., Bourrat, F., Ederya, M ., 2004. Effects of microcystin-LR on development of medaka fish embryos (*Oryzias latipes*). *Toxicon*, 43, 141–147.

James, K.J., Furey, A., Sherlock, I.R., Stack, M.A., Twohig, M., Caudwell, F.B., *et al.,* 1998. Sensitive determination of anatoxin-a, homoanatoxin-a and their degradation products by liquid chromatography with fluorimetric detection. *J. Chromatogr. A.,* 798(1–2),1 47–57.

Jang, M.H., Ha, K., Takamura, N., 2007. Reciprocal allelopathic responses between toxic cyanobacteria (*Microcystis aeruginosa*) and duckweed (*Lemna japonica*). *Toxicon*, 49, 727-733.

Jann, K., Jann, B., 1984. Structure and biosynthesis of 0-antigens. In: *Handbook of endotoxin. Vol.1. Chemistry of endotoxin.* (ed. E.T. Rietschel).Elsevier,Amsterdam. pp.138-186.

Jonasson, S., Vintila, S., Sivonen, K., El-Shehawy, R., 2008. Expression of the nodularin synthetase genes in the Baltic Sea bloom-former cyanobacterium *Nodularia spumigena* strain AV1. *FEMS Microbiology Ecology*, 65, 31–39.

Johnson, C.H., Mori, T. and Xu, Y., 2008. A Cyanobacterial Circadian Clockwork. *Current Biology*, 18, R816-R825.

Kaebernick, M., Neilan B.A., 2001. Ecological and molecular investigations of cyanotoxin production. *FEMS Microbiology Ecology*, 35, 1-9.

Karamyan V.T., Speth R.C., 2008. Animal models . *Life Sciences*, 82, 233-246.

Kaya, K., 1996. Toxicology of microcystins. In: Watanabe, M.F., Harada, H., Carmichael, W.W., Fujiki (Eds.), Toxic *Microcystis*. H. Press. Inc., CRC, pp. 175–202.

Keil, C., Forchert, A., Fastner, J., Szewzyk, U., Rotard, W., Chorus, I., Krätke, R., 2002. Toxicity and microcystin content of extracts from a *Planktothrix* bloom and two laboratory strains. *Water Research*, 36 (8), 2133-2139.

Kent, M.L., Dawe, S.C., Hilaire, S.St., Anderson, R.J., 1996. Effects of feeding rate, seawater entry, and exposure to natural biota on the severity of net-pen liver disease among pen-reared Atlantic salmon, *Progress. Fish Cult.*, 58, pp. 43–46.

Kinnear, S.H.W., Duivenvoorden, L.J., Fabbro, L.D., 2007. Sublethal responses in *Melanoides tuberculata* following exposure to *Cylindrospermopsis raciborskii* containing cylindrospermopsin. *Harmful Algae*, 6, 642-650.

Korpinen, S., Karjalainen, M., Viitasalo, M., 2006. Effects of cyanobacteria on survival and reproduction of the littoral crustacean *Gammarus zaddachi* (Amphipoda), *Hydrobiologia*, 559, pp. 285–295.

Kotak, B.G., Zurawell, R.W., Prepas, E.E., Holmes, C.F.B., 1996. Microcystin–LR concentration in aquatic food web compartments from lakes of varying trophic status. *Can. J. Fish Aquat. Sci.,* 53(9), 1974–1985.

Krienitz, L., Ballot, A., Kotut, K., Wiegand, C., Pütz, S., Metcalf, J.S., Codd, G.A., Pflugmacher, S., 2003.Contribution of hot spring cyanobacteria to the mysterious deaths of Lesser Flamingos at Lake Bogoria, Kenya. *FEMS Microbiology Ecology*, 43(2), 141-148.

Kubo, T., Kato, N., Hosoya, K., Kaya, K., 2008. Effective determination method for a *cyanobacterial* neurotoxin, beta-N-methylamino-l-alanine. *Toxicon*, 51,1264–1268.

Kuiper-Goodman, T., Falconer, I., Fitzgerald, J.,1999. Human health aspects. In: Chorus, I., Bartram, J. (Eds.), Toxic cyanobacteria in water: A guide to their public health consequences, monitoring and management. World Health Organization, Taylor and Francis, London and New York, pp. 113–153.

Kurmayer, R., 2001. Competitive ability of Daphnia under dominance of nontoxic filamentous cyanobacteria. *Hydrobiologia,* 442, 279-289.

Lance E., Neffling M.R., Gérard C., Meriluoto J., Bormans M., 2010. Accumulation of free and covalently bound microcystins in tissues of *Lymnaea stagnalis* (Gastropoda) following toxic cyanobacteria or dissolved microcystin-LR exposure. *Environmental Pollution*, 158, 674-680.

Lawton, L. A., Beattie, K. A., Hawser, S. P., Campbell, D. L., Codd, G. A., 1994. Evaluation of assay methods for the determination of cyanobacterial hepatotoxicity. In *Detection Methods for Cyanobacterial Toxins.* (G. A. Codd, T. M. Jefferies, C. W. Keevil and E. Potter, ed.), pp. 111–116. The Royal Society of Chemistry, Cambridge, UK.

LeBlanc, S., Pick, F.R., Aranda-Rodriguez, R., 2005. Allelopathic Effects of the Toxic Cyanobacterium *Microcystis aeruginosa* on Duckweed, *Lemna gibba* L. *Environ. Toxicol.*, 20, 67.

Lehtonen, K.K., Kankaanpää, H., Leiniö, S., Sipiä, V.O., Pflugmacher, S., Sandberg-Kilpi, E., 2003. Accumulation of nodularin-like compounds from the cyanobacterium *Nodularia spumigena* and changes in acetylcholinesterase activity in the clam *Macoma balthica* during short-term laboratory exposure. *Aquatic Toxicology*, 64 (4), 461-476.

Li, R., Carmichael, W.W., Brittain, S, Eaglesham, G.K., Shaw, G.R., Mahakhant, A., Noparatnaraporn, N., Yongmanitchai, W., Kaya, K., Watanabe, M.M., 2001. Isolation and identification of the cyanotoxin cylindrospermopsin and deoxy-cylindrospermopsin from a Thailand strain of *Cylindrospermopsis raciborskii* (Cyanobacteria). *Toxicon*, 39, 973-980.

Li, X.Y., Chung, I.K., Kim, J.I., Lee, J.A., 2004. Subchronic oral toxicity of microcystin in common carp (*Cyprinus carpio* L.) exposed to *Microcystis* under laboratory conditions. *Toxicon,* 44, 821-827.

Li, L., Xie, P., Chen, J., 2007. Biochemical and ultrastructural changes of the liver and kidney of the phytoplanktivorous silver carp feeding naturally on toxic *Microcystis* blooms in Taihu Lake, China. *Toxicon,* 49(7), 1042-1053.

Li, L., Xie, P., Li, S., Qiu, T., Longgen, Guo, L. 2007. Sequential ultrastructural and biochemical changes induced in vivo by the hepatotoxic microcystins in liver of the phytoplanktivorous silver carp *Hypophthalmichthys molitrix. Comparative Biochemistry and Physiology, Part C*, 146, 357–367.

Li, L., Xie, P., Guo L., 2008. Antioxidant response in liver of the phytoplanktivorous bighead carp (*Aristichthys nobilis*) intraperitoneally- injected with extracted microcystins. *Fish. Physiol Biochem.*, 36 (2), 165-172, DOI: 10.1007/s10695-008-9228-z.

Li, D., Li, G., Chen, W., Liu, Y., 2009. Interactions between a cyanobacterial bloom (*Microcystis*) and the submerged aquatic plant *Ceratophyllum oryzetorum* Kom. Chinese *Journal of Oceanology and Limnology,* 27, 38-42.

Li, D., Xie, P., Zhang, Xa., Zhao, Y., 2009. Intraperitoneal injection of extracted microcystins results in hypovolemia and hypotension in crucian carp (*Carassius auratus*). *Toxicon,* 53, 638–644.

Li, A., Tian, Z., Li, J., Yu, R., Banack, S.A., Wang, Z., 2010. Detection of the neurotoxin BMAA within cyanobacteria isolated from freshwater in China.*Toxicon*, 55, (5), 947-953.

Lindholm, T., Ôhman, P., Kurkihelasmo, K., Kincaid, B., Meriluoto, J., 1999. Toxic algae and fish mortality in a brackish-water lake in Âland, SW Finland. *Hydrobiologia,* 397,109-120.

Liu, Y.D., Song, L.R., Li, X.Y., Liu, T.M., 2002. The toxic effects of microcystin-LR on embryo-larval and juvenile development of loach, *Misguruns mizolepis* Gunthe. *Toxicon*, 40(4), 395-399.

Liu, Y., Xie, P., Wu, X.P., 2006. Effects of toxic and nontoxic *Microcystis aeruginosa* on survival, population-increase, and feeding of two small cladocerans. Bull. Environ. Contam. Toxicol. 77, 566–573.

Lugomela, C., Pratap, B.H., Mgaya, D. Y., 2006. Cyanobacteria blooms—A possible cause of mass mortality of Lesser Flamingos in Lake Manyara and Lake Big Momela, Tanzania. *Harmful Algae,* 5, 534–541.

MacKintosh, C., Beattie, K.A., Klumpp, S., Cohen, P., Codd, G.A., 1990. Cyanobacterial microcystin-LR is a potent and specific inhibitor of protein phosphatases 1 and 2A from both mammals and higher plants. *FEBS Lett*., 264, 187–192.

Magalhães, V.F., Soares, R.M., Azevedo, S., 2001. Microcystin contamination in fish from the Jacarepagua Lagoon (Rio de Janeiro, Brazil): ecological implication and human health risk. *Toxicon,* 39, 1077-1085.

Mahmood, N.A., Carmichael, W. W., 1986. Paralytic shellfish. *Toxicon*, 24, 175-186.

Malbrouck, C., Trausch, G., Devos, P., Kestemont, P., 2003. Hepatic accumulation and effects of microcystin-LR on juvenile goldfish *Carassius auratus* L. *Comp. Biochem. Physiol. Cytoxicol. Pharmacol*., 135(1), 39-48.

Malbrouck, C., Kestemont, P., 2006. Effects of microcystins on fish. *Environ. Toxicol. Chem.* 25(1), 72-86.

Mares, J., Palikova, M., Kopp, R., Navratil, S., Pikula, J., 2009. Changes in the nutritional parameters of muscles of the common carp (*Cyprinus carpio*) and the silver carp (*Hypophthalmichthys molitrix*) following environmental exposure to cyanobacterial water bloom. *Aquac. Res*. 40(2), 148-156.

Mayer, H., and Weckesser J., 1984. 'Unusual' lipid A's: structures, taxonomical relevance and potential value for endotoxin research, p. 221-247. In E. T. Rietschel (ed.), Handbook of endotoxin, vol. 1: chemistry of endotoxin, Elsevier Science Publishing, Inc., New York.

McDermott, C. M., Feola, R., Plude, J., 1995. Detection of cyanobacterial toxins (microcystins) in waters of northeastern wisconsin by a new immunoassay technique. *Toxicon*, 33, (11), 1433-1442.

McElhiney, J., Lawton, L.A., 2005. Detection of the cyanobacterial hepatotoxins microcystins. *Toxicol. Appl. Pharmacol.*, 203, 219–230.

Metcalf, J.S., Lindsay, J., Beattie, K.A., Birmingham, S., Saker, M.L., Törökné, A.K., Codd, G.A., 2002. Toxicity of cylindrospermopsin to the brine shrimp *Artemia salina*: comparisons with protein synthesis inhibitors and microcystins. *Toxicon*, 40 (8), 1115-1120.

Metcalf, J.S., Banack, S.A., Lindsay, J., Morrison, L.F., Cox, P.A., Codd, G.A., 2008. Cooccurrence of b-N-methylamino-L-alanine, a neurotoxic amino acid with other cyanobacterial toxins in British waterbodies, 1990–2004. *Environ. Microbiol.,* 10, 702–708.

Miller, M.A., Kudela, R.M., Mekebri, A., Crane, D., Oates, S.C., Tinker, M.T., Staedler, M., Miller, W.A., Toy-Choutka, S., Dominik, C., Hardin, D., Langlois, G., Murray, M., Ward, K., Jessup, D.A. 2010. Evidence for a novel marine harmful algal bloom: cyanotoxin (microcystin) transfer from land to sea. *Plos One*, 5(9), e12576.

Mitrovic, S.M., Allis, O., Furey, A., James, K.J., 2005. Bioaccumulation and harmful effects of microcystin-LR in the aquatic plants *Lemna minor* and *Wolffia arrhiza* and the filamentous alga *Chladophora fracta*. *Ecotoxicology and Environmental Safety*, 61, 345-352.

Mohamed, Z.A., Carmichael, W.W., Hussein, A.A., 2003. Estimation of microcystins in the freshwater fish *Oreochromis niloticus* in an Egyptian fish farm containing a Microcystis bloom. *Environmental Toxicology,* 18, 137-141.

Mohamed, Z.A., 2008. Polysaccharides as a protective response against microcystin induced oxidative stress in *Chlorella vulgaris* and *Scenedesmus quadricauda* and their possible significance in the aquatic ecosystem. *Ecotoxicology*, 17, 504–516.

Monson CS, Banack SA, Cox, PA., 2003. Conservation Implications of Chamorro Consumption of Flying Foxes as a Possible Cause of Amyotrophic Lateral Sclerosis–Parkinsonism Dementia Complex in Guam. *Conservation Biology*, 17(3), 678–686.

Montagnolli, W., Zamboni, A., Luvizotto-Santos, R., Yunes, J.S., 2004. Acute effects of *Microcystis aeruginosa* from the Patos Lagoon estuary, southern Brazil, on the microcrustacean *Kalliapseudes schubartii* (Crustacea: Tanaidacea), *Arch. Environ. Contam. Toxicol.*, 46, pp. 463–469.

Murch, S.J., Cox, P.A., Banack, S.A., 2004. A mechanism for slow release of biomagnified cyanobacterial neurotoxins and neurodegenerative disease in Guam. *Proc. Natl. Acad. Sci. U.S.A.* 101, 12228–12231.

Msagati, T.A.M., Siame, B.A., Shushu, D.D., 2006. Evaluation of methods for the isolation, detection and quantification of cyanobacterial hepatotoxins. *Aquatic Toxicology*, 78 (4), 382-39.

Nam, S., Joo, S., Kim, S., Baek, N.-I., Choi, H.-K., Park, S., 2008. Induced Metabolite Changes in *Myriophyllum spicatum* during Co-existence Experiment with the Cyanobacterium *Microcystis aeruginosa. Journal of Plant Biology*, 51, 373-378.

Namikoshi, M., Murakami, T.,Watanabe, M.F., Oda, T., Yamada, J., Tsujimura, S., *et al.,* 2003. Simultaneous production of homoanatoxin-a, anatoxin-a, and a new nontoxic 4-hydroxyhomoanatoxin-a by the cyanobacterium *Raphidiopsis mediterranea* Skuja. *Toxicon*, 42(5), 533–8.

Nasri H., El Herry S., Bouaïcha N., 2008. First reported case of turtle. *Ecotoxicology and Environmental Safety*, 71, 535-544.

Neffling, M.R., Spoof, L., Quilliam, M., Meriluoto, J., 2010. LC–ESI-Q-TOF-MS for faster and accurate determination of microcystins and nodularins in serum. *Journal of Chromatography B*, 878 (26), 2433-2441.

Negri, A.P., Jones, G.J., 1995. Bioaccumulation of paralytic shellfish poisoning (Psp) toxins from the cyanobacterium *Anabaena circinalis* by the fresh-water mussel *Alathyria condola. Toxicon,* 33, 667- 678.

Nishiwaki-Matsushima, R., Ohta, T., Nishiwaki, S. et al., 1992. Liver tumor promotion by the cyanobacterial cyclic peptide toxin microcystin-LR. *J. Cancer Res. Clin. Oncol.,* 118, 420–424.

Nogueira, I.C.G., Pereira,P., Dias, E., Pflugmacher, S., Wiegand, C., Franca, S., Vasconcelos, V., 2004. Accumulation of paralytic shellfish. *Toxicon*, 44, 773-780.

Norris, R.L., Eaglesham, G.K., Pierens, G., Shaw, G.R., Smith, M.J., Chiswell, R.K., Seawright, A.A. and Moore, M.R., 1999. Deoxycylindrospermopsin, an analog of cylindrospermopsin from *Cylindrospermopsis raciborskii. Environmental Toxicology,* 14, 163-165.

Oberemm, A., Becker, J., Codd, G.A., Steinberg, C.E.W., 1999. Effects of cyanobacterial toxins and aqueous crude extrakt of cyanobacteria on the development of fish and amphibians. *Environ. Toxicol.*, 14, 77–88.

Ohtani, I., Moore, R.E., Runnegar, M.T.C., 1992. Cylindrospermopsin, a potent hepatotoxin from the blue-green alga *Cylindrospermopsis raciborskii. J. Amer. Chem. Soc.,* 114, 7941-7942.

Osswald, J., Rellán, S., Gago, A., Vasconcelos, V., 2007. Toxicology and detection methods of the alkaloid neurotoxin produced by cyanobacteria, anatoxin-a. *Environment International*, 33, (8), 1070-1089.

Ott, J.L., Carmichael, W.W., 2006. LC/ESI/MS method development for the analysis of hepatotoxic cyclic peptide microcystins in animal tissues. *Toxicon* 47, 734–741.

Palikova, M., Navratil, S., Marsalek, B., Blaha,, L. 2003. Toxicity of crude extract of cyanobacteria for embryos and larvae of carp (*Cyprinus carpio* L.). *Acta. Vet. BRNO,* 72(3), 437-443.

Palikova, M., Navratil, S., Tichy, F., Sterba, F., Marsalek, B., Blaha, L., 2004. Histopathology of carp (*Cyprinus carpio* L.) larvae exposed to cyanobacteria extract. *Acta. Vet. BRNO,* 73(2), 253-257.

Papapetropoulos, S., 2007. Is there a role for naturally occurring cyanobacterial toxins in neurodegeneration? The beta-N-methylamino-L-alanine (BMAA) paradigm. *Neurochem. Int.* 50, 998–1003.

Park, H.-D., Watanabe, M.F., Harada, K.-I., Suzuki, M., Hayashi, H. and Okino, T. 1993. Seasonal variations of *Microcystis* species and toxic heptapeptide microcystins in Lake Suwa. *Environ. Toxicol. Water Qual.,* 8, 425-435.

Pereira P., Onodera H., Andrinolo D., Franca S., Araújo F., Lagos N., Oshima Y., 2000. Paralytic shellfish. *Toxicon*, 38, 1689-1702.

Pereira, P., Dias, E., Franca, S., Pereira, E., Carolino, M., 2004. Accumulation and depuration of cyanobacterial paralytic shellfish toxins by the freshwater mussel *Anodonta cygnea. Aquatic Toxicology,* 68, 339-350.

Peng L., Liu Y., Chen W., Liu L., Kent M., Song L., 2010. Health risks . *Ecotoxicology and Environmental Safety*, 73, 1804-1811.

Pflugmacher, S., Wiegand, C., Beattie, K.A., Krause, E., Steinberg, C.E.W., Codd, G.A., 2001. Uptake, effects and metabolism of cyanobacterial toxins in the emergent reed plant *Phragmites australis* (CAV.) Trin. Ex Steud. *Environmental Toxicology and Chemistry,* 20, 846–852.

Pflugmacher, S., 2002. Possible Allelopathic Effects of Cyanotoxins, with Reference to Microcystin-LR, in Aquatic Ecosystems. *Wiley InterScience,* 17, 407–413.

Pietsch, C., Wiegand, C., Amé, M.V., Nicklisch, A., Wunderlin, D., Pflugmacher, S., 2001. The Effects of a Cyanobacterial Crude Extract on Different Aquatic Organisms: Evidence for Cyanobacterial Toxin Modulating Factors. *John Wiley and Sons, Inc. Environ. Toxicol.*, 16, 535–542.

Pinho, G.L.L., Moura da Rosa, C., Yunes, J.S., Luquet, C.M., Bianchini, A., Monserrat, J.M., 2003. Toxic effects of microcystins in the hepatopancreas of the estuarine crab *Chasmagnathus granulatus* (Decapoda, Grapsidae). Comparative Biochemistry and Physiology Part C: *Toxicology and Pharmacology,* 135, 459-468.

Pomati, F., Manarolla, G., Rossi, O., Vigetti, D., Rossetti, C., 2001.The purine degradation . *Environment International*, 27, 463-470.

Poon, G.K., Griggs, L.J., Edwards, C., Beattie, K.A., Codd, G.A., 1993. Liquid chromatography- electrospray ionization-mass spectrometry of cyanobacterial toxins. *J. of Chromatogr.,* 628(2), 215–33.

Prepas, E.E., Kotak, B.G., Campbell, L.M., Evans, J.C., Hrudey, S.E. *et al.,* 1997. Accumulation and elimination of cyanobacterial hepatotoxins by the freshwater clam *Anodonta grandis simpsoniana*. *Can. J. Fish Aquat. Sci.* 54(1), 41–46.

Prieto, A., I., Pichardo, S., Jos, A ., Moreno, I., Cameàn, A.,M., 2007. Time-dependent oxidative stress responses after acute exposure to toxic cyanobacterial cells containing microcystins in tilapia fish (*Oreochromis niloticus*) under laboratory conditions. *Aquatic Toxicology,* 84,337–345.

Qiu, T., Xie, P., Guo, L.G., Zhang, D.W., 2009. Plasma biochemical responses of the planktivorous filter-feeding silver carp (*Hypophthalmichthys molitrix*) and bighead carp (*Aristichthys nobilis*) to prolonged toxic cyanobacterial blooms in natural waters. *Environ. Toxicol. Pharmacol.,* 27(3), 350-356.

Rapala, J., Lahti, K., Sivonen , K., Niemelä, S.I., 1994. Biodegradability and adsorption on lake sediments of cyanobacterial hepatotoxins and anatoxin-a. *Lett. App. Microbiol.,* 19, 423-8.

Rellán, S., Osswald, J., Vasconcelos, V., Gago-Martinez, A., 2007. Analysis of anatoxin-a in biological samples using liquid chromatography with fluorescence detection after solid phase extraction and solid phase microextraction. *Journal of Chromatography A.,*1156, 134–40.

Rodger, H.D., Turnbull, T., Edwards, C., Codd, G.A., 1994. Cyanobacterial (blue-green algal) bloom associated pathology in brown trout, Salmo trutta L., in Loch Leven, Scotland. J. Fish Dis. 17, 177–181.

Rohrlack, T., Christoffersen, K., Dittmann, E., Nogueira, I., Vasconcelos, V., 2005. Ingestion of microcystins by *Daphnia*: Intestinal uptake and toxic effects. *Limnology and Oceanography* 50, 440-448.

Rosen, J., Hellenäs, K.E., 2008. Determination of the neurotoxin BMAA (beta-Nmethylamino- L-alanine) in cycad seed and cyanobacteria by LC-MS/MS (liquid chromatography tandem mass spectrometry). *Analyst. (Lond),* 133, 1785–1789.

Runnegar, M. T., Xie, C., Snider, B. B., Wallace, G. A., Weinreb, S. M., and Kuhlenkamp, J., 2002. In vitro hepatotoxicity of the cyanobacterial alkaloid cylindrospermopsin and related synthetic analogues. *Toxicol. Sci.,* 67, 81–87.

Saker, M.L., Eaglesham, G.K., 1999. The accumulation of cylindrospermopsin from the cyanobacterium *Cylindrospermopsis raciborskii* in tissues of the Redclaw crayfish *Cherax quadricarinatus*. *Toxicon*, 37 (7), 1065-1077.

Saker, M. L., Metcalf, J.S., Codd, G.A., Vasconcelos, V., 2004. Accumulation and depuration of the cyanobacterial toxin cylindrospermopsin in the freshwater mussel *Anodonta cygnea. Toxicon*, 43 (2), 185-194.

Saker, M.L., Jungblut, A.D., Neilan, B.A., Rawn, D.F.K., Vasconcelos, V.M., 2005. Detection of microcystin synthetase genes in health food supplements containing the freshwater cyanobacterium *Aphanizomenon flosaquae*. *Toxicon*, 46, 555-562.

Saqrane, S., Elghazali, I., Ouahid, Y., ElHassni, M., ElHadrami, I., Bouarab, L., del Campo, F.F., Oudra, B., Vasconcelos, V., 2007. Phytotoxic effects of cyanobacteria extract on the aquatic plant *Lemna gibba*: Microcystin accumulation, detoxication and oxidative stress induction. *Aquatic Toxicology,* 83, 284-294.

Saqrane, S., El ghazali, I., Oudra, B., Bouarab, L., Dekayir, S., Mandi, L., Ouazzani, N., Vasconcelos, V.M., 2009. Detection of microcystin contamination by the measurement of the variability of the in vivo chlorophyll fluorescence in aquatic plant *Lemna gibba. Toxicon*, 53, 9-14.

Sasner Jr., J. J., Ikawa, M., Foxall, T. L., 1984. Studies on *Aphanizomenon* and *Microcystis* toxins. In: E. P. Ragelis (ed.) Seafood Toxins. *American Chemical Society*: Washington, D.C., pp. 391-406.

Seawright, A.A., Nolan, C.C., Shaw, G.R., Chiswell, R.K. Norris, R.L., Moore, M.R. and Smith, M.J., 1999. The oral toxicity for mice of the tropical cyanobacterium *Cylindrospermopsis raciborskii* (Woloszynska). *Environmental Toxicology*, 14, 135-142.

Sedmak, B., Kosi, G., 1998. The role of microcystins in heavy cyanobacterial bloom formation. *Journal of Plankton Research*, 20, 691–708.

Sedmak, B., Elersek, T., 2005. Microcystins Induce Morphological and Physiological Changes in Selected Representative Phytoplanktons. *Microbial Ecology*, 50, 298–305.

Shen P. P., Shi Q., Hua Z. C., Kong F. X., Wang Z. G., Zhuang S. X., Chen D. C., 2003. Analysis of microcystins in cyanobacteria blooms and surface water . *Environment International*, 29, 641-647.

Singh, D.P., Tyagi, M.B., Kumar, A., Thakur, J.K., Kumar, A., 2001. Antialgal activity of a hepatotoxin-producing cyanobacterium, *Microcystis aeruginosa. World Journal of Microbiology and Biotechnology*, 17, 15-22.

Sipiä, V.O., Kankaanpaa, H.T., Flinkman, J., Lahti, K., Meriluoto, J.A.O., 2001. Time dependent accumulation of cyanobacterial hepatotoxins in flounders (*Platichthys flesus*) and mussels (*Mytilus edulis*) from the northern Baltic Sea. *Environmental Toxicology*, 16, 330-336.

Sivonen, K., Jones, G.J., 1999. Cyanobacterial toxins. In: Chorus, I., Bartram, J. (Eds.), Toxic cyanobacteria in water: A guide to their public health consequences, monitoring and management.World Health Organization, Taylor and Francis, London and New York, pp. 41–111.

Smith, J.L., Haney, J.F., 2006. Foodweb transfer, accumulation, and depuration of microcystins, a cyanobacterial toxin, in pumpkinseed sunfish (*Lepomis gibbosus*). *Toxicon,* 48, 580–589.

Smith, J.L., Boyer, G.L., Zimba P.V., 2008. A review of cyanobacterial odorous and bioactive metabolites: Impacts and management alternatives in aquaculture. *Aquaculture,* 280, 5–20.

Smith, J.L., Boyer, G.L., 2009. Standardization of microcystin extraction from fish tissues: A novel internal standard as a surrogate for polar and non-polar variants. *Toxicon*, 53 (2), 238-245.

Smith J. L., Schulz K. L., Zimba P. V., Boyer G. L., 2010. Possible mechanism for the foodweb transfer of covalently bound microcystins. *Ecotoxicology and Environmental Safety,* 73, 757-761.

Soares, R.M., Magalhães, V.F., Azevedo, S.M.F.O., 2004. Accumulation and depuration of microcystins (cyanobacteria hepatotoxins) in *Tilapia rendalli* (Cichlidae) under laboratory conditions. *Aquatic Toxicology*, 70 (1), 1-10.

Song L., Chen W., Peng L., Wan N., Gan N., Zhang X., 2007. Distribution and bioaccumulation. *Water Research*, 41, 2853-2864.

Stewart, I., Schluter, P.J., and Shaw, G.S. 2006. Cyanobacterial lipopolysaccharides and human health–a review. *BioMed. Central,* 1-23.

Sueoka, E., Sueoka, N., Okabe, S. *et al.,* 1997. Expression of the tumor necrosis factor alpha gene and early response genes by nodularin, a liver tumor promoter, in primary cultured rat hepatocytes. *J. Cancer Res. Clin. Oncol.,* 123, 413–419. PMID: 9292703.

Terao, K., Ohmori, S., Igarashi, K., Ohtani, I., Watanabe, M. F., Harada, K. I., Ito, E., and Watanabe, M., 1994. Electron microscopic studies on experimental poisoning in mice induced by cylindrospermopsin isolated from the blue-green alga *Umezakia natans. Toxicon*, 32, 833–844.

Tester, P.A., Turner, J.T., Shea, D., 2000. Vectorial transport of toxins from the dinoflagellate *Gymnodinium breve* through copepods to fish. *Journal of Plankton Research*, 22, 47-62.

Thostrup, L., Christoffersen, K., 1999. Accumulation of microcystin in *Daphnia magna* feeding on toxic *Microcystis. Arch. Hydrobiol.* 145(4), 447–467.

Tricarico, E., Bertocchi, S., Brusconi, S., Casalone, E., Gherardi, F., Giorgi, G., Mastromei, G., Parisi, G., 2008. Depuration of microcystin-LR from the red swamp crayfish *Procambarus clarkii* with assessment of its food quality. *Aquaculture*, 285, 90–95.

Turner, A.D., Hatfield, R.G., Rapkova, M., Higman, W., Algoet, M., Suarez-Isla, B.A., Cordova, M., Caceres, C., Van de Riet, J., Gibbs, R., Thomas, K., Quilliam, M., Lees, D.N. 2011. Comparison of AOAC 2005.06 LC official method with other methodologies for the quantitation of paralytic shellfish poisoning toxins in UK shellfish species. *Anal Bioanal. Chem.* (2011) 399:1257–1270.

Valdor, R., Aboal, M., 2007. Effects of living cyanobacteria, cyanobacterial extracts and pure microcystins on growth and ultrastructure of microalgae and bacteria. *Toxicon*, 49, 769-779.

Vasconcelos, V.M., 1999. Cyanobacterial toxins in Portugal: effects on aquatic animals and risk for human health. Braz. *J. Med. Biol. Res.,* 32, 249–254.

Vinagre, T.M, Alciati, J.C., Yunes, J.S., A., Richards, J., Bianchini, A. Monserrat J.M., 2002. Effects of extracts from the cyanobacterium *Microcystis aeruginosa* on ion- regulation and gill Naq, Kq-ATPase and Kqdependent phosphatase activities of the estuarine crab Chasmagnathus granulate (Decapoda, Grapsidae). *Physiological and Biochemical Zoology*. 75(6), 600-608.

Wang, P.J., Chien, M.S., Wu, F.J., Chou, H.N., Lee, S.J., 2005. Inhibition of embryonic development by microcystin-LR in zebrafish, *Danio rerio*. *Toxicon*, 45(3), 303-308.

Wang L.J., Tian Z.H 2009. Analysis on urban residents consumption characteristics of aquatic products in China. *Econ. Anal.,* 1, 8-11.

Wang, M., Wang, D., Lin, L., Hong, H., 2010. Protein profiles in zebrafish (*Danio rerio*) brains exposed to chronic microcystin-LR. *Chemosphere,* 81, 716–724.

Ward, C.J., Codd, G.A., 1999. Comparative toxicity of four microcystins of different hydrophobicities to the protozoan, *Tetrahymena pyriformis*. *J. Appl. Microbiol.* 86, 874–882.

Watanabe, M.F., Park, H.D., Kondo, F., Harada, K.I., Hayashi, H., Okino, T., 1997. Identification and estimation of microcystins in freshwater mussels. *Nat. Toxins* 5, 31–35.

White, S.H., Duivenvoorden, L.J., Fabbro, L.D., Eaglesham, G.K. 2007. Mortality and toxin bioaccumulation in Bufo marinus following exposure to *Cylindrospermopsis raciborskii* cell extracts and live cultures. *Environ. Pollution* 147, 158-167.

Whitton, B.A., Potts, M., 2000. The Ecology of Cyanobacteria: Their Diversity in Time and Space (*Hardcover*), 704p.

WHO 2007. http://www.euro.who.int/document/fos/news57e.pdf.

Wiegand, C., Peuthert, A., Plugmacher, S., Carmeli, S., 2002. Effects of microcin SF608 and microcystin-LR, two cyanobacterial compounds produced by *Microcystis* sp., on aquatic organisms. *Environmental Toxicology*, 17, 400-406.

Williams, D. E., Craig, M., Dawe, S. C., Kent, M. L., Andersen, R. J. and Holmes, C. F. B. 1997*a*. 14C-labeled microcystin-LR administered to Atlantic salmon via intraperitoneal injection provides *in vivo* evidence for covalent binding of microcystin-LR in salmon livers. *Toxicon,* 35, 985–989.

Williams, D.E., Craig, M., Dawe, S.C., Kent, M.L., Holmes, C.F.B., 1997*b*. Evidence for a covalently bound form of microcystin-LR in salmon liver and Dungeness crab larvae. *Chemical Research in Toxicology*, 10 (4), 463-469.

Williams, D.E., Dawe, S.C., Kent, M.L., Andersen, R.J., Craig, M., Holmes, C.F.B., 1997*c*. Bioaccumulation and clearance of microcystins from salt water mussels, *Mytilus edulis*, and *in vivo* evidence for covalently bound microcystins in mussel tissues. *Toxicon*, 35, 1617-1625.

Wu, X., Xiao, B., Li, R., Wang, Z., Chen, X., Chen, X., 2009. Rapid quantification of total microcystins in cyanobacterial samples by periodate-permanganate oxidation and reversed-phase liquid chromatography. *Analytica Chimica Acta*, 651 (2), 241-247.

Xie, L., Xie, P., Ozawa, K., Honma, T., Yokoyama, A., Park, H.D., 2004. Dynamics of microcystins-LR and -RR in the phytoplanktivorous silver carp in a sub-chronic toxicity experiment. *Environmental Pollution*, 127 (3), 431-439.

Xie, L., Yokoyama, A., Nakamura, K., Parka, H., 2007. Accumulation of microcystins in various organs of the freshwater snail *Sinotaia histrica* and three fishes in a temperate lake, the eutrophic Lake Suwa, Japan. *Toxicon*, 49, 646–652.

Xiao, F.-G., Zhao, X.-L., Tang, J., Gu, X.-H., Zhang, J.-P., Niu, W.-M., 2009. Necessity of Screening Water Chestnuts for Microcystins After Cyanobacterial Blooms Break Out. *Arch. Environ. Contam. Toxicol*., 57, 256–263.

Yamakasi, S., 1993. Probable effects of algal bloom on the growth of *Phragmites australis* (Cav.) Trin. ex Steud. *Botanical Society of Japan*, Tokyo, Japan.

Yoshizawa, S., Matsushima, R., Watanabe, M.F., Harada, K., Ichihara, A., Carmichael, W.W., Fujiki, H., 1990. Inhibition of protein phosphatases by microcystins and nodularin associated with hepatotoxicity. *J. Cancer Res. Clin. Oncol.,* 116, 609–614.

Zhang, D.W., Xie, P., Liu, Y.Q., Chen, J., Liang, G.D., 2007. Bioaccumulation of the hepatotoxic microcystins in various organs of a freshwater snail from a subtropical Chinese lake, Taihu Lake, with dense toxic *Microcystis* blooms. *Environ. Toxicol. Chem*., 26, 171–6.

Zhang, X.Z., Xie, P., Wang, W.M., Li, D.P., Li, L., Tang, R., *et al.,* 2008. Dose-dependent effects of extracted microcystins on embryonic development, larval growth and histopathological changes of southern catfish (*Silurus meridionalis*). *Toxicon,* 51(3), 449-456.

Zhang, D.W., Xie, P., Liu, Y.Q., Qiu, T., 2009. Transfer, distribution and bioaccumulation of microcystins in the aquatic food web in Lake Taihu, China, with potential risks to human health. *Sci. Total Environ*., 407, 2191–2199.

Zimba, P.V., Grimm, C.C., Dionigi, C.P., 2001a. Phytoplankton community structure, biomass, and off flavor: pond-size relationships in Louisiana catfish ponds. *J. World Aquac. Soc.,* 32, 96–104.

Zimba, P.V., Khoo, L., Gaunt, P.S., Brittain, S., Carmichael, W.W., 2001b. Confirmation of catfish, *Ictalurus punctatus* (Rafinesque), mortality from *Microcystis* toxins. *J. Fish Dis.*, 24, 41–47.

In: Zooplankton and Phytoplankton
Editor: Giri Kattel, pp. 35-61

ISBN 978-1-61324-508-8

Chapter 2

INVESTIGATIONS ON THE USE OF MICROALGAE FOR AQUACULTURE

José Antonio López Elías[1], Luis Rafael Martínez Córdova[1] and Marcel Martínez Porchas[2]

[1]Departamento de Investigaciones Científicas y Tecnológicas de la Universidad de Sonora, Blvd. Luis Donaldo Colosio s/n entre Reforma y Sahuaripa, Edificio 7G, Hermosillo, Sonora, 83000 México

[2]Centro de Investigación en Alimentación y Desarrollo. Hermosillo, Sonora, México

ABSTRACT

Microalgaes are probably the most commonly used organisms as live feed during the larviculture and nursery of aquacultural organisms, and many researches have been made worldwide in this field. This chapter is a summary of the investigations conducted in the University of Sonora, México, on the use of microalgae for aquaculture of diverse organisms, mainly crustaceans and fishes. The document is divided in five sections related to the most important aspects of the culture such as: production systems; evaluation of particular species and their nutritional value for selected organisms (mollusks and crustaceans), effect of physical and chemical environmental factors on the production and nutritional value, evaluation of commercial laboratories of microalgaes, and evaluation of alternative mediums for their culture. The results of our investigations are discussed when compared to similar studies on the same or different species and in the same or different regions.

INTRODUCTION

Despite of its high cost, the use of live feed is essential and frequently irreplaceable in the aquaculture of mollusks, fishes, crustaceans, and some other aquatic organisms, especially in the larviculture and nursery phases (Lin et al. 2009). The use of microalgae during these phases seems to be a universal practice, because some microalgae species have adequate physical and nutritional characteristics for the early development of aquatic organisms, and the operative costs for their production is commonly lower compared to the production of other organisms or formulated feeds (Martínez Córdova et al. 1999; Lovatelli et al. 2004).

A great number of microalgae species have been used worldwide for the larviculture of aquatic organisms, and the selection of a suitable species to feed a particular organism (mollusk, fish or crustacean), depends on many factors such as: size, habitat (plankton or benthos), nutritional value, replication and growth rates, nutrimental requirements, tolerance to environmental factors, etcetera.

Many experimental researches have been done worldwide, including our institution (University of Sonora), related to the use of microalgae for the farming of diverse aquatic organisms. Those researches have covered many aspects of the culture process, including: production systems, evaluation of particular species and their nutritional value for selected organisms (mollusks and crustaceans), effect of physical and chemical environmental factors on the production and nutritional value, evaluation of commercial laboratories of microalgae, evaluation of alternative mediums for their culture, etcetera. All these aspects are widely discussed in the five sections of this chapter.

The authors hope this information can be useful for researchers, aquaculturists, students, and other people related to the microalgae culture.

1. A Brief History

Soeder (1986) hypothesized that our ancestors used to collect the cyanobacteria *Nostoc* as a gelatinous mass from the soil, as well as *Spirulina* from lakes in order to complement their nutrition. He also mentioned that *Nostoc* was consumed in Mongolia and China, and that *Spirulina* was an essential part of Africans and Aztecs (Mexico ancestors) diets; moreover, Africans still at present, consuming the microalgae.

The investigations on microalgae began around 1890, being Beijerinck the pioneer microbiologist who purified the freshwater microalgae *Chlorella vulgaris*. After that, Otto Warburg (1919) produced the same species at higher concentrations, and conducted some studies regarding to its photosynthesis process (Abalde et al. 1995). However, the first massive production of microalgae under controlled conditions occurred in Germany in the early 1920´s and had the objective to produce great microalgae biomasses rich in lipids. Such research was subsequently recuperated in a document entitled “Algal Culture: From Laboratory to Pilot Plant” (Burlew 1953), and was the first evidence of the possibility to produce microalgae at commercial scale (Pipes and Gotaas 1960).

In the 1940´s the massive culture of microalgae was initiated; the first cultures were performed outdoor, because of the high costs of artificial light. In this type of culture, sunlight, carbon dioxide (CO_2), and environmental temperature were used, so the production depended on the natural variation of these parameters (Richmond 1986).

The technical bases for the industrial culture were established in the early 1950´s in Japan, Germany, USA, Israel, and other countries. By that time, Oswald and colleagues from the University of California, produced massively microalgae with the objectives of treating wastewaters and producing protein (Abalde et al. 1995).

The biotechnology for the massive culture of microalgae has been developed successfully in many countries. At first, the production for human consumption was an interesting expectative, however, the high production costs decreased the interest in this type of production, which is now focused on animal nutrition, soil treatment, bioremediation, (Becker, 1994), production of renewable biofuels (Chisti 2008), and extraction of bioactive compounds for human use, such as: antioxidants, lipids, proteins, vitamins, and others (Cleber Bertoldi et al. 2006; Seon Jin and Melis 2003; Matsunaga et al. 2006).

2. Production Systems and Routine Modifications

The phytoplankton culture has been traditionally done in three main types of systems: static, semi-continuous and continuous (Vonshak 1988). The most commonly used in commercial hatcheries is the static system arranged in steps, which consists in starting the culture at low volumes (assay tubes or Erlenmeyer flasks), passing to bottles (19 L or more), then to columns, and finally to pools, big plastic bags or tanks. The commercial medium f/2 is the most frequently used for microalgae culture (Guillard andandand Ryther, 1962), which is prepared with highly purified chemicals. However, there are several other mediums based on highly purified chemical or commercial products such as plant fertilizers, which diminish productive costs.

The selection of containers for production of microalgae is a key aspect. The form (rounded, oval, squared, rectangular or cylindrical), depth (from 60 cm to 2 m) and material are the most important characteristics. These factors could have an important influence on the water column dynamics and thus on the microalgae contact with light, which is strongly related to the cell development and reproduction (Richmond 1986; Oswald 1988).

In aquaculture, the stock strains of the different species are maintained under controlled conditions (continuous illumination, and constant temperature, using air conditioners), using assay tubes, Erlenmeyer flasks or bigger bottles for great volumes. However, the massive production is commonly done outdoor and rarely indoor (López-Elías et al. 2003). Despite photobioreactors can produce much higher amounts of microalgae (sometimes by one order of magnitude), they also require greater economical investments; therefore, many farms cover their microalgae requirements using the traditional outdoor production systems (Zhang and Richmond 2003; López-Elías et al. 2005a). Thus, microalgae culture for the aquaculture industry does not always require highly technified systems; however, in some particular cases, the use of technology could be necessary.

Regarding to the production routines, a relatively constancy has been observed among commercial laboratories (López-Elías et al. 2005a); such laboratories usually report their highest productions during warm months due to the better environmental conditions, while during cold months the production is almost null. Despite the simplicity of the traditional systems for microalgae production, several adjustments and modifications can be done in order to increase the quantity and/or quality of microalgae. For instance, slight modifications in environmental factors or nutrients can result in high changes in the productive response of microalgae (Medina-Reyna and Cordero-Esquivel 1998).

Some experimental studies have been made in our institution to evaluate the different production systems and the effectiveness of the routines used in those systems, including the evaluation of some modifications on the traditional systems.

2.1.

An experimental study focused on the evaluation of an alternative system for the outdoor culture of *Chaetoceros muelleri* and *Dunaliella* sp. during winter and spring, was done in Northwest Mexico (Becerra-Dorame et al. 2010). The alternative system had the same structure than the traditional outdoor systems, but it also had a recirculation cascade in which the water was pumped up from the bottom of the plastic containers by means a pump of 1/8HP at a flux rate of 3.68 L·min^{-1}, and thereafter flowed down over a transparent plastic sheet arranged in steps, to improve the exposure of microalgae to the light (Figure 1). The hydraulic retention time (h) was 87 min, and the hydraulic loading rate (flow/horizontal area of cascade structure) was 11.5 L·mm^{-2}.

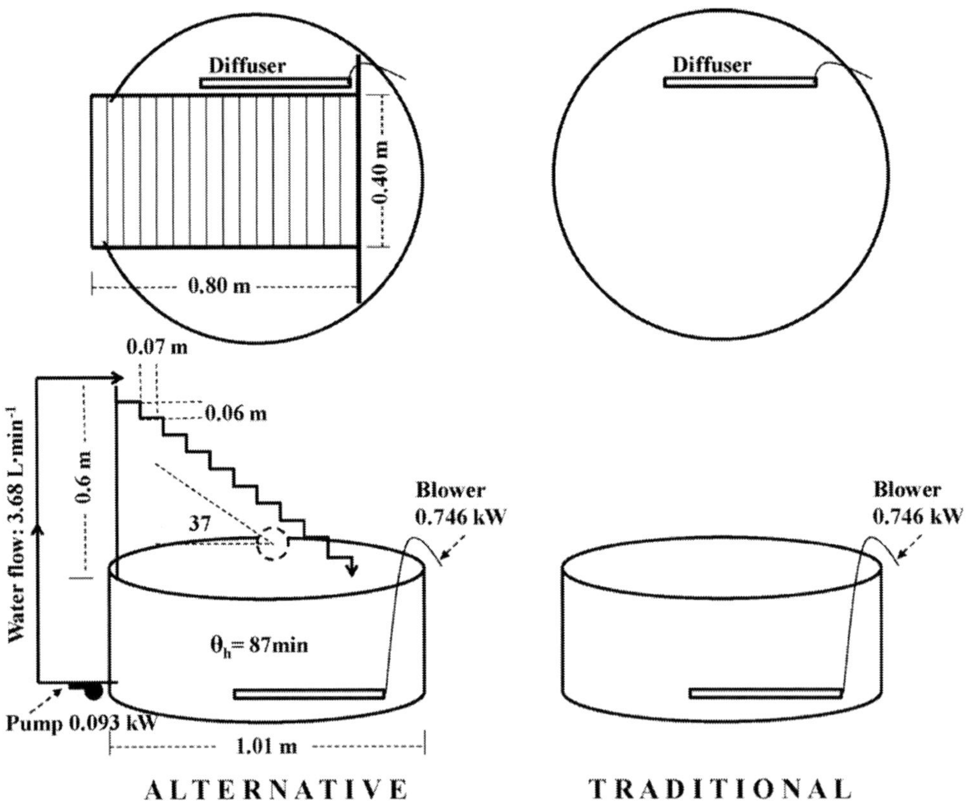

Figure 1. Schemes of the traditional and alternative systems used to culture *C. muelleri* and *Dunaliella* sp. The traditional system is commonly used in shrimp farms or commercial laboratories to fed shrimp larvae. Source: Becerra-Dórame et al. (2010), *Aquacultural Engineering*.

The modification on the structure increased the microalgae exposure to light and the movement of microalgae within the system. The diatom *C. muelleri* showed a greater final cell density with such modifications compared to those cultured in the traditional system (Figure 2).

No differences on final biomass of *C. muellerii* were found among systems in the winter trial, but *Dunaliella* sp. had higher biomass in the alternative system. Contrarily, in the spring trial, *C. muellerii* had a greater biomass in the alternative system, but *Dunaliella sp.* did not show differences.

It was concluded that the new system is a good alternative for the culture of these two microalgae, especially during spring.

2.2.

An experimental study was performed to evaluate the effect of management routines such as: inoculation hour, inoculum concentration, and culture medium on the growth and final biomass of *C. muelleri, Dunaliella* sp. and *Tetrasemis chuii* cultured outdoor during spring, in 250-L plastic tanks (López-Elías, et al., 2008; López-Elías, 2010a, Becerra-Dórame, et al., 2010).

The inoculation for the three species was done at 0600 and 1200 h; the inoculum concentrations were 0.2 and 0.4 x 10^6 cells·mL^{-1} for *C. muelleri*; 0.04 and 0.08 x 10^6 cells·mL^{-1} for *Dunaliella* sp. and *T. chuii*; the f and the f/2 mediums (Guillard, 1975) were used for *C. muelleri*, while f/2 and 2f were used for the two other species.

Results showed that the inoculation time had an effect on the productive response of microalgae, while some were also affected by the culture medium and/or the inoculums density. Final cellular density of *C. muelleri* at 72 h varied from 1.21 to 2.83 x 10^6 cells·mL^{-1} with the best results obtained when inoculation was made at 0600 h, at a concentration of 0.4 x 10^6 cells·mL^{-1}, and using the f medium. The maximum growth rate was similar among treatments (0.82 and 1.27 divisions·day^{-1}), but the cumulated growth was higher with an inoculum concentration of 0.2 x 10^6 cells·mL^{-1}, in which 2.59 to 3.09 total divisions were obtained (Table 1).

For *Dunaliella* sp., the final cellular density was greater when the inoculum concentration was 0.08 x 10^6 cells·mL^{-1} and the 2f medium was used (0.75 and 0.83 x 10^6 cells·mL^{-1} at the end of the culture) (Figure 3). The inoculation hour did not have an effect on the final cellular density.

Tetraselmis chuii, showed a higher final cellular density when inoculated at 0600 h with a concentration of 0.65 x 10^6 cells·mL^{-1}, as compared when inoculated at 1200 h with 0.56 x 10^6 cells·mL^{-1} (Table 2). The greater duplication rates were 3.92 and 3.99 divisions·day^{-1}, when the inoculum concentration was 0.04 x 10^6 cells·mL^{-1} and was done at 0600 h. The culture medium did not have a significant effect on the cellular density. As conclusion of the study it was established that the higher cellular densities of microalgae were obtained in the cultures initiated at 0600 h with an inoculum concentration of 0.08 x 10^6 cells·mL^{-1}. For *C. muellerii*, the greater biomasses were achieved in general with the higher inoculum concentration, initiated at 0600 h and the f medium (Table 1). For *Dunaliella* sp., the biomass and organic matter were greater in the cultures initiated at 1200 h, using the inoculum of highest concentration, and independently of the culture medium (Table 3). *Tetraselmis chuii*, had a greater final cellular density when inoculated at 0600 h, with a concentration of 0.08×10^6 cells·mL^{-1} and using the 2f medium. (Table 2).

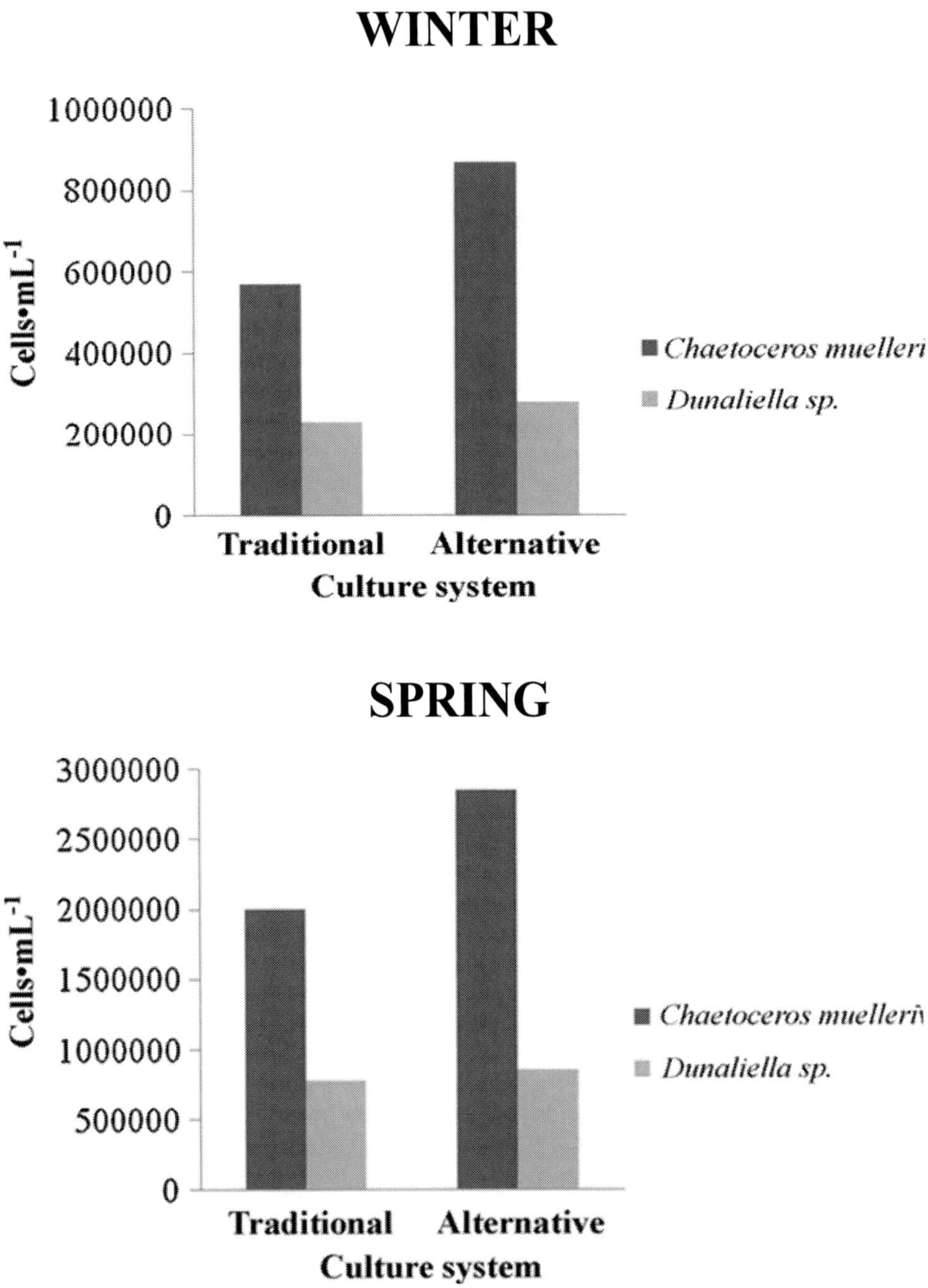

Figure 2. Cell densities of *C. muellerii* and *Dunaliella* sp. outdoor produced in the traditional and alternative systems, during winter and spring (Becerra-Dorame et al. 2010).

Table 1. Mean ± SD of final cellular density, maximum growth rate (MGR), and cumulated growth rate (CGR) of *C. muellerii* at 48 y 72 hours, inoculated at 0600 and 1200 h with inoculums of 0.2 y 0.4 x 10^6 cells.mL^{-1} in the f and f/2 medium

$\alpha = 0.05$; a < b < c.

Treatment	Cells·mL^{-1} x 10^6		MGR (Divisions·day^{-1})	CGR (Total Divisions)	Dry matter (g·m^{-3})	Organic matter (g·m^{-3})
	48 h	72 h				
Inoculum 0600 h						
0.2 x 10^6 f/2	0.75 ± 0.01^a	1.52 ± 0.11^a	1.24 ± 0.17^b	2.93 ± 0.10^{bc}	0.087 ± 0.025^a	0.055 ± 0.003^a
0.2 x 10^6 f	0.82 ± 0.04^a	1.71 ± 0.17^a	0.90 ± 0.07^a	3.09 ± 0.14^c	0.131 ± 0.017^b	0.060 ± 0.029^a
0.4 x 10^6 f/2	1.31 ± 0.13^b	1.75 ± 0.02^a	1.15 ± 0.16^{ab}	2.13 ± 0.01^a	0.085 ± 0.006^a	0.064 ± 0.004^a
0.4 x 10^6 f	1.55 ± 0.04^c	2.83 ± 0.07^b	0.96 ± 0.05^{ab}	2.80 ± 0.04^b	0.186 ± 0.010^c	0.099 ± 0.007^b
Inoculum 1200 h						
0.2 x 10^6 f/2	0.73 ± 0.18^a	1.21 ± 0.11^a	0.82 ± 0.36^a	2.59 ± 0.04^b	0.121 ± 0.012^{ab}	0.069 ± 0.016^a
0.2 x 10^6 f	0.92 ± 0.05^{ab}	1.41 ± 0.17^b	1.10 ± 0.12^a	2.82 ± 0.08^c	0.155 ± 0.033^{bc}	0.065 ± 0.011^a
0.4 x 10^6 f/2	1.13 ± 0.12^{bc}	1.39 ± 0.02^{ab}	1.21 ± 0.21^a	1.79 ± 0.01^a	0.110 ± 0.003^a	0.066 ± 0.002^a
0.4 x 10^6 f	1.28 ± 0.03^c	2.78 ± 0.07^c	1.27 ± 0.07^a	2.80 ± 0.06^c	0.173 ± 0.023^c	0.099 ± 0.015^b

Table 2. Means ± SD of cellular density and biomass of *T. chuii* at 72 hours in mass culture, with different inoculum hours, concentration, and culture mediums

	Inoculation hour	Inoculum concentration (x10^6 cells·mL^{-1})	Culture medium	Cells·mL^{-1} x 10^5	Biomass (g·L^{-1})
A	*0600*	*0.04*	*f/2*	6.128^{abc} ± 0.360	0.35^d ±0.12
B	*1200*	*0.04*	*f/2*	4.202^d ± 0.350	0.50^{bcd} ±0.03
C	*0600*	*0.04*	*2f*	6.401^{abc} ± 0.652	0.63^b ±0.17
D	*1200*	*0.04*	*2f*	6.345^{abc} ± 1.087	0.39^{cd} ±0.11
E	*0600*	*0.08*	*f/2*	7.489^a ± 0.311	0.40^{cd} ±0.09
F	*1200*	*0.08*	*f/2*	6.821^a ± 0.360	0.57^{bcd} ±0.05
G	*0600*	*0.08*	*2f*	5.920^{abcd} ± 1.390	0.94^a ±0.03
H	*1200*	*0.08*	*2f*	4.938^{cd} ± 1.123	0.41^{cd} ±0.15

Different letters means significant differences (p<0.05).

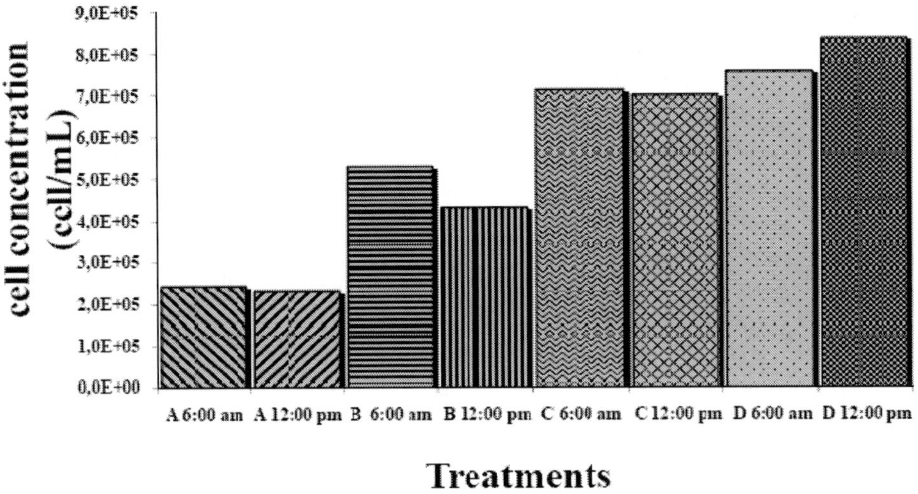

Figure 3. Cellular density of *Dunaliella* sp. at 72 h cultured outdoor with inoculums at 0600 and 1200 (A: inoculum 0.04 x 10^6 cell·mL^{-1} in medium f/2; B: inoculum 0.04 x 10^6 cell·mL^{-1} in medium 2f; C: inoculum 0.08 x 10^6 cell·mL^{-1} in medium f/2; B: inoculum 0.08 x 10^6 cell·mL^{-1} in medium 2f).

3. Evaluation of Nutritional Value of Microalgae for the Culture of Particular Species

Despite the wide use of microalgae in aquaculture and the high nutritional value of such natural feed, the nutritional profiles of microalgae are different not only between different species, but also among the same species (Volkman et al. 1989). As mentioned above, sight changes in the environmental conditions could modify the nutritional composition of microalgae (Sánchez-Saavedra and Voltolina 1996).

Biochemical composition of microalgae varies widely because of diverse factors inherent to the culture systems, such as culture age, pH, salinity, light, temperature and management (Voltolina and López Elías 2002). In addition, it is almost impossible to identify all the factors influencing the microalgae composition in an outdoor system (Voltolina et al. 1999). However, some modifications in the system or management procedures may improve the quantity and quality of microalgae. The nutritional quality of microalgae is a mains aspect for the success of aquaculture.

Some experiments are mentioned below, regarding to the use of microalgae cultured in different mediums and their effectiveness for zooplankton production.

3.1.

The effectiveness of *Nanochloropsis oculata* and *C. muellerii*, as feed for rotifers and copepods, was evaluated in a 7-week experiment, at mesocosms level. Both microalgae were cultured in the conventional f/2 medium, and two alternative mediums, one agricultural (mono ammonium phosphate) and one aquacultural (Nutrilake). *Nanochloropsis oculata* was used as feed for the rotifer *Brachionus rotundiformis* and *C. muellerii* for the copepod *Calanus pacificus.*

Table 3. Mean dry weight, organic matter, and ash of *Dunaliella* sp. in the four treatments

Treatment	Day	Dry weight ($g{\cdot}L^{-1}$)	Organic matter ($g{\cdot}L^{-1}$)	Ash ($g{\cdot}L^{-1}$)
0600 f/2	2	0.079[a]	0.054[ab]	0.025[a]
0600 f/2	3	0.094[ab]	0.065[bc]	0.029[ab]
0600 2f	2	0.098[bc]	0.070[c]	0.028[ab]
0600 2f	3	0.102[bc]	0.071[c]	0.031[ab]
1200 f/2	2	0.078[a]	0.050[a]	0.028[ab]
1200 f/2	3	0.104[bc]	0.069[c]	0.035[b]
1200 2f	2	0.093[ab]	0.064[bc]	0.029[ab]
1200 2f	3	0.114[c]	0.067[c]	0.047[c]

Different letters mean significant differences at $P < 0.05$.

Table 4. Production parameters and chemical proximate composition of *B. rotundiformis* fed *N. oculata* cultured on the alternative mediums and the control

Medium	Final density Rotifers mL^{-1}	Carbohydrate (%)	Lipid (%)	Protein (%)
MAP	96.75±10.56a	25.92±1.72c	26.34±2.39b	47.74±3.47a
NLK	132.5±6.35c	12.95±1.10a	22.79±3.33a	64.27±4.42b
f/2	111.5±9.29b	18.79±2.15b	23.08±2.62a	58.13±2.82b

Results showed that the culture mediums had an effect on the nutritional quality of microalgae; such microalgae quality had a subsequent effect on the productive response and chemical composition of rotifers and copepods when was used as feed source. Rotifer recorded differences on the production parameters as well as on the chemical proximate composition when fed on *N. oculata* cultured in the alternative mediums, as compared to the control. The best final density of rotifers and eggs, as well as the higher content of protein, were found in the organisms fed on microalgae produced in the aquacultural fertilizer, followed by those fed on microalgae produced in f/2 and MAP (Table 4). Carbohydrate and lipid contents were higher in the MAP treatment.

The chemical proximate composition of the rotifers showed significant differences on the carbohydrate content among treatments, with higher levels on the organisms fed on microalgae cultured in the agricultural fertilizer compared to those fed cultured in the aquacultural one. The lipid content was higher in the rotifers fed microalgae from the agricultural medium as compared to the two other. The protein content was greater in the organisms fed microalgae farmed in the aquacultural and conventional f/2 medium (Table 4).

The copepod *C. pacificus* fed on *C. muellerii* which was previously cultured in the agricultural fertilizer, showed a higher final density (Table 5). In addition, significant differences were found on the protein content of the copepods fed microalgae produced in the different mediums. The higher levels were found in those fed on microalgae from the conventional medium (f/2), compared to the agricultural and aquacultural medium. No differences in carbohydrate and lipid content were recorded.

Table 5. Production parameters and chemical proximate composition of *C. pacificus* fed on *C. muellerii* cultured on alternative and conventional mediums

Mediums	Final density copepods mL^{-1}	Carbohydrate (%)	Lipid (%)	Protein (%)
MAP	1.41±1.79b	18.68±1.44a	25.80±2.56a	55.52±1.21a
NLK	1.62±1.22c	15.51±0.80ab	25.81±1.16a	58.68±1.60a
GF/2	0.97±2.6a	15.91±2.31a	24.83±0.48a	62.26±1.92b

3.2.

The nutritional value of microalgae *Isochrysis* sp. (Diet 1), *Chaetoceros muelleri* (Diet 2), and the combination of both (Diet 3), was evaluated as feed source for larviculture and chemical proximate composition of *Litopenaeus vannamei*. The protein levels of shrimp larvae were similar when for all treatments (~43 – 44%), (Table 6). Polyunsaturated fatty acids (PUFA) 20:5w3 and 22:5w3, were more abundant in larvae fed on Diet 1; 22:6w3 was more abundant in organisms fed on Diet 2. However, the profile of fatty acids was more complete in those larvae fed on Diet 3. No significant differences on larvae survival were found among diets (> 60%) (Rodríguez, et al., 2010). The best growth was observed in larvae feed Diet 2 and 3 (Table 7).

These results differed from those reported by D´Souza and Loneragan (1999) who found that a monoalgal diet based on *Isochrysis* was unsatisfactory for shrimp larvae nutrition. Also, Piña, et al. (2006), found that monoalgal diet with *Isochrysis* sp. did not improve survival rate and rate of development in *L. vannamei* protozoea larvae. In this study however, *Isochrysis* was cultivated in a greenhouse, being able to synthesize a high amounts of DHA and other important cell constituents, which had a positive effect on the larval culture. Thus, microalgae species that are not adequate for shrimp or fish rearing, can be modified in their nutritional quality in order to cover the nutritional requirements of any aquatic organism.

Table 6. Nutrient fraction (%) of shrimp larvae (*L. vannamei*) fed monospecific diets (*C. muelleri* and *Isochrysis* sp.) and the mixture of both. Two similar experiments were done (Exp I andandand Exp II)

Larvae	Protein (%)		*Carbohydrate* (%)		Lipid (%)	
	Exp. I	Exp. II	Exp. I	Exp. II	Exp. I	Exp. II
Nauplii	40.58 (0.00)	37.06 (9.74)	31.64 (0.00)	32.13 (1.07)	12.64 (0.00)	10.16 (0.08)
Chaetoceros Zoea III	39.94[a] (6.53)	45.94[a] (6.73)	30.77[a] (0.09)	30.84[a] (0.49)	15.87[a] (4.75)	15.12[a] (9.80)
Isochrysis Zoea III	43.68[a] (7.69)	42.34[a] (4.73)	31.37[a] (0.39)	30.82[a] (0.94)	18.54[a] (3.18)	12.05[a] (5.78)
Mixture Zoea III	38.94[a] (2.36)	48.63[a] (6.95)	31.20[a] (0.11)	30.57[a] (0.11)	26.28[b] (3.11)	13.75[a] (6.72)

Table 7. Fatty acid profile (%) of *L. vannamei* larvae fed with Diet I (*Chaetoceros muelleri*), Diet II (*Isochrysis* sp.) and Diet III (mixture) in two experimental runs

Fatty Acids	Diet I	Diet II	Diet III
14:0	31.62	11.70	66.41
16:0	41.30	----	----
16:1	----	----	6.46
18:0	----	25.75	2.24
18:1w9	----	13.45	----
18:1w7	----	----	1.52
18:2w6	----	7.16	1.52
20:1w9	----	----	6.4
20:5w3	11.01	7.53	4.01
22:1w11	----	----	2.48
22:1w9	----	----	4.83
22:5w3	10.26	11.59	8.89
22:6w3	----	13.84	----
24:1w9	5.79	15.69	----

4. Effect of Physical, Chemical and Management Factors, on Production and Composition of Microalgae

As mentioned, the production response and biochemical composition of microalgae, both in natural or aquaculture environments, are strongly influenced by abiotic factors such as: light intensity, temperature, salinity, pH, and nutrients availability.

The salinity is one of the most important environmental factors in the culture of microalgae, because it affects their growth and replication rate. In the natural environment, this parameter determines the distribution and abundance of phytoplankton species in the aquatic ecosystems. Additionally, it has a determinant influence on the density, viscosity and solubility of diverse gases in the water column.

Variations on salinity cause physiological responses in microalgae related to the cellular volume and osmotic adjustment. These responses imply the release of ions, synthesis of organic compounds, diminution of CO_2 fixation, and modification on the nitrogen metabolism (Kirst, 1989).

Most of the microalgae have a high tolerance to salinity variations, and can growth well in a wide range (Brown et al. 1996). The hypo osmotic conditions tend to be more adverse than the hyper osmotic ones. Growth rate and metabolic activity of many marine microalgae decline significantly when salinity drop under 20 ppt (Kirst 1989).

The optimum growth and tolerance to variations on salinity, depends mainly on the species and its natural habitat. For instance, Thessen et al. (2005) argued that the optimal salinity for microalgae growth depends on the species, and documented that *Pseudo-nitzschia delicatissima* had a maximum growth rate at low salinities (10-30 ppt), while *Pseudo-nitzschia pseudodelicatissima* had the best growth performance at higher salinities (25-40 ppt) and *Pseudo-nitzschia multiseries* showed an intermediate behavior (25-30 ppt). In addition, the estuarine species are more tolerant to low salinities than the oceanic species (Kirst, 1989). Renaud and Parry (1994), reported that salinities from 10 to 35 ppt do not have any effect on the growth rates of *Isochrysis* sp. and *Nitzschia frustulum*, however, the growth rate of *N. oculata*, decreased at 35 ppt. In some aquaculture hatcheries, the salinity may be difficult to control, because the water sources are estuaries or the ocean, thus, the salinity levels are 35 ppt or higher if evaporation is considered.

Temperature is an abiotic factor that affects the cellular metabolism and, in consequence, the growth and chemical proximate composition of microalgae (Richmond, 2004). Most of the species used for aquaculture purposes growth well on the range of 10 °C to 35 °C, with an optimum within 16 °C and 24 °C (Voltolina et al. 1989). In commercial hatcheries, temperature is controlled by air conditioned apparatus for indoor cultures but, outdoor systems are exposed to the environmental variation of the region.

The pH is one of the most important factors to consider for microalgae production, because their membranes are completely permeable to H^+ and OH^- ions, and the concentration of such ions may affect some cellular functions, causing death at extreme concentrations of H^+ (very low pH). High levels of OH^- (high pH), can affect microalgae development, but are not as lethal as a low pH condition. The optimum pH for most of the microalgae ranges from 7 to 8 (Abalde et al. 1995).

Another aspect to consider for microalgae culture is the nitrogen: phosphorus ratio. Changes in biochemical functions can be attributed to N:P ratio; however, the ratio may also depend upon the N or P sources. Some of the mediums used for microalgae culture can be modified in their N:P or C:N ratios in order to improve production and quality of particular microalgae species.

4.1.

A study conducted under laboratory conditions, evaluated the effect of four salinities (25, 30, 35, 40, 45 and 50 ppt) on the growth and chemical proximate composition of the microalgae *T. weissflogii* at three culture phases. The best growth rates and final cellular densities were observed at 25 ppt. Diminutions in size, as well as morphological changes, were observed at high salinities (≥ 40 ppt). The most drastic changes occurred at 50 ppt (Table 8). Protein and carbohydrate contents were greater at 25 and 30 ppt, and the higher levels were recorded at the stationary phase. Lipid contents were higher at low salinities, but did not change during any of the three growth phases (López-Elías, et al., 2009).

Lipids and carbohydrates are considered as stored energy products, and their decrease can negatively affect the growth and metabolic activities of microalgae (Brown et al. 1997). In addition, the osmotic challenge at higher or lower salinities increase the energy demand of cells, leading to an enhanced use of lipids and carbohydrates to cope with the energy demand.

Table 8. Means and standard deviations of maximum cellular density, specific growth rate and cell volume of *Thalassiosira weissflogii* at different salinities

Salinity (ppt)	Maximum cellular density ($x10^4$ mL^{-1})		Specific growth rate (μ) day^{-1}		Cell volume ($m\mu^3$)
25	43 ±	0.5^{ab}	1.24 ±	0.025^{a}	1594.3 ± 145.1^{a}
30	42 ±	1.2^{b}	1.12 ±	0.023^{b}	1489.0 ± 151.6^{a}
35	44 ±	1.2^{a}	1.16 ±	0.039^{b}	1401.4 ± 57.2^{a}
40	40 ±	0.9^{c}	0.99 ±	0.030^{c}	1013.9 ± 145.0^{b}
45	37 ±	0.6^{d}	0.82 ±	0.038^{d}	700.0 ± 14.8^{c}
50	35 ±	1.0^{e}	0.81 ±	0.007^{d}	563.7 ± 29.9^{c}

Different letters show significant differences (α=0.05).

Several species of microalgae are tolerant to great variations in salinity, although, their chemical proximate composition may be affected by such factor (Renaud and Parry 1994; Brown et al. 1996). Protein, lipids and carbohydrates in most of the microalgae species seem to be affected by high salinities (Richmond 1986). In some species, a considerable increase in ash and lipid contents have been documented when salinity increase (Kirst 1989).

4.2.

An other laboratory study was conducted to evaluate the growth rate of *Thalassiosira pseudonana* in a static system, at salinities of 15, 25 and 35 ppt, under two conditions: continuous lighting or photoperiod of 12L:12D; the f/2 medium was used. It was found that salinity and light affected the final cellular density and the growth rate of the microalgae (Table 9). The highest growth rate (1.71 divisions·day^{-1}) and the greatest cellular density (5.6 x 10^5 cells·mL^{-1}) were recorded at 35 ppt, in the cultures using continuous illumination. Those values decreased when photoperiod was used (1.0 divisions·day^{-1} and 4.8 x 10^5 cells·mL^{-1}), but decreased even more at 15 ppt, in either continuous lighting or photoperiod. The longer light period increased the microalgae production, such results may explain the lower efficiency of outdoor systems compared to those indoor ones. Tzovenis et al. (2003), documented that light has an important roll in the development of microalgae. Similar results were reported by Brown et al. (1996), who recorded, a growth rate of 1.9 divisions·day^{-1} at continuous illumination and 1.0 divisions·day^{-1}, using a 12:12 photoperiod for the same species (López-Elías, et al., 2009). Moreover, it was observed that the salinity can potentiate or diminish the positive effect of light on the growth performance of the microalgae.

Table 9. Means ± SD of final cellular density, duplication rate (μ maximum, μ mean, μ cummulated) of *T. pseudonana* at three salinities, using photoperiod (P) or continuous lighting (CL)

Salinity (ppt)	Light condition	Final cellular density (Cells·mL^{-1} x10^5)	Specific growth rate (Divisions·day^{-1})		
			μ maximum	μ mean	μ cummulated
15	P	3.0 ± 1.5 a	0.68 ± 0.26a	0.33 ± 0.05a	2.30 ± 0.36a
25	P	4.2 ± 3.0 b	0.95 ± 0.18a	0.38 ± 0.01ab	2.74 ± 0.16ab
35	P	4.5 ± 3.5 bc	1.00 ± 0.20ab	0.46 ± 0.04bc	3.23 ± 0.29b
15	CL	3.9 ± 1.8 ab	0.71 ± 0.10a	0.35 ± 0.06a	2.50 ± 0.34a
25	CL	4.5 ± 1.8 bc	1.39 ± 0.09bc	0.43 ± 0.03abc	3.21 ± 0.28b
35	CL	5.5 ± 1.4c	1.71 ± 0.15c	0.49 ± 0.05c	3.43 ± 0.37b

Different letters mean significant differences at $P < 0.05$.

4.3.

A similar investigation was done to evaluate the growth, cellular density, and biomass production of *Isochrysis* sp. under controlled laboratory conditions, at salinities of 20, 30, 40, 55 and 60 ppt, in a static system. The number of cells was counted every 12 h and the biomass production was evaluated every 24 h. The higher cellular densities and growth rates were recorded at the lower salinities (20 and 30 ppt), with final values of 9.0 and 8.8 x10^6 cells·mL^{-1}, respectively. The pH increased over the culture period at the salinities of 20, 30 and 40 ppt, and remained without changes at 55 and 60 ppt. The lowest values of dry weight, organic matter and ash were observed at salinities of 20 and 30 ppt, and increased as salinity did (Table 10). It was concluded that low salinities favored duplication rate of the species, and high salinities increase cellular volume and diminish the growth rate (López-Elías et al., 2004a).

Richmond (1986) and Abalde et al. (1995), documented that microalgae equilibrate the osmotic pressure by enhancing the synthesis of organic compounds osmoprotectants or incorporating inorganic salts. That suggests that *Isochrysis* sp. in our study used one or both of these strategies.

4.4.

In a related investigation, the growth of *T. chuii* at five salinities (20, 30, 40, 50 and 60 ppt), and two nitrogen:phosphorous rates (15:1 and 30:1), was evaluated. An inoculum of 1.0 x 10^5 cells·mL^{-1} was used for the treatments. The pH increased over time from 7.2 at the beginning to 8.8 at the end of the trial. The greatest final cellular density was obtained at 30 and 40 ppt, with values of ~5.0 x 10^6 cells·mL^{-1} at the N:P ratio of 30:1. The final cellular concentrations were also higher at 30 and 40 ppt, at N:P ratio of 15:1, however, they were lower (~4.2 x 10^6 cells·mL^{-1}) than those observed for microalgae cultured in 30N:1P. A greater decrease in growth was observed in microalgae cultured in 15N:1P at higher salinities (50 and 60 pps) (< 3.7 x 10^6 cells·mL^{-1}), (Figure 4) (López-Elías et al., 2010b). Independently of salinity, the higher means of daily duplication rate were observed at the higher N:P ratio (30:1; 0.6 divisions·day^{-1}) compared to the 15:1 (0.5 divisions·day^{-1}). Domínguez and Guevara (1994) and López-Elías et al. (2006) also found even higher growth performances for the same species at a higher N:P ratios.

Table 10. Means ± SD of final cellular density dry weight, organic matter and ash of *Isochrysis* sp. at different salinities

Salinity (‰)	Density (cells·mL^{-1} x 10^6)	Dry weight (pg·cell^{-1})	Organic matter (pg·cell^{-1})	Ash (pg·cell^{-1})
20	9.02± 0.85 ^{d}	27.2±17.56 ^{a}	15.76±7.53 ^{a}	11.45±2.94 ^{a}
30	8.77±2.06 ^{d}	40.97±19.39 ^{b}	23.28±16.73 ^{a}	17.69±8.42 ^{b}
40	3.45±0.36 ^{c}	47.96±14.45 ^{b}	26.31±3.62 ^{b}	23.90±14.01 ^{b}
55	1.84±0.17 ^{b}	97.07±14.12 ^{c}	38.03±8.54 ^{c}	59.04±12.95 ^{c}
60	0.87±0.06 ^{a}	226.92±30.92 ^{d}	89.66±2.42 ^{d}	137.56±29.03 ^{d}

Different letters in a column jeans significant differences at $\alpha = 0.05$.

4.5.

In shrimp hatcheries, is important to feed the first larvae phases of aquatic organisms with phytoplankton, however, cultures of green flagellated are commonly not efficient, because of the inadequate formulation and use of mediums, which subsequently leads to lower productions. The growth of *T. chuii* was evaluated in outdoor mass cultures using different culture mediums. These cultures were maintained for two days in tanks of 250 L. The mediums used were f/2, f, 2f, 4f and two alternative mediums with modifications in their N:P ratio [N_{30}:P_1(1) and N_{30}:P_1(2)]. N_{30}:P_1(1) medium was formulated with sodium nitrate (150 g·L^{-1}) and monobasic sodium phosphate (8.09 g·L^{-1}), while N_{30}:P_1(2) was prepared with sodium nitrate (183.6 g·L^{-1}) and monobasic sodium phosphate (10 g·L^{-1}). Final cell density varied from 2.7 to 4.3 x 10^5 cells·mL^{-1} and dry weight ranged between 32 and 51 g.m^3 (Table

11). The highest growth was observed both modified mediums [N_{30}:P_1(1) and N_{30}:P_1(2)] compared to those considered as conventional. It was concluded that and adequate management of N:P ratio canenhance cell density of outdoor mass culture of *T.chuii* at less cost (López-Elías et al. 2006). In commercial hatcheries form Sinaloa, Mexico, mean cellular densities around 0.2 x 10^6 cél mL^{-1} were obtained in 3-days culture (López-Elías et al., 2003), which means that the modification of the culture medium has a positive effect on the growth of the species as reported by Lourenco et al. (1997).

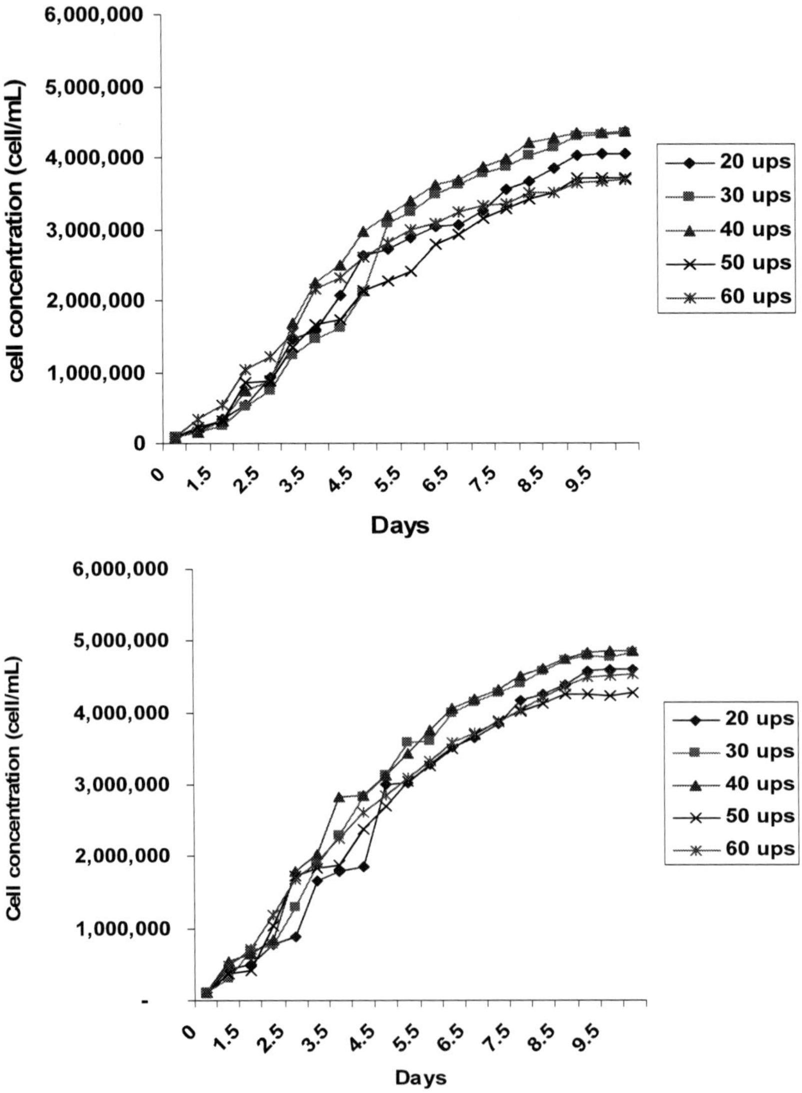

Figure 4. Growth of T. chuii at salinities of 20, 30, 40, 50 y 60 pps a and N:P rates of 15:1 (A) and 30:1 (B).

Table 11. Means ± SD of cellular density, growth rate and dry weight production in the outdoor culture of *T. chuii* in the mediums f/2, f, 2f, 4f, N_{30}:P_1(1) and N_{30}:P_1(2)

Day	Cells·mL^{-1} x 10^5	Variation (%)	μ (Divisions·day^{-1})	Final biomass (g·m^3)
f/2 medium				
0	0.60	0.00	0.92	34.42 ± 5.69 ^{a}
1	1.13 ± 0.33	29.20		
1.5	0.93 ± 0.29	31.18	1.44	
2	3.08 ± 0.65^a	21.10		
f medium				
0	0.60	0.00	0.56	32.44 ± 3.24^a
1	0.89 ± 0.21	23.60		
1.5	0.93 ± 0.15	16.13	1.60	
2	2.99 ± 0.21^a	7.02		
2f medium				
0	0.60	0.00	1.14	51.24 ± 5.67^b
1	1.32 ± 0.30	2.27		
1.5	1.18 ± 0.50	4.24	1.17	
2	2.97 ± 0.48^a	16.16		
4f medium				
0	0.60	0.00	0.93	49.80 ± 2.96 ^{b}
1	1.14 ± 0.16	14.04		
1.5	11.4 ± 0.11	9.65	1.31	
2	2.84 ± 0.26^a	9.15		
30N:1P (1)				
0	0.60	0.00	0.81	47.97 ± 6.04 ^{b}
1	1.05 ± 0.11	10.48		
1.5	1.29 ± 0.08	6.20	1.62	
2	3.24 ± 0.05^a	1.54		
30N:1P (2)				
0	0.60	0.00	1.18	48.37 ± 5.58 ^{b}
1	1.36 ± 0.10	7.35		
1.5	1.39 ± 0.03	2.16	1.66	
2	4.30 ± 0.42^b	9.77		

5. Evaluation of Production Protocols in Commercial Laboratories

Although the traditional protocols for microalgae production are similar among laboratories, some routines can vary in diverse aspects such as: type of containers (size, form, material and etcetera), size and hour of inoculum, illumination intensity, culture medium, time for harvest, and others. Those differences may have a significant influence in the production results in terms of cellular concentration, duration of each one of the development phases, final biomass and chemical proximate composition.

5.1.

The protocols for microalgae outdoor production were evaluated in different commercial hatcheries from Sinaloa and Sonora states, Northwest Mexico, during three years. It was found that in Sinaloa, the containers for microalgae production were pools from 2 to 4 m^3 and fiberglass transparent cylinders of 1 m^3; the harvests were made in a period from 1 to 4 days. In Sonora, some hatcheries used pools of 4 m^3 and the microalgae were harvested in 2 to 3 days; other laboratories used opaque cylinders of 0.8 m^3, and harvested at 3 days, while others used pools of 2.5 m^3 and harvested microalgae at days 2 and 3 (López-Elías et al. 2003). The total of evaluated laboratories performed outdoor cultures to reduce costs, and because the biomass obtained was greater than that needed.

5.2.

The growth rate, final biomass, and proximate composition of massive outdoor culture of *C. muelleri*, were evaluated in two commercial hatcheries of Sonora, Mexico: El Camarón Dorado (CD), and Aqualarvas (AL) (Table 12). The first one maintained its intermediate cultures in a laboratory multi step system at temperatures between 20 and 22 °C, and a continuous photon flux of 7.9 $mol{\cdot}m^{-2}{\cdot}d^{-1}$; and for higher volume cultures, they used outdoor ponds of 2.5 m^3 and 1 m depth. The second hatchery maintained similar conditions but used tanks of 3.3 m^3 and 0.5 m depth for the last phase.

In CD hatchery, the indoor culture (300 L) achieved a cellular density of 0.77 x 10^6 $cells{\cdot}mL^{-1}$ and a duplication rate of 0.74 $divisions{\cdot}day^{-1}$ in three days; however, the growth performance significantly increased when microalgae were transferred to outdoor tanks (2.04 x 10^6 $cells{\cdot}mL^{-1}$; 1.4 $divisions{\cdot}day^{-1}$).

In AL hatchery, microalgae reached in two days a mean cellular density of 1.38 x 10^6 $cells{\cdot}mL^{-1}$, but during the warm months the harvest was done after 1 day with a density of 1.84 x 10^6 x 10^6 $cells{\cdot}mL^{-1}$, while in the cold months microalgae were harvested after 3 days with a concentration of only 0.72 x 10^6 $cells{\cdot}mL^{-1}$.

The final biomass was significantly greater in the pools from CD hatchery (0.5 m depth), with values of 40 $g{\cdot}m^{-3}$ after 1 day and 49 $g{\cdot}m^{-3}$ with the normal routine of 2 days. In the tanks of AL hatchery (1 m depth), the production ranged from 32.8 to 33.6 $g{\cdot}m^{-3}$ independently of the time of harvest.

The harvest time, had a significant effect on the chemical proximate composition of the microalgae (Table 13). In AL hatchery, the protein content in dry basis was 75.0% at day 1 and 60.5% at day 2. The content of lipids plus carbohydrates ranged from 5.1 to 5.4 $g{\cdot}m^{-3}$ after 1 day and 10.1 $g{\cdot}m^{-3}$ after 2 days. In CD hatchery, the protein obtained was 17.6 (53.6 %) at day 2, but increased to 26.4 (78.8 %) in the following 24 hours (day 3) (López-Elías, et al., 2005a).

Table 12. Mean cell concentrations (10^6 cells·mL^{-1}) and growth rates (divisions·day^{-1}) of *C. muelleri* cultured in 300-L cylinders (A) and 2.5-3.3 m^3 tanks (B) in two commercial hatcheries

Day	Acualarvas (n=26)		Camarón Dorado (n=25)	
	A			
	cells·mL^{-1} x10^6	Growth rate μ	cells·mL^{-1} x10^6	Growth rate μ
0	0.36 ± 0.19	-	0.21 ± 0.07	-
1	1.34 ± 0.40	2.01 ± 0.45	0.38 ± 0.10	0.98 ± 0.39
2	2.04 ± 0.46	0.65 ± 0.54	0.56 ± 0.17	0.54 ± 0.19
3	-	-	0.77 ± 0.21	0.46 ± 0.25
$\bar{X}\mu$	-	1.39 ± 0.36	-	0.74 ± 0.24

Day	B (n=6) N	μ	(n=12) N	μ	(n=16) N	μ	(n=6) N	μ
0	0.51	-	0.44	-	0.22 ± 0.13	-	0.17	-
1	1.84	1.89 ± 0.32	1.08	1.36 ± 0.52	0.58 ± 0.26	1.32 ± 0.45	0.27	0.70 ± 0.48
2	-	-	1.38	0.41 ± 0.28	0.88 ± 0.33	0.54 ± 0.65	0.48	0.81 ± 0.24
3	-	-	-	-	-	-	0.72	0.59 ± 0.27
$\bar{\mu}$	-	1.89 ± 0.32	-	0.88 ± 0.30	-	0.94 ± 0.35	-	0.70 ± 0.19

Table 13. Mean of ash-free biomass (AFW), proteins, carbohydrates, and lipids on the outdoor culture of *C. muelleri* after 1, 2, and 3 days in two commercial hatcheries

Hatchery	Days	n	AFW g·m^{-3}	Protein g·m^{-3}	Carbohydrate g·m^{-3}	Lipid g·m^{-3}
AL	1	19	39.98 ± 10.7b	29.54 ± 7.3b	5.40 ± 2.45a	5.06 ± 1.76a
	2	65	48.65 ± 16.3c	29.47 ±12.8b	10.15 ± 6.38b	10.13 ± 5.25b
CD	2	65	32.84 ± 6.9a	17.60 ± 3.7a	5.57 ± 1.96a	9.66 ± 3.57b
	3	14	33.62 ± 6.5a	26.36 ±6.5ab	4.85 ± 1.31a	2.40 ± 0.61a

Different letters indicate significant differences between values in the same column (two way nonparametric analysis of variance and Dunn's multiple comparison tests. α = 0.05). a≤ab≤b and a<b.

5.3.

The microalgae production routines of 7 shrimp hatcheries of Sonora (3) and Sinaloa (4) were described, using the field data of each one of the hatcheries. The cellular density of inoculum and during each one of the growth phases was evaluated, considering the environmental conditions. The inoculum density varied in both, within the same hatchery and among different hatcheries (0.08–0.60 x 10^6 cells·mL^{-1}). In general, it was found that production routines were carried out in long periods, leading to late harvests when microalgae were at the stationary phase; i.e., they lost time trying to obtain higher productions (without results) instead of make a harvest and perform a semi-continuous culture. In the last step of production, the cellular density varied from 0.82 to 2.35 x 10^6 cells·mL^{-1} (Table 14). The conclusion of the study was that it is necessary to reduce the routine times and standardize the concentration of inoculums at all levels of production (López-Elías, et al., 2005b).

5.4.

In a similar investigation, the cellular concentration and the final biomass of *C. muelleri* cultured indoor and outdoor at levels of 300 and 3000 L, during spring, summer, and winter, were evaluated (Table 15). Temperature varied from 8 °C during the winter to more than 45 °C at summer. Outdoor systems showed a higher variability in terms of production, such results were associated to environmental conditions and water quality. In general, the cellular density and biomass production were significantly better in the outdoor systems (except for indoor culture at 300 L) (López-Elías, et al., 2005a). Regarding to indoor systems, the production was constant in all seasons and hatcheries because of the higher level in the control of variables; such productions were similar to those obtained at laboratory conditions (López-Elías et al. 1999). It was concluded that producers can choice to maintain their microalgae cultures outdoor with a higher but variable production, or indoor with a constant but lower production. Otherwise, they can practice outdoor cultures during warm months and indoor during cold months.

Table 14. Production parameters of *C. muelleri* at inoculation and harvest (x 10^6 cells·mL^{-1}) cultured outdoor in hatcheries of Sinaloa and Sonora. Cellular variation coefficient (CVC), hour of inoculum, replication rate (RR, divisions/day), days of culture (DC), and growth rate (GR. 24 h) outdoor cultured in pools in hatcheries of Sinaloa and Sonora

Hatchery	Inoculum concentration	CVC (%)	RR	DC	GR	Harvest
Tanks, year 1999						
Maricultura Lab. 1	0.157 (0.065)	41.4	1.21	3	2.43	0.798 (0.195)
Maricultura Lab. 2	0.080 (0.039)	48.9	1.59	3	3.02	0.597 (0.096)
Generación 50	0.419 (0.039)	9.3	1.95	1	1.95	1.620 (0.145)

Hatchery	Inoculum concentration	CVC (%)	RR	DC	GR	Harvest
Tanks, year 1999						
Camarón Dorado	0.237 (0.071)	30.2	1.50	3	2.49	1.187 (0.077)
Tanks, year 2000						
Maricultura Lab. 1	0.288 (0.119)	41.3	1.08	3	1.84	1.065 (0.148)
Maricultura Lab. 2	0.230 (0.025)	10.9	0.93	3	1.84	0.835 (0.167)
Maricultura Lab. 2 (Tinas de 1.0 m^3)	0.486 (0.223)	45.9	1.19	2	2.06	1.836 (0.173)
Generación 50	0.452 (0.056)	12.4	1.03	2-3	1.76	1.534 (0.367)
Camarón Dorado	0.187 (0.066)	35.4	1.08	2	1.45	0.822 (0.406)
Acualarvas	0.372 (0.268)	72.1	1.01	2	1.61	1.068 (0.821)
Aremar (Cylinders 800 L)	0.305 (0.071)	23.4	1.27	3	2.79	2.093 (0.130)
DICTUS (Tanks 200 L)	0.434 (0.038)	8.8	1.95	2	2.42	2.350 (0.467)

Table 15. Mean cell density and organic yield of indoor and outdoor (CREMES), indoor and greenhouse (UEK) cultures of *C. muelleri* during spring, summer and winter

	Cell density (cells·ml^{-1} x10^6)		*Organic yield (g·m^{-3})*	
CREMES(300 L)	Indoor	Outdoor	Indoor	Outdoor
Spring	2.2 ± 0.1	4.9 ± 0.2	0.101 ± 0.001	0.159 ± 0.012
Summer	1.4 ± 0.2	1.8 ± 0.2	0.076 ± 0.003	0.124 ± 0.007
Winter	1.8 ± 0.1	1.1 ± 0.7	0.066 ± 0.012	0.036 ± 0.100
CREMES (3000 L)				
Spring	1.0 ± 0.1	1.3 ± 0.0	0.036 ± 0.011	0.046 ± 0.007
Summer	0.7 ± 0.1	1.5 ± 0.3	0.049 ± 0.003	0.076 ± 0.002
Winter	0.6 ± 0.1	1.1 ± 0.1	0.045 ± 0.002	0.050 ± 0.002
UEK (300 L)				
Spring	1.3 ± 0.5	1.4 ± 0.4	0.057 ± 0.012	0.068 ± 0.026
Summer	1.1 ± 0.6	0.9 ± 0.2	0.060 ± 0.007	0.084 ± 0.018
Winter	1.2 ± 0.5	1.4 ± 0.4	0.084 ± 0.031	0.072 ± 0.027

Different letters indicate significant differences (two way ANO VA and Tukey multiple comparison test; a=0.05). Standard deviations in parenthesis.

6. Evaluation of Alternative Culture Mediums

The mediums commonly used to culture microalgae are those considered chemically complete, such as f, f/2, Walne and others (Tzovenis et al. 1997). However some alternative mediums have been proposed and evaluated in order to improve production or decrease costs (Piña et al., 2007; Lopez-Elias et al. 2008). The use of agricultural fertilizers seems to be an adequate alternative since they have nutrients that can be used by microalgae; in addition, their costs are commonly lower compared to the conventional mediums. The feasibility of using such alternative has been demonstrated; moreover it was observed that microalgae cultures in these mediums had similar production and quality parameters compared to those cultured in f/2 medium (López-Elías and Voltolina 1993; Piña et al. 2007). Thus, the conventional mediums can be substituted by alternative and cheaper mediums.

6.1.

An experimental study was conducted to evaluate effectiveness of two alternative mediums for the culture of microalgae *N. oculata* and *C. muellerii*, in terms of production response and nutritional value. The mediums evaluated were: mono ammonium phosphate (MAP, used in agriculture) and Nutrilake (NLK, used in aquaculture). The control consisted in the conventional medium Guillard f/2 (Guillard 1975). The cultures were initiated in 80-L fiberglass columns at a density of 4 x 10^6 cells mL^{-1}; a light intensity of 128-192 $\mu mol \cdot m^{-2}s^{-1}$ was maintained during the experiment, by using white light lamps, and an environmental temperature of 18-20 °C as suggested by Lopez-Elías et al. (2004).

Some differences on the production parameters of *N. oculata* (Table 16) and *C. muellerii* were recorded (Table 17). For both species, the best final cell density was obtained with the agricultural medium. *N. oculata* had similar protein content when farmed in the three mediums (28-32%), while *C. muelleri* ranged from 24-27% in protein content. Carbohydrate content increased in the alternative medium MAP, but the lipid content decreased. Results suggested that the medium used for agricultural purposes (MAP) was an adequate alternative to substitute the conventional medium, not only because it increased the production of both microalgae, but also because it had a quite lower price than the conventional medium.

Table 16. Production parameters and chemical proximate composition of *N. oculata* cultured in the two alternative mediums and the control

Medium	K	TD	Final density (cells mL^{-1})	Carbohidrates (%)	Lípids (%)	Proteins (%)
MAP	0.69±0.81a	1.23±0.57a	2.05±0.67b	36.83±1.49b	34.54±1.49a	28.63±2.00a
NLK	0.46±0.29a	1.38±0.67a	1.26±0.49a	31.63±0.69b	37.58±0.69a	32.12±0.74a
GF/2	0.71±0.65a	1.35±0.75a	1.76±0.80b	28.96±1.83a	42.01±1.83b	29.02±2.25a

K: growth rate (divisions day^{-1}), DT: duplication time (days). The proximate composition is expressed in organic fraction. Different letters in a row means significant differences at $P < 0.05$.

Table 17. Production parameters and chemical proximate composition of *C. muelleri* in the two alternative mediums and the control

Medium	K	DT	Final density (cells mL^{-1})	Carbohidrates (%)	Lípids (%)	Proteins (%)
MAP	1.00±0.27b	0.89±0.46a	3.69±0.76b	35.52±2.64b	40.45±2.10a	24.03±2.70a
NLK	0.61±0.22a	1.33±0.45b	2.28±0.67a	34.31±1.32ab	41.67±0.69a	24.02±1.00a
GF/2	0.86±0.29ab	0.83±0.31a	2.94±0.09ab	30.76±1.43a	41.68±2.78a	27.56±1.48b

K: growth rate (divisions day^{-1}), DT: duplication time (days). The proximate composition is expressed in organic fraction. Different letters in a row means significant differences at $P < 0.05$.

6.2.

An investigation at laboratory and massive level was performed to evaluate the effectiveness of two agricultural fertilizers (FERTIMEX, México, and Peters, U.S.A.) on the growth and chemical proximate composition of *I. galbana* (Table 18).

At laboratory level, culture was done in 2.0-L Erlenmeyer flasks, and it was observed that final cellular density was significantly higher in the stationary phase using FERTIMEX (7.47 x 10^6 cells·mL^{-1}), compared to that using Peters (5.21 x 10^6 cells·mL^{-1}), however no differences were found for dry biomass among treatments (~112 g.L^{-1}). Chemical proximate composition varied depending on the culture age and the medium used. The lipid content tended to increase over time in both treatments; contrarily, the protein content decreased over time. Apparently, the microalgae at their low growth phase, had a more balanced nutrient content; however, it also depends on the requirements of the species that is pretended to feed with the microalgae (López-Elías et al., 2008). The conclusion was that the agricultural fertilizers were adequate for the culture of *I. galbana* microalgae, and that higher contents of protein are found in the exponential phase of growth.

Table 18. Cellular density, dry biomass and other constituents of *I. galbana* cultured with two agricultural ferilizers in different growth phases (EXP= exponential, LG= low growth and STAT= stationary) in a static system

	Fertimex			**Peters (FP)**		
	EXP	*LG*	*STAT*	*EXP*	*LG*	*STAT*
Cellular density	1.56 ^{b}	3.90 ^{a}	7.47 ^{b}	0.98 ^{a}	3.47 ^{a}	5.21 ^{a}
(x 10^6 cél·mL^{-1})	(0.50)	(0.68)	(1.01)	(0.32)	(0.54)	(1.12)
Dry biomass	0.049^a	0.093 ^{a}	0.122 ^{b}	0.055 ^{a}	0.099 ^{a}	0.112 ^{a}
(g.L^{-1})	(0.008)	(0.015)	(0.028)	(0.005)	(0.015)	(0.013)
Proteins	4.80 ^{a}	4.45 ^{a}	3.12 ^{a}	9.47 ^{b}	4.22 ^{a}	3.48 ^{a}
(pg $cell^{-1}$)	(1.79)	(0.73)	(0.18)	(2.60)	(0.54)	(0.64)
Carbohydrate	2.22 ^{a}	2.44 ^{ab}	2.42 ^{ab}	2.90 ^{b}	2.55 ^{ab}	2.44 ^{ab}
(pg $cell^{-1}$)	(0.32)	(0.51)	(0.20)	(0.46)	(0.29)	(0.46)
Lipids	2.23 ^{a}	2.26 ^{a}	3.83 ^{c}	3.01 ^{b}	2.68 ^{ab}	3.97 ^{c}
(pg $cell^{-1}$)	(0.40)	(0.16)	(0.23)	(0.49)	(0.23)	(0.07)

6.3.

A similar study was conducted to evaluate the aquacultural fertilizer Nutrilake, compared to urea and the traditional f medium, as nitrogen source for the culture of the microalgae *C. muelleri*, *T. weissflogii*, *Isochrysis* sp. and *T. suecica*. Final cellular concentration, duplication rate, dry weight, biomass, and chemical proximate composition were monitored. *C. muelleri* showed the highest cellular concentrations and biomass using urea as culture medium, and no significant differences among mediums were found with respect to the chemical proximate composition. *T. weissflogii* observed greater cellular concentrations with Nutrilake. For *Isochrysis* sp. y *T. suecica* no significant differences were found in cellular concentrations among mediums. Their greatest protein contents were obtained with urea. Carbohydrate content of *T. weissflogii* was significantly higher with Nutrilake and f medium, while higher limpid contents were found using urea. No changes in dry weight were recorded for *T. weissflogii* and *T. suecica,* but *Isochrysis* sp. showed higher values with Nutrilake. The replication rate was higher using urea and Nutrilake for *T. weissflogii, T. suecica* and *Isochrysis* sp., but not for *C. muelleri*. (Piña et al., 2007).

REFERENCES

Abalde, J., Cid, A., Hidalgo, J. P., Torres, E. andandand Herrero, C. *1995*. Microalgas: Cultivo y Aplicaciones. Universidad de Coruña, España, 209 pp.

Arroyo-Pacheco, L.E. y Martínez-Baldenebro F. *1994*. Producción de biomasa y composición química de dos especies de microalgas marinas a diferentes salinidades. Tesis de licenciatura. Depto. de Ciencias Químico Biológicas, Universidad de Sonora. 78 pp.

Becerra-Dórame, M., López-Elías, J.A., Enríquez-Ocaña, F., Huerta-Aldaz, N., Voltolina, D., Osuna-López, I. andandand Izaguirre-Fierro, G. *2010*. The effect of initial cell and nutrient concentrations on the growth and biomass production of outdoor cultures of *Dunaliella* sp. Ann. Bot. Fennici., 47:109-112.

Becker, E.W. *1994*. Microalgae. *Biotechnology and microbiology*. Cambridge University Press. Great Britain. 293 pp.

Becker, W. *2004*. Microalgae for Aquaculture: The Nutritional value of Microalgae for Aquaculture. En Richmond A. Handbook of Microalgal culture: *Biotechnology and Applied Phycology*. Blackwell Publishing. U.S.A. 380-391 pp.

Brown R.M., Dunstan, A.G., Norwood, J.A. andandand Miller, A.K.. *1996*. Effects of harvest stage and light on the biochemical composition of the diatom Thalassiosira pseudonana *Journal of Phycology*., 32, 64-73.

Brown M.R., Jeffrey, S.W., Volkman, J.K. andandand Dunstan, G.A. *1997*. Nutritional properties of micro algae for mariculture. *Aquaculture,* 151, 315-331.

Burlew, J.S. 1953. Current status of the large-scale culture of algae. In: Burlew, J.S. (ed.). Algal Culture form Laboratory to Pilot Plant. Carnegie Institution or Washington, USA. No. 600, Pp 3-33.

Chisti Y. *2008*. Biodiesel from microalgae beats bioethanol. *Trends Biotechnol.* 26:126–131.

Cleber Bertoldi, F., Sant'Anna, E., Villela da Costa Braga, M. andandand Barcelos Oliveira, J.L. *2006.* Lipids, fatty acids composition and carotenoids of *Chlorella vulgaris* cultivated in hydroponic wastewater. *Grasas y Aceites.* 57 (3): 270-274.

Domínguez, R.L. M. andandand Guevara E.B.L. *1994.* Optimización de la razón nitrógeno:fósforo sobre el crecimiento y biomasa de dos especies de microalgas. *Tesis de licenciatura.* Universidad de Sonora, México, 91 pp.

D'Souza, F.M. andandand Loneragan, N.R.. *1999.* Effects of monospecific and mixed-algae diets on survival, development and fatty acid composition of penaeid prawn (*Penaeus* sp.) larvae. *Marine Biology.* 133: 621-633.

Gallegos-Simental, G., Voltolina, D., López-Elías, J.A., Enríquez-Ocaña, F. y P., Piña. *2002.* Producción a la intemperie y en el laboratorio de la diatomea Chaetoceros muelleri en Bahía Kino, Sonora, México. *Oceánide.* 17 (2): 85-91.

García-Lagunas, N., López-Elías, J.A., Miranda-Baeza, A., Martínez-Porchas, M., Huerta-Aldaz, N. andandand García-Triana A. *2010.* Effect of salinity on growth and chemical composition of the diatom *Thalassiosira weissflogii* in three phases of culture. (Enviada).

Guillard, R.L. andandand Ryther, J. H. *1962.* Studies on marine planktonic diatoms I. *Cyclotella nana* Husted and Detonula confervacea (Cleve) Gran. *Canadian Journal of Microbiology*, 8: 229-239.

Guillard, R.R.L. *1975.* Culture of the phytoplankton for feeding marine invertebrate larvae.

In: Smith, W.L., Chanley, M.E. (Eds.), *Culture of Marine Invertebrate Animals.* Plenum Publishing Co., New Cork, pp. 296-360.

Kirst, G.O. *1989.* Salinity tolerance of eukaryotic marine algae. *Annu. Rev. Plant Physiol. Plant Mol. Biol.* 40: 21-53.

López Elías, J.A. y Voltolina, D. *1993.* Cultivos semicontinuos de cuatro especies de microalgas con un medio no convencional. *Ciencias Marinas* 19 (2): 169-180.

López-Elías, J.A., Encinas-Arreola, A. R., García-Valenzuela, A.C., Váldez, J. andandand Hoyos-Chaires, F. *1999.* Producción anual de dos especies de microalgas en un centro acuícola en Bahia Kino, Sonora. OCEÁNIDES, 14 (1): 59-65.

López-Elías, J.A., Voltolina, D., Chavira-Ortega, C.O., Rodríguez-Rodríguez, B.B., Sáenz-Gaxiola, L.M., Cordero-Esquivel, B. y Nieves-Soto, M. 2003. Mass production of microalgae in six commercial shrimp hatcheries of the Mexican northwest. *Aquacultural engineering* 29: 155-164.

López Elías J.A., Huerta Aldaz N, Estrada Durán G.J., Celis Salgado M. P., De la Re Vega E., Quintero Arredondo N., Estrada Quintero J. A., Niebla Larreta J. L., Miramontes Higuera N., García Quiroz K. M., Niebla Rodríguez S. J., Carvajal Sánchez I. S. y Velazco Rameños J. *2004a.* Efecto de la salinidad en el crecimiento de *Isochrysis sp.* Bajo condiciones de cultivo estático. Biotécnia Vol. 6 No. 3: 10-15.

López-Elías, J. A., Voltolina, D., Nieves-Soto, M. y Figueroa-Ortiz, L. *2004b.* Producción y Composición de Microalgas en Laboratorios Comerciales del Noroeste de México. In: Cruz Suárez, L.E., Ricque Marie, D., Nieto López, M.G., Villarreal, D., Scholz, U. y González, M. Avances en Nutrición Acuícola VII. Memorias del VII Simposium Internacional de Nutrición Acuícola.16-19 Noviembre, 2004. Hermosillo, Sonora, México.

López Elías, J.A., Voltolina, D., Enríquez Ocaña, F. and Gallegos Semental, G. *2005a.* Indoor and outdoor mass production of the diatom *Chaetoceros muelleri* in Mexican commercial hatchery. *Aquacultural Engineering* 33 (3): 181-191.

López Elías, J.A., Cortes González, I., Nieves Soto, M., Enríquez Ocaña, F., Piña Valdéz, P., Voltolina, D. y Pablos Mitre, N.M. *2005b*. Seguimiento de cultivos masivos de microalgas en siete laboratorios productores de larvas de camarón. *Biotécnia* Vol. 7 No. 3: 36.43.

López-Elías, J.A., Badilla-Flores, K.D., Estrada-Raygoza, L.A., Fimbres Olivarría, D., Ochoa-Castillo, L. de J., Ramos-Brito, L. y Huerta-Aldaz, N.. *2006*. Manejo de medios de cultivo en el crecimiento de *Tetraselmis chuii* en cultivos al exterior. *Biotecnia.* Vol 8 No. 3: 14-21.

López-Elías, J.A., Carvallo-Ruiz, G., Enríquez-Ocaña, F., Huerta-Aldaz, N. y Aguirre-Rosas, J.C.. *2008.* Crecimiento y composición bioquímica de *Isochrysis galbana* cultivada en fertilizantes agrícolas. *Biotecnia,* vol. X No. 2: 22-32.

López-Elías, J.A., Enríquez-Ocaña, F., Pablos-Mitre, M.N., Huerta-Aldaz, N., Leal, S., Miranda-Baeza, A., Nieves-Soto, M. and Vásquez- Salgado, I. *2008.* Growth and biomass production of *Chaetoceros muelleri* in mass outdoor cultures: Effect of the hour of the inoculation, size of the inoculum and culture medium. *Rev. Invest. Mar.* 29 (2): 171-177.

López-Elías, J.A., García-Lagunas, N., Jiménez-Gutiérrez, L.R. y Huerta-Aldaz, N. *2009.* Crecimiento de la diatomea *Thalassiosira pseudonana* en cultivos estáticos con iluminación continua y fotoperiodo a diferentes salinidades. *Biotecnia,* Vol. XI No.1: 11-18.

López-Elías, J.A, Esquer-Miranda, E., Martínez-Porchas, M., Garza-Aguirre, M. del C., Rivas-Vega, M., Huerta-Aldaz, N., Aguirre-Hinojosa, E. y Nieves-Soto, M. *2010a.* Evaluación del efecto de la hora y la concentración de inóculo en el desarrollo de cultivos estáticos masivos de *Tetraselmis chuii* (Butcher,1958*)* en los medios f/2 y 2 f. (en preparación).

López-Elías, J.A., Gonzalez-Bello, G. E.., Huerta-Aldaz, N., Murguia-López, A., Mercado-Castillo, L. y Miranda-Baeza, A. *2010b.* Efecto de cinco salinidades y dos razones nitrógeno-fósforo sobre el crecimiento de la microalga *Tetraselmis chuii.* XII Congreso de la Asociación de Investigadores del Mar de Cortés y Simposium Internacional sobre el Mar de Cortés. Marzo 2 a 5. 2010. Guayamas, Son., Mex.

Lourenco, S.O., Lanfer Marquez, U.M., Mancini-Filho, Barbarino, E. and Aidar, E. *1997.* Changes in biochemical profile of *Tetraselmis gracilis* I. Comparison of two culture media. *Aquaculture*. 148: 153-168.

Lovatelli, A. *2004*. *Hatchery Culture of bivalves*. A practical manual. FAO Technical paper 471. 177 pp.

Martínez-Córdova. *1999*. Cultivo de Camarones Peneidos. *Principios y Prácticas.* AGT Editor, México, D.F. 102 p.

Matsunaga, T., Takeyama, H., Miyashita, H. andandand Yokouchi, H. *2006.* Marine microalgae. Adv. Biochem. Engin. Biotechnol. 96:165 – 188.

Medina-Reyna C.E. andandand Cordero-Esquivel, B. *1998.* Crecimiento y composición bioquímica de la diatomea *Chaetoceros muelleri* (Lemerman), mantenida en cultivo estático con un medio comercial. *Ciencia y Mar 6*, 19-26.

Oswald, W.J. *1988*. Large-scale algal culture systems (engineering aspects). En Borowitzka, M.A. y Borowitzka, L. J. (eds.). *Microalgal Biotechnology*. Cambridge University Press. pp. 357-394.

Piña, P., Voltolina, D., Nieves, M. andandand Robles, M. *2006.* Survival, development and growth of the Pacific White Shrimp *Litopenaeus vannamei* protozea larvae, fed with monoalgal and mixed diets. *Aquaculture* 253:523-530.

Piña, P. Medina A.M., Nieves, M., Leal, S., López-Elías, J.A.y Guerrero, M.A. *2007***.** Cultivo de cuatro especies de microalgas con diferentes fertilizantes utilizados en acuicultura. *Rev. Invest. Mar.* 28 (3): 225-236.

Pipes, W.O. andandand Gotaas, H.B. *1960*. Utilization of organic matter by *Chlorella* grown in seawage. *Appl. Microbiol.* 8:163-169.

Richmond, A. *1986.* Handbook of microalgal mass culture. CRC Press, Boca Raton Florida. 528 pp.

Richmond, A, *2004.* Handbook of Microalgal Culture. *Biotechnology and Applied Phycology. Blackwell* Publishing, USA, 566 pp.

Renaud S.M. and Parry D.L. *1994.* Microalgae for use in tropical aquaculture II: Effect of salinity on growth, gross chemical composition and fatty acid composition of three species of marine microalgae. *Australian Journal of Botany*. 51: 703-713.

Rodríguez, E. O., López-Elías, J.A., Aguirre-Hinojosa, E., Garza-Aguirre, M. del C., Constantino-Franco, F., Miranda-Baeza, A. andandand Nieves-Soto, M. *2010.* Evaluation of the nutritional quality of *Chaetoceros muelleri* Schütt (Chaetocerotales: Chaetocerotaceae) and *Isochrysis* sp. (Isochrysidales: Isochrysidaceae) grown outdoors for the larval development of *Litopeneaus vannamei* (Boone, 1931) (Decapoda: Penaeidae). (En preparación).

Rosales, N., Ortega, J., Mora, R. andandand Morales, E. *2005.* Influencia de la salinidad sobre crecimiento y composición bioquímica de la cianobacteria *Synechococcus* sp. Ciencias Marinas. 31(2): 349-355.

Seon Jin. E. andandand A., Melis. *2003.* Microalgal biotechnology: Carotenoid production by the green algae *Dunaliella salina. Biotechnology and Bioprocess Engineering.* 8: 331-337.

Soeder, C.J. *1986.* An historical outline of applied algology. Pp. 25-41. In Richmond, A. (ed.). Handbook of Microalgal Mass Culture. CRC.

Thessen, A., Dortch, Q., Parsons, M. andandand Morrison, W. *2005*. Effect of salinity of *Pseudo-nitzschia* species (bacillariophyceae) growth and distribution. *Journal of Phycology*. 41: 21-29.

Tzovenis, I., De Pauw N. andandand Sorgeloos P. *2003.* Optimization of T-ISO biomass production rich in essential fatty acids I. Effect of different light regimes on growth and biomass production. *Aquaculture*, 216;203-222.

Tzovenis, I., DePauw, N. y Sorgeloos, P. *1997.* Effect of different light regimes on the docohexaenoic acid (DHA) content of *Isochrysis* aff. *galbana* (clone T-ISO). *Aquaculture International*, 5: 489-507.

Sánchez-Saavedra, M.P. and Voltolina, D. *1996*. Effects of blue-green light on growth rate and chemical composition of three diatoms. *Journal of Applied Phycology*. 8:131-137.

Voltolina, D., Bückle Ramírez, L.F. y Morales Guerrero, E. L. *1989.* Manual de metodologías y alternativas para el cultivo de microalgas (2ª Ed.) Centro de Investigación Científica y de Educación Superior de Ensenada, B. C., México. *Informe Especial* OC-89-01, 67 pp.

Voltolina, D. and López-Elías, J.A. *2002*. Cultivos de apoyo para la acuacultura: Tendencias e innovaciones. Pp. 23-41. En Martínez-Córdova L.R. (ed.). Camaronicultura. Avances y tendencias. AGT Editor, S.A., Mex., 167 pp.

Voltolina, D., Piña, P. and Nieves, M. *1999.* Fertilizers as cheap growth media: a mexican point of view. *Rivista Italiana di Acquacoltura,* 34: 43-45.

Volkman, J.K., Jeffrey, S.W., Nichols, P.D., Rogers, G.I. andandand Garland, C.D. *1989.* Fatty acid composition of 10 species of microalgae used in mariculture. *Journal of Experimental Marina Biology and Ecology*. 128:219-240.

Vonshak, A. *1988.* Microalgas: técnicas de cultivo en laboratorio y para producción de biomasa a la intemperie. p. 155- 165. En Coombs, J., May, D.O., Long, S.P. y Scurlock, J.M. (eds.) Técnicas en Fotosíntesis y Bioproductividad. Colegio de Graduados, Chapingo, Estado de México, México, 258 pp.

Zhang C.W., Richmond A. *2003.* Sustainable, high-yielding outdoor mass cultures of *Chaetoceros muelleri* var. *subalsum* and *Isochrysis galbana* vertical plate reactors. *Marine Biotechnology*. 5:302-310.

In: Zooplankton and Phytoplankton
Editor: Giri Kattel, pp. 63-88
ISBN 978-1-61324-508-8

Chapter 3

ANNUAL CYCLE OF THE PLANKTON BIOMASS IN THE NATIONAL PARK SISTEMA ARRECIFAL VERACRUZANO, SOUTHWESTERN GULF OF MEXICO

Yuri B. Okolodkov**, *José A. Aké-Castillo, María G. Gutiérrez-Quevedo, Horacio Pérez-España, and David Salas-Monreal
Instituto de Ciencias Marinas y Pesquerías (ICIMAP),
Universidad Veracruzana,
Calle Hidalgo No. 617, Col. Río Jamapa, C.P. 94290,
Boca del Río, Veracruz, México

ABSTRACT

From April 2007 through May 2008, a total of 12 monthly surveys were carried out at nine oceanographic stations in the National Park Sistema Arrecifal Veracruzano, southwestern Gulf of Mexico. Horizontal tows were taken with a net of 120 μm mesh size for zooplankton, and with a net of 30 μm mesh size for phytoplankton. In total, 216 samples were analyzed. ANOVA and a principal component analysis showed temporal variation in biomass and parameters influencing it but no spatial pattern. Phytoplankton species richness was highest in June and October 2007 and lowest in April, November and December 2007. Zooplankton taxa richness was highest in April and June 2007 and lowest in March 2008. A list of 36 major zooplankton taxa and 275 phytoplankton species and infraspecific taxa is given. Plankton biomass ranged from 23 to 3637 mg wet weight m^{-3} (mean 552.6 mg m^{-3}), equaling 1.3-202.8 mg C m^{-3} (mean 30.7 mg C m^{-3}). The biomass within the park was unimodal in annual dynamics, with peak production in September-October at the end of the rainy season when the currents changed direction. Taxa richness contributed to the seasonal variation, and the calanoid copepods (mostly copepodite stages) and centric diatoms were the most conspicuous plankton components. Annual plankton biomass values showed a rather homogeneous spatial distribution.

* Fax: 52 (229) 956 70 70; e-mail: yuriokolodkov@yahoo.com.

Keywords: annual cycle, biomass, coral reefs, Gulf of Mexico, Mexico, national park, plankton, seasonal changes.

INTRODUCTION

Zooplankton has been shown to be an important component of coral reef ecosystems (Emery, 1968; Moore and Sander, 1976; Ferraris, 1982; Alldredge and King, 1985; Sorokin, 1990; McKinnon and Thorrold, 1993; Morales and Murillo, 1996; Castellanos-Osorio and Suárez-Morales, 1997; Alvarez-Cadena et al., 1998; Suárez-Morales and Rivera-Arriaga, 1998). The role of the so-called demersal (or distance demersal) zooplankton has begun to be understood only recently, when nocturnal migrations of some zooplankters in reef waters were taken into account and quantitative data on zooplankton abundance were obtained (Emery, 1968; Tranter and George, 1972; Sale et al., 1976; Alldredge and King, 1977, 1980, 1985; Porter and Porter, 1977; Porter et al., 1977; Hobson and Chess, 1979; Sorokin, 1990; Lewis and Boers, 1991; Heidelberg et al., 2010). Most of the reef-dwelling zooplankters associated with the sea bottom belong to demersal zooplankton. This is composed of adult invertebrates (mainly crustaceans) that usually live in or on solid substrates in the daytime and rise at night to swim near the bottom or in the water column. Sorokin (1990) distinguishes three ecological groups of demersal zooplankton, and meroplankton is one of them; the other two include adult individuals that go up into the water for a short or a long time at night. Most migrating demersal zooplankters are much larger in size than daytime holoplanktonic species. They form swarms, sheltering during the daytime. Demersal plankton appears especially adapted to avoid fish predation; the highest mean density of zooplankters emerged from living coral out of six studied substrate types (Alldredge and King, 1977). Branching corals provide the most demersal plankton and sand the least (Porter and Porter, 1977). The reef zooplankton is different from the zooplankton populations in the open waters in species composition, behavior (swarming and schooling) and abundance. It is a component of the reef community that uses its resources, and its main specific feature is its domination by populations of species associated with the bottom (Sorokin, 1990). In natural samples, zooplankton includes reef resident species, migrating demersal species, open water holoplanktonic species advected onto reefs and meroplankton from both open sea and reef sources (Emery, 1968; Heidelberg et al., 2004).

In the Mexican Atlantic, the ecology and community structure of the reef zooplankton have been studied more in the Caribbean Sea than in the Gulf of Mexico (Ferraris, 1982; Suárez and Gasca, 1990; Morales and Murillo, 1996; Suárez-Morales and Gasca, 1996, 2000; Castellanos-Osorio and Suárez-Morales, 1997; Alvarez-Cadena et al., 1998; Suárez-Morales and Rivera-Arriaga, 1998; Suárez-Moralez et al., 1999; Tovar et al., 2009).

The reefs of the National Park Sistema Arrecifal Veracruzano (NPSAV), located in the southwestern part of the Gulf of Mexico, are composed mainly of hermatypic corals that behave as planktivores. In the NPSAV, zooplankton is one of the main food sources for larger organisms. In addition to corals, planktivorous fish such as herring, anchovies, halfbeaks, damselfishes, sergeants, gobies and juveniles of many species are abundant. Other taxonomic groups such as sponges, barnacles, polychaetes and bivalves are also common planktivores in the area (Domínguez-Castanedo et al., 2007; Gómez, 2007; Martínez-Hernández and

Camacho-Olivares, 2007). Zooplankton must play an important role as the basis of the food web, and the majority of the ~1267 animal species living in the NPSAV (CONANP, 2007) have a planktonic stage as a part of their life cycles. Additionally, the reef zooplankton resources are actively used by pelagic fishes of the open sea, and these food resources are likely as or more important than that of the benthos because of their accessibility for grazing by economically important fish and fish larvae (Sorokin, 1990).

Some zooplankton biomass values have been previously reported from the southern Gulf of Mexico. These values ranged from 0.100 to 2532 g m^{-3} (Khromov, 1965a, 1965b, 1967; Bogdanov et al., 1968; De la Cruz, 1972, Flores-Coto et al., 1988; Gasca et al., 1995; g wet weight m^{-3} here and further in the article). Plankton biomass values in the Gulf of Mexico and the Caribbean Sea were summarized by Suárez and Gasca (1994), Gasca et al. (1995) and Okolodkov (2003). In the 1960-1970s, Flores-Coto (1965, 1974) studied the morphology of appendicularians of the NPSAV and reported the first zooplankton (appendicularians) abundance estimate in Veracruz waters. Other groups such as copepods, pteropods and chaetognaths were also studied (*e.g.*, Aguayo-Saviñón, 1965; Leal-Rodríguez, 1965; Vega-Rodríguez, 1965; Campos, 1980; Suárez, 1991, 1992; Campos-Hernández and Suárez-Morales, 1994). In all cases, local spatial distributions and organism morphology were reported. Information on seasonal changes in zooplankton biomass and its horizontal patterns is still scarce. The aim of this study was to determine the spatial-temporal distribution of the net plankton biomass in the NPSAV and to evaluate the partial contributions of the major zooplankton taxa and phytoplankton species.

Material and Methods

Study Area

The NPSAV is located in the southwestern Gulf of Mexico (Figure 1). It is formed of 23 coral reefs, 11 of which are located off the port of Veracruz in the northwestern part of the national park, while another 12 are located in its southeastern part, off Antón Lizardo village. The Jamapa River divides the two coral reef areas within the park. La Antigua River is its northern natural limit, and the Papaloapan River is its southern limit (Lara et al., 1992). The Jamapa River is a sediment source, and together with a pair of cyclonic and anticyclonic eddies formed off the river, inhibits the formation of coral reefs in this area; however, a relatively high concentration of plankton biomass has been suggested to occur over the cyclonic eddy (Salas-Monreal et al., 2009). The reefs are 0.3 to 3.2 km long, covering an area of 1 to 19 km^2 with a mean height varying between 20 to 45 m above the seafloor. The coastline and bathymetry of the state of Veracruz are characterized by a narrow continental shelf 6 to 33 km wide, with a mean depth of about 70 m. These features explain the diurnal tidal characteristics of the area (Salas-Perez et al., 2008), which result in one high and one lower zooplankton peak near the surface every 24 h (Salas-de-Leon et al., 2004a, 2008). The bathymetry is complicated due to the shallow-water sandy banks, reefs and islands that form a series of canals and sills of variable size and depth. This type of relief results from the large-scale circulation within the Gulf of Mexico (Salas-de-Leon et al., 2004b, 2008).

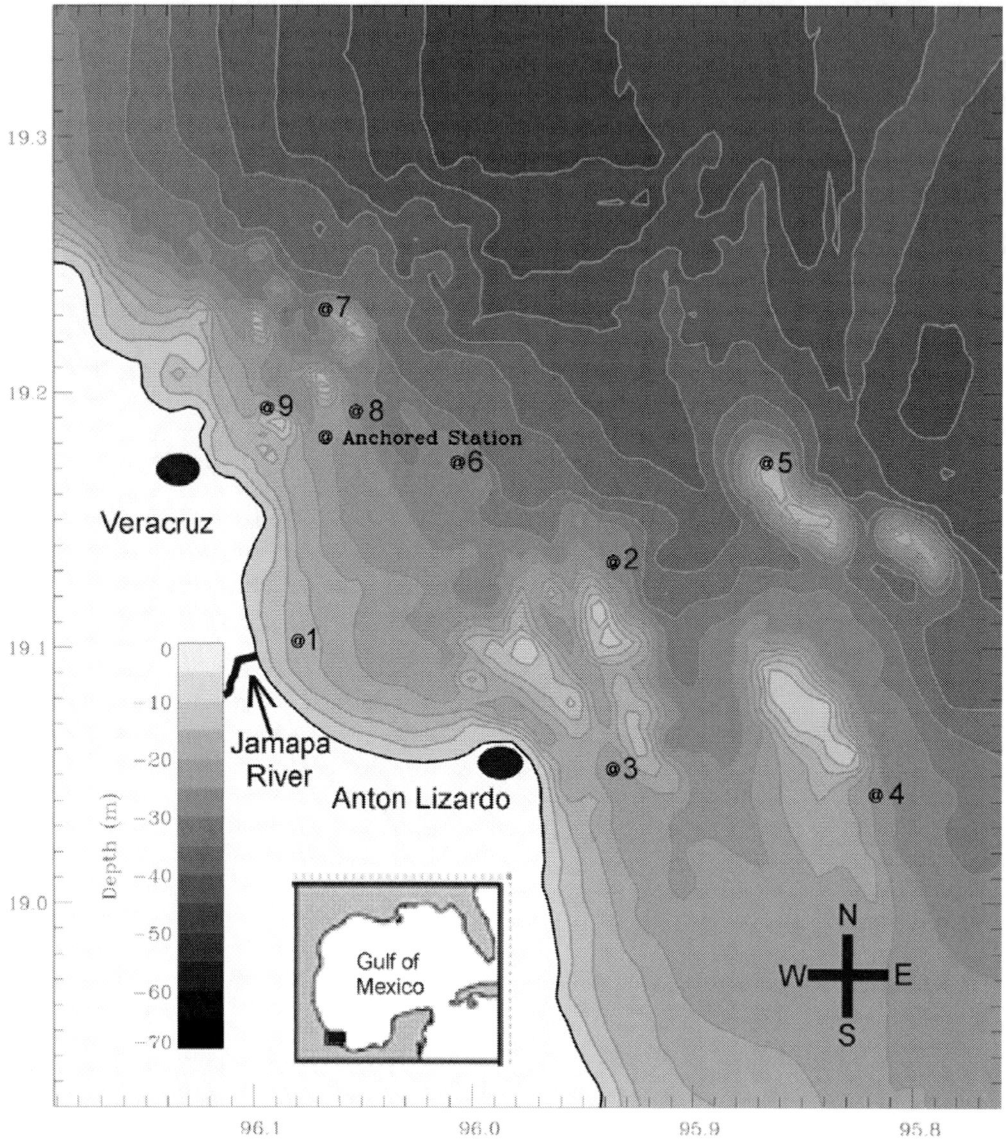

Figure 1. Location of the nine plankton sampling stations and the anchored ADCP station within the National Park Sistema Arrecifal Veracruzano, in 2007-2008.

The climate is hot and humid with abundant rains in summer and autumn and a short dry period in April-May. The mean annual air temperature is 26°C, with minimum temperatures close to 18°C in January-February (García-Abdeslem, 1987). The predominant northerlies are commonly observed during autumn and winter. During spring and summer, tropical storms occur and sometimes become hurricanes (Gutiérrez-de-Velasco and Winant, 1996) generating southerly winds. The southerly wind season may last until November. The diurnal tidal range reaches 95 to 105 cm (Salas-Pérez et al., 2008; González et al., 2009). Average water temperature is 26.6°C and usually fluctuates between 22.9°C and 32°C, while the salinity offshore is about 36 to 38 psu (Carrillo et al., 2007; Okolodkov et al., 2007; Salas-Monreal et al., 2009). The residual temperature (the mean averaged temperature after a diurnal tidal cycle) within the area is approximately 28°C with variations of less than 1°C (Salas-Monreal et al., 2009).

Data Collection

A total of 12 monthly surveys were carried out at 9 oceanographic stations from the 33-foot boat *CEPIA* (Figure 1, Table 1). Sampling was performed from April 2007 through May 2008, except for July and February, always between 9:00 to 15:00 hours. A total of 216 samples were taken with a plankton hand net of 120 μm mesh size with a mouth of 30 cm and a phytoplankton hand net of 30 μm mesh size. Horizontal samples of the superficial 30-cm layer were taken at a speed of 2.5 km h^{-1} for 5 min. They were immediately fixed with a 37% formalin stock solution to a final concentration of 4% in 500-ml plastic bottles. At each station temperature, salinity and water depth were measured *in situ* using a thermometer, refractometer and an eco-sounder, respectively. The volume of water filtered through the plankton net was calculated based on the net mouth and the boat speed.

In addition to net samplings, a 1 kHz Acoustic Doppler Current Profiler (ADCP) was anchored (Table 1) to obtain water velocities from surface to bottom near one of the net stations (Figure 1). The ADCP was calibrated to obtain one data set every 15 min. with a vertical resolution of 1 m. These data were used to correlate plankton biomass with current velocities to reveal any relationship between the direction and magnitude of the current and plankton concentrations within the study area. The data used here were averaged once every three days (72 h) before net samplings to obtain a single velocity averaged from the surface to 6 m depth, the depth of the seasonal pycnocline reported by Salas-Monreal et al. (2009).

Table 1. Geographic coordinates and site depth of oceanographic stations and the anchored ADCP station in the National Park Sistema Arrecifal Veracruzano in 2007-2008

Station	Site	Latitude °N	Longitude °W	Depth (m), mean±SD
1	Bocana de Jamapa	19°06'09.5"	96°04'57.2"	14.5±2.0
2	Enmedio	19°07'51.6"	95°56'41.4"	31.0±5.0
3	Antón Lizardo	19°03'02.7"	95°56'28.7"	24.5±4.4
4	Cabezo	19°02'43.1"	95°49'25.4"	28.0±5.9
5	Anegada de Afuera	19°10'39.5"	95°52'14.3"	24.0±5.9
6	Afuera Jamapa	19°10'26.9"	96°01'01.3"	34.0±4.2
7	Anegada de Adentro	19°14'06.5"	96°04'16.1"	32.0±7.3
8	Verde	19°11'27.3"	96°03'24.0"	21.0±2.9
9	Sacrificios	19°10'07.1"	96°05'33.9"	15.0±1.7
ADCP	Sacrificios	19°11'01.5"	96°06'06.8"	15.0±0.5

Laboratory Analysis

All plankton net samples were divided into two equal parts using a polycarbonate Folsom plankton splitter of 1-liter capacity: one half was for wet-weight biomass determination using a suction filtration technique (Zavala-García and Flores-Coto, 1989), and the other half was for taxonomic analysis. Half of each sample for wet-weight biomass measurement was processed using previously mounted equipment that included a vacuum pump, a graduated plastic cylinder of 250 ml, a Büchner funnel of 10 cm diameter, a glass conical Kitasato flask of 500 ml, a plastic cylinder of 9.5 cm diameter and 5 cm height, an 80 μm nylon mesh disc and a pre-weighed Whatman paper filter (Zavala-García and Flores-Coto, 1989). An Adam Equipment ADP 110L top loading balance (±0.001 g) was used to measure wet biomass. To convert it into carbon units, the factor of 0.0555, which was a mean value for phytoplankton (0.047) and crustaceans (0.064), was used (Vinogradov and Shushkina, 1987). The other halves of the net plankton samples were analyzed in a Petri dish to study the taxa composition using a stereo microscope Olympus SZ40 or Motic SMZ-168. Phytoplankton net samples were observed using a photonic microscope Motic BA300 for studying microalgae species composition.

Statistical Analyses

To determine temporal or spatial patterns, separate analyses of variance (ANOVAs) were conducted for each variable (wet biomass, number of phytoplankton species, number of major zooplankton taxa, temperature and salinity), using first the "month" as a source of variation and then "site" in different analyses. When considering month as a source of variation the variable biomass failed the homogeneity of variances test, so a non-parametric ANOVA by ranks (Kruskal Wallis) was performed for this variable. *Post hoc* comparison tests were performed (the Bonferroni test) in order to find groups that differed.

A principal component analysis (PCA) was performed using all variables to integrate information and to determine the contribution of the plankton community to the characterization of the NPSAV (Clarke and Warwick, 2001).

Results

Plankton Biomass

Biomass estimates ranged from 23 to 3637 mg m^{-3} during the study period (Table 2). The Kruskal-Wallis ANOVA showed significant differences among the sum of ranks ($H_{(11, N=116)}=55.85702$, $p<0.001$). Multiple comparisons of means of ranks indicated that September and October were the months when the biomass values differed from the rest (Figure 2A), with the highest median values. No specific month had a minimal value. These were also the months when currents changed direction from northward to southward. The ANOVA considering site as the variation source showed no significant differences among mean values of biomass (F=0.51, p=0.86); therefore no differences in plankton biomass among sampling stations were observed (Figure 2B). The highest plankton biomass peak was observed in September, when the highest mean value of precipitation, from January 1943 to December 1993 (Carrillo et al., 2007), had been recorded in the study area (Figure 3).

Table 2. Plankton biomass reported for the Gulf of Mexico and the western Caribbean Sea

Region	Biomass, mg m^{-3} (wet weight)	Biomass, mg C m^{-3}*	Net mesh size (μm), gear	References
SW and SE Gulf of Mexico (coastal zone)	100-1000	5.6-55.5	168	Khromov, 1967 (as pictured in Figure 2); Bogdanov et al., 1968 (as pictured in Figure 4)
SE Gulf of Mexico	200-1000	11.1-55.5	170	De la Cruz, 1972
SE Gulf of Mexico	100-1500	5.6-83.3	500	Flores-Coto et al., 1988
W Caribbean Sea	20-125	1.1-6.9	500	Suarez and Gazca, 1994
NE Gulf of Mexico, Florida Keys	No data	3.1-21.4	Plankton pump	Heidelberg et al., 2010
NPSAV, SW Gulf of Mexico	23-3637 (mean 552.6)	1.3-202.8 (mean 30.7)	120	This study

* Calculated using conversion factor of 0.0555 averaged from phytoplankton (0.047) and crustaceans (0.064) from Vinogradov and Shushkina (1987).

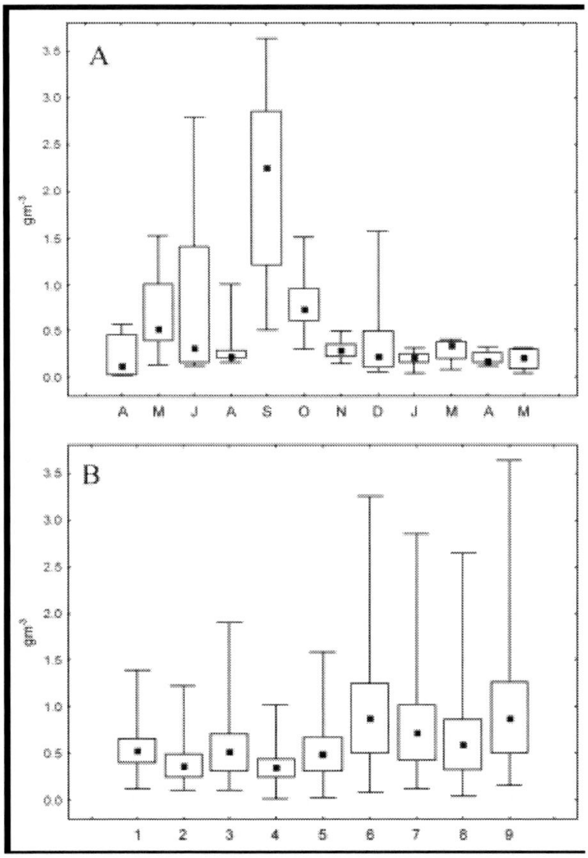

Figure 2. Mean values of wet-weight net-plankton biomass: A – per month; B – per station; boxes represent the standard error, and whiskers represent the minimum and the maximum values.

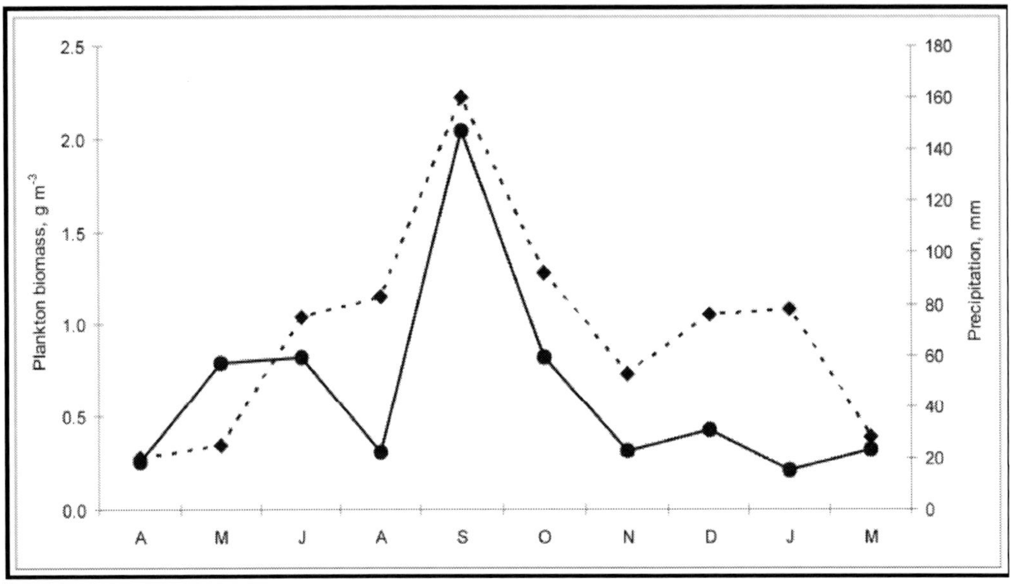

Figure 3. Mean values of wet-weight net-plankton biomass recorded from April 2007 to March 2008 (continuous line) and mean values of monthly precipitation from January 1945 to December 1993 (dashed line).

Plankton Taxa Richness

The number of species or upper level taxa per station and per month fluctuated between 6 to 45 for both phytoplankton and zooplankton. The number of planktonic species differed significantly in mean values among months ($F_{phytoplankton}=6.83$, $F_{zooplankton}=11.21$, $p<0.001$). Phytoplankton species richness was highest in June and October 2007 and lowest in April, November and December 2007 (Figure 4A). Zooplankton taxa richness was highest in April and June 2007 and lowest in March 2008 (Figure 4C). The spatial analysis of plankton data did not show any significant differences among sites ($F_{phytoplankton}=0.38$, $F_{zooplankton}=0.838$, $p>0.5$) (Figure 4B, D).

Temperature and Salinity

Temperature differed significantly in mean values among months ($F=44.7$, $p<0.005$), with the highest temperatures in May and October 2007 and May 2008 (Figure 5A). The lowest temperatures were recorded from November 2007 to March 2008. Salinity varied significantly over time ($F=2.6$, $p=0.006$), October being the month with the lowest values (Figure 5C).

The spatial analysis of temperature showed no pattern ($F=0.8$, $p=0.57$); temperature was homogenous among sampling stations (Figure 5B). In contrast, salinity differed significantly ($F=5.1$, $p<0.001$), with Stn 1 different from the rest and also having the lowest value (Figure 5D).

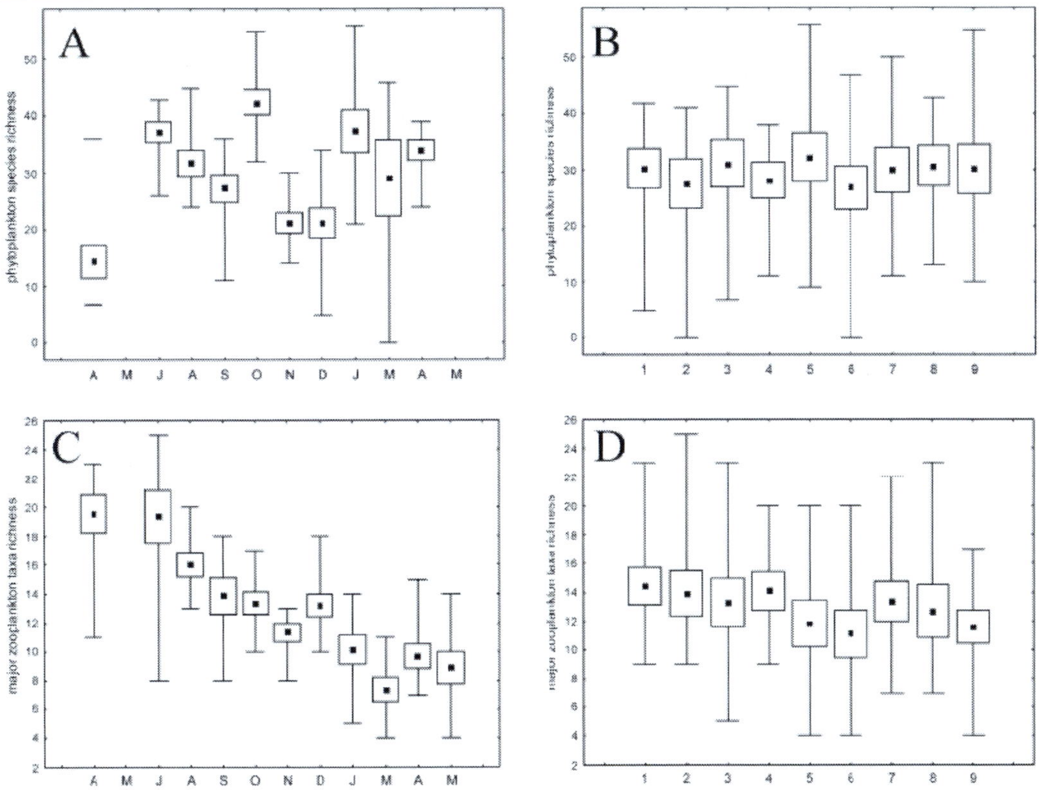

Figure 4. Mean values of phytoplankton species richness (A – per month and B – per station) and major zooplankton taxa richness (C – per month and D – per station); boxes represent the standard error, and whiskers represent the minimum and the maximum values.

Principal Component Analysis

Two first components explained 56.4% of data variation (36.20% and 20.22% for each component). Contribution of the variables in the first (X) axis was higher for temperature and plankton biomass, which were inversely correlated to salinity. For the second (Y) axis, phytoplankton richness was the variable with a higher contribution (Figure 6). Ordination data showed a separation of some sampling sites mainly corresponding to April 2007, indicating an important decrease in phytoplankton species richness during this month (Figure 7).

Taxa Composition

During the sampling period, 36 higher taxa of zooplankton and 275 phytoplankton species were identified, including 152 diatoms, 118 dinoflagellates, three cyanophytes and two silicoflagellates. These occurred with different frequencies, mainly seasonally.

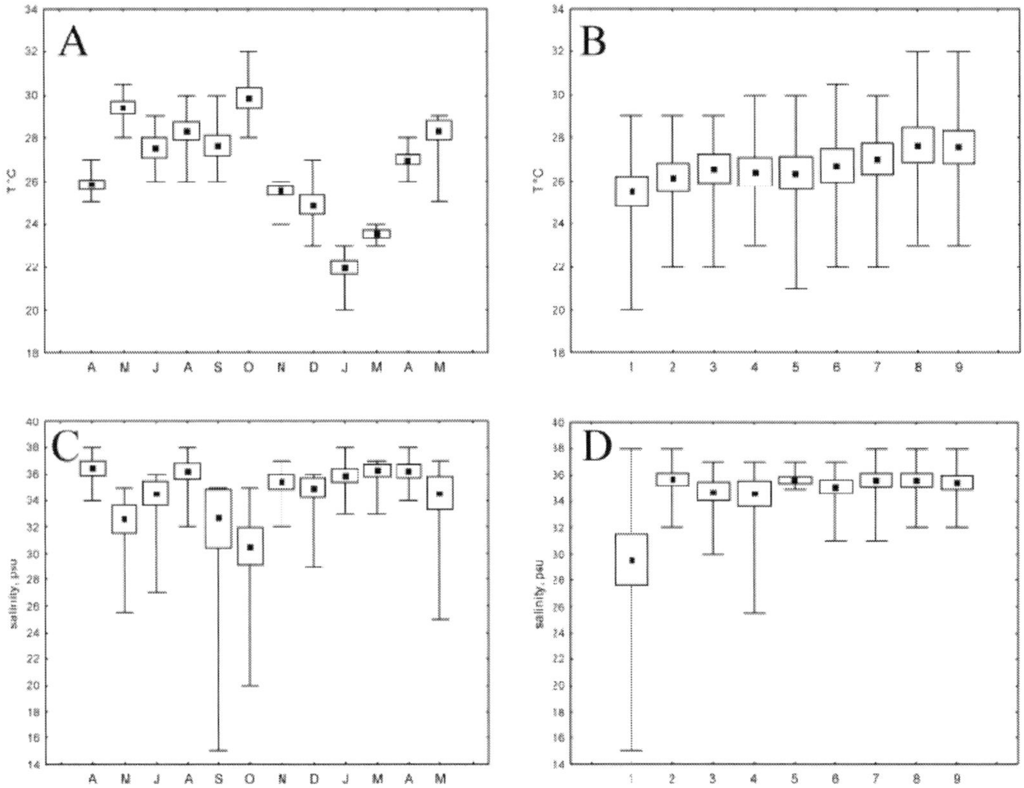

Figure 5. Mean values of temperature (A – per month and B – per station) and salinity (C – per month and D – per station); boxes represent the standard error, and whiskers represent the maximum and the minimum values.

Ten zooplankton taxa were found to be dominant (10% or more of total abundance in a sample) or subdominant (6-9%): calanoid Copepoda, Bivalvia (larvae), Chaetognatha, Appendicularia, Polychaeta (larvae), Gastropoda (larvae), Decapoda (larvae), Echinodermata (larvae), Hydrozoa (hydromedusae) and Cladocera. Calanoid copepods (mostly copepodite stages) were observed at most sampling stations. The most frequent phytoplankton species were mainly the diatoms *Chaetoceros affinis*, *C. diversus*, *C. lorenzianus*, *C. peruvianus*, *Guinardia stolterfothii*, *Hemiaulus hauckii*, *H. membranaceus*, *Lioloma pacificum*, *Proboscia alata*, *Pseudosolenia calcar-avis*, *Thalassionema nitzschioides*, *Thalassiothrix frauenfeldii*; the dinoflagellates *Ceratium trichoceros* and *Protoperidinium oviforme*; and the cyanophyte *Trichodesmium erythraeum* (Table 3).

Plankton and Current Velocity Analysis

The relatively high biomass observed in September 2007 was consistent with high current speed values obtained from the anchored station (Figure 8) as noted from the speed values near Stn 8 (Table 1, Figure 1). Figure 8 shows the net-plankton biomass at Stn 8 and the mean current velocity throughout the first 6 m of the water column averaged for the three days before sampling. Positive current velocity values indicate a northward averaged velocity, while a negative value indicates a southward one. During southerlies, from April to October of 2007, the plankton biomass was highly correlated with the vertical integrated speed at Stn 8 ($r^2 = 0.86$). As the wind speed increases, the biomass value increases accordingly. The same pattern was observed during northerlies from November 2007 to March 2008. The minimum value during the period of observations was in November 2007, while the maximum value was in September 2007.

Table 3. List of plankton taxa from the National Park Sistema Arrecifal Veracruzano in 2007-2008 (some dinoflagellate species found during the study period at other oceanographic stations within the NPSAV were also included)

Major zooplankton taxa

Amphipoda
Appendicularia
Bivalvia larvae
Brachyopoda
Bryozoa
Chaetognatha
Cirripedia larvae
Cladocera

- *Evadne*
- *Penilia*

Copepoda:

- Calanoida
- Cyclopoida
- Harpacticoida
- Poecilostomatoida
- copepodites and nauplii

Ctenophora
Decapoda

- nauplii
- mysis
- zoea Anomura
- zoea Brachyura
- postlarvae
- megalopa
- Luciferidae

Doliolida
Echinodermata larvae:

- Ophiuroidea (ophiopluteus)
- Echinoidea (pluteus, echinopluteus)
- Asteroidea (bipinnaria, brachiolaria)
- Holothuroidea (doliolaria)

Enteropneusta (tornaria)
Euphausiacea
Foraminifera
Gastropoda juv.
Heteropoda
Hydromedusae juv.
Invertebrata eggs
Mysidacea
Nemertini larvae
Ostracoda
Pisces eggs and larvae
Phoronida
Polychaeta larvae and juv.
Pteropoda
Radiolaria
Scyphomedusae
Siphonophora
Tintinnida (Ciliata)

Phytoplankton taxa

Diatoms

Achnanthes cf. *sublaevis* Hust.
Actinocyclus octonarius Ehrenb.
Actinoptychus senarius Ehrenb.
Amphipleura sp.
Asterionellopsis glacialis (Castrac.) Round
Asterolampra marylandica Ehrenb.
Asteromphalus cleveanus Grunov
Asteromphalus flabellatus (Bréb.) Grev.
Aulacoseira granulata (Ehrenb.) Simonsen
Azpeitia nodulifera (A. W. F. Schmidt) G. A. Fryxell et P. A. Sims
Bacillaria paxillifera (O. F. Müll.) M. Marsson
Bacteriastrum delicatulum Cleve
Bacteriastrum furcatum Shadbolt
Bacteriastrum hyalinum Lauder
Biddulphia pulchella S. F. Gray
Biddulphia tuomeyi (Bailey) Roper
Campylodiscus sp.
Cerataulina bergonii (H. Perag.) F. Schütt

Cerataulina pelagica (Cleve) Hendey
Cerataulus californicus A. W. F. Schmidt
Cocconeis sp.
Corethron hystrix Hensen
Chaetoceros affinis Lauder
Chaetoceros atlanticus Cleve
Chaetoceros brevis F. Schütt
Chaetoceros castracanei G. Karst.
Chaetoceros coarctatus Lauder
Chaetoceros compressus Lauder
Chaetoceros costatus Pavill.
Chaetoceros curvisetus Cleve
Chaetoceros dadayi Pavill.
Chaetoceros debilis Cleve
Chaetoceros decipiens Cleve f. *decipiens*
Chaetoceros decipiens f. *singularis* Gran
Chaetoceros denticulatus Lauder
Chaetoceros didymus Ehrenb. var. *anglica* (Grunov) Gran
Chaetoceros didymus Ehrenb. var. *protuberans* (Lauder) Gran et K. Yendo
Chaetoceros diversus Cleve
Chaetoceros laciniosus F. Schütt
Chaetoceros lorenzianus Grunov
Chaetoceros messanensse Castrac.
Chaetoceros okamurae J. Ikari
Chaetoceros pendulus G. Karst.
Chaetoceros peruvianus Brightw.
Chaetoceros cf. *pseudocrinitum* Ostenf.
Chaetoceros pseudocurvisetus L. Mangin
Chaetoceros radicans F. Schütt
Chaetoceros rostratus Lauder
Chaetoceros seychellarus G. Karst.
Chaetoceros tortissimus Gran
Climacodium frauenfeldianum Grunov
Climacosphenia moniligera Ehrenb.
Coscinodiscus centralis Ehrenb.
Coscinodiscus granii Gough
Coscinodiscus cf. *marginatus* Ehrenb.
Coscinodiscus radiatus Ehrenb.
Coscinodiscus wailessi Gran et Angst
Cyclotella stylorum Brightw.
Cyclotella sp. 1
Cyclotella sp. 2
Cyclotella sp. 3
Cylindrotheca closterium (Ehrenb.) Reimann et J. C. Lewin
Cymbella sp.
Dactyliosolen antarcticus Castrac.

Dactyliosolen fragilissimus (Bergon) Hasle
Diploneis sp.
Ditylum brightwelli (T. West) Grunov
Entomoneis alata (Ehrenb.) Ehrenb.
Eucampia cornuta (Cleve) Grunov
Fragilaria goulardii (Bréb. *ex* Grunov) Lange-Bert.
Fragilaria ulna (Nitzsch) Lange-Bert.
Fragilariopsis cf. *kerguelensis* (O'Meara) Hust.
Gomphonema sp. 1
Gomphonema sp. 2
Guinardia cylindrus (Cleve) Hasle
Guinardia flaccida (Castrac.) H. Perag.
Guinardia striata (Stolterf.) Hasle
Gramatophora marina (Lyngb.) Kütz.
Gyrosigma sp.
Haslea trompii (Cleve) Simonsen
Haslea wawrickae (Hust.) Simonsen
Hemiaulus hauckii Grunov *ex* Van Heurck
Hemiaulus membranaceus Cleve
Hemiaulus sinensis Grev.
Isthmia minima Harvey et Bailey
Isthmia nervosa Kütz.
Lauderia annulata Cleve
Leptocylindrus danicus Cleve
Leptocylindrus minimus Gran
Licmophora comunnis (Heib.) Grunov
Licmophora sp.
Lioloma delicatulum (Cupp) Hasle
Lioloma pacificum (Cupp) Hasle
Lithodesmium undulatum Ehrenb.
Mastogloia sp.
Melosira numoloides C. Agardh
Melosira sp.
Meuniera membranacea (Cleve) P. C. Silva
Nitzschia longissima (Bréb.) Ralfs
Nitzschia sigma (Kütz.) W. Smith
Nitzschia sp.
Neocalyptrella robusta (G. Norman *ex* Ralfs) Hernández-Becerril et Meave del Castillo
Odontella longicruris (Grev.) Hoban
Odontella mobiliensis (Bailey) Grunov
Palmeria hardmanniana Grev.
Paralia sulcata (Ehrenb.) Cleve
Pinnularia sp.
Plagiotropis sp.
Pleurosigma normanii Ralfs
Pleurosigma hamuliferum Brun

Pleurosigma rigidum W. Smith
Pleurosigma sp.
Podocystis sp.
Proboscia alata (Brightw.) Sundström f. *alata*
Proboscia alata f. *gracillima* (Cleve) Gran
Pseudoeunotia sp.
Pseudoguinardia recta Stosch
Pseudo-nitzschia "pseudodelicatissima complex" spp.
Pseudo-nitzschia "americana complex" sp.
Pseudo-nitzschia "seriata complex" sp.
Pseudo-nitzschia cf. *subcurvata* (Hasle) Fryxell
Pseudosolenia calcar-avis (Schultze) Sundström
Rhabdonema sp.
Rhizosolenia acuminata (H. Perag.) H. Perag.
Rhizosolenia bergonii H. Perag.
Rhizosolenia clevei Ostenf.
Rhizosolenia clevei var. *comunis* Sundström
Rhizosolenia castracanei H. Perag.
Rhizosolenia formosa H. Perag.
Rhizosolenia hebetata Bailey var. *semispina* (Hensen) Gran
Rhizosolenia hyalina Ostenf.
Rhizosolenia imbricata Brightw.
Rhizosolenia pungens A. Cleve
Rhizosolenia setigera Brightw.
Rhizosolenia temperei H. Perag.
Rhizosolenia sp.
Skeletonema costatum (Grev.) Cleve
Skeletonema pseudocostatum L. K. Medlin
Skeletonema tropicum Cleve
Stephanopyxis palmeriana (Grev.) Grunov
Streptotheca tamesis Shrubsole
Striatella unipunctata (Lyngb.) C. Agardh
Surirella spp.
Thalassionema nitzschioides (Grunov) Mereschkowsky
Thalassiosira decipiens (Grunov) Jörg.
Thalassiosira eccentrica (Ehrenb.) Cleve
Thalassiosira sp. 1
Thalassiosira sp. 2
Thalassionema frauenfeldii (Grunov) Tempère et H. Perag.
Triceratium favus Ehrenb.
Toxarium hennedyanum (W. Greg.) Pelletan
Tropidoneis sp.

Dinoflagellates

Akashiwo sanguinea (Hirasaka) G. Hansen et Moestrup

Alexandrium sp.
Amphisolenia bidentata Schröd.
Amphidinium bifurcata J. Murray et M. C. Whitting
Amphisolenia schauislandi Lemmermann
Ceratium arietinum Cleve
Ceratium azoricum Cleve
Ceratium belone Cleve
Ceratium bigelowii Kof.
Ceratium candelabrum (Enrenb.) F. Stein
Ceratium carriense Gourret
Ceratium contortum (Gourret) Cleve
Ceratium contrarium (Gourret) Pavill.
Ceratium declinatum (G. Karst.) Jörg.
Ceratium dens Ostenf. et J. Schmidt
Ceratium euarcuatum Jörg.
Ceratium extensum (Gourret) Cleve
Ceratium furca (Ehrenb.) Clap. et J. Lachm. var. *furca*
Ceratium furca var. *hircus* (Schröd.) Margalef
Ceratium fusus (Ehrenb.) Dujard.
Ceratium gibberum Gourret var. *dispar* (Pouchet) Sournia
Ceratium hexacanthum Gourret
Ceratium horridum (Cleve) Gran
Ceratium inflatum (Kof.) Jörg.
Ceratium karstenii Pavill.
Ceratium kofoidii Jörg.
Ceratium limulus Gourret
Ceratium longirostrum Gourret
Ceratium lunula (Schimper *ex* G. Karst.) Jörg.
Ceratium macroceros (Ehrenb.) Vanhöffen var. *gallicum* (Kof.) Sournia
Ceratium massiliense (Gourret) Jörg. var. *armatum* (G. Karst.) Jörg.
Ceratium pentagonum Gourret var. *tenerum* Jörg.
Ceratium symetricum Pavill.
Ceratium teres Kof.
Ceratium trichoceros (Ehrenb.) Kof.
Ceratium tripos (O. F. Müll.) Nitzsch
Ceratium vultur Cleve f. *vultur Ceratocorys horrida* F. Stein
Dinophysis amandula Sournia
Dinophysis argus (F. Stein) T. H. Abé
Dinophysis caudata Saville-Kent
Dinophysis doryphora (F. Stein) T. H. Abé
Dinophysis favus (Kof. et J. R. Michener) T. H. Abé
Dinophysis fortii Pavill.
Dinophysis hastata F. Stein
Dinophysis mitra (F. Stein) T. H. Abé
Dinophysis rapa (F. Stein) T. H. Abé
Dinophysis rotundata Clap. et Lachm.

Dinophysis schuettii J. Murray et M. C. Whitting
Gonyaulax polygramma F. Stein
Gonyaulax spinifera (Clap. et Lachm.) Diesing
Heterocapsa sp.
Histioneis isselii Forti
Karenia brevis (Davis) G. Hansen et Moestrup
Noctiluca scintillans (Macartney) Kof. et Sw.
Ornithocercus heteroporoides T. H. Abé
Ornithocercus heteroporus Kof.
Ornithocercus magnificus F. Stein
Ornithocercus quadratus F. Schütt
Ornithocercus steinii F. Schütt
Ornithocercus thumii (A. Schmidt) Kof. et Skogsb.
Ostreopsis cf. *ovata* Fukuyo
Oxytoxum scolopax F. Stein
Oxytoxum constrictum (F. Stein) Buetschli
Peridinium quinquecorne T. H. Abé
Podolampas bipes F. Stein
Podolampas elegans F. Schütt
Podolampas palmipes F. Stein
Podolampas spinifera Okamura
Prorocentrum compressum (Bailey) T. H. Abé
Prorocentrum concavum Fukuyo
Prorocentrum gracile F. Schütt
Prorocentrum lima (Ehrenb.) F. Stein
Prorocentrum cf. *mexicanum* B. F. Osorio
Prorocentrum micans Ehrenb.
Prorocentrum cf. *triestinum* J. Schill.
Protoperidinium abei (Paulsen) Balech
Protoperidinium brochii (Kof. et Sw.) Balech
Protoperidinium claudicans (Paulsen) Balech
Protoperidinium compressum (T. H. Abé) Balech
Protoperidinium cristatum Balech
Protoperidinium concinnum Faust
Protoperidinium conicum (Gran) Balech
Protoperidinium crassipes (Kof.) Balech
Protoperidinium depressum (Bailey) Balech
Protoperidinium divaricatum (Meunier) Parke et J. D. Dodge
Protoperidinium divergens (Ehrenb.) Balech
Protoperidinium excentricum (Paulsen) Balech
Protoperidinium cf. *hirobis* (T. H. Abé) Balech
Protoperidinium latissimum (Kof.) Balech
Protoperidinium longipes Balech
Protoperidinium obtusum (G. Karst.) Parke et J. D. Dodge
Protoperidinium oceanicum (Vanhöffen) Balech
Protoperidinium oviforme (P. A. Dang.) Balech

Protoperidinium ovum (J. Schill.) Balech
Protoperidinium pacificum (Kof. et J. R. Michener) Balech
Protoperidinium pellucidum Bergh
Protoperidinium pentagonum (Gran) Balech
Protoperidinium persicum (J. Schill.) Okolodkov
Protoperidinium quarnerense (Schröd.) Balech
Protoperidinium solidicorne (L. Mangin) Balech
Protoperidinium sphaericum (J. Murray et M. C. Whitting) Balech
Protoperidinium cf. *subinerme* (Paulsen) A. R. Loebl.
Protoperidinium cf. *subsphaericum* (Balech) Balech
Protperidinium thorianum (Paulsen) Balech
Protoperidinium thulesense (Balech) Balech
Protoperidinium venustum (Matzen.) Balech
Protoperidinium sp. C ortho-hexa (see Okolodkov, 2008)
Protoperidinium sp. D meta-hexa (see Okolodkov, 2008)
Pyrocystis fusiformis (Wyville-Thompson *ex* Haeckel) V. H. Blackmann
Pyrocystis robusta Kof.
Pyrophacus horologium F. Stein
Pyrophacus steinii (J. Schill.) D. Wall et Dale
Scrippsiella cf. *trochoidea* (F. Stein) A. R. Loebl.
Spyraulax jolliffei (J. Murray et M. C. Whitting) Kof.

Cyanophytes

Trichodesmium erythraeum Ehrenb.
Trichodesmium thiebautii Gomont
Oscillatoria sp.

Silicoflagellates

Dictyocha fibula Ehrenb.
Dictyocha crux Ehrenb.

Discussion

Plankton biomass in the NPSAV varied seasonally with no distinction among sampling stations. The peak of plankton biomass occurred at the end of the rainy season with the highest values in September-October, which corresponded to one of the three peaks of water temperature. This unique biomass peak characterizes the NPSAV as having a unimodal production cycle similar to that of enclosed coastal ecosystems in temperate regions (Cebrián and Valiela, 1999) where the peak occurs in summer.

This peak occurred in September, when the highest multi-year mean value of precipitation has been recorded in the study area (Figure 3). As known from the literature, tropical zooplankton may show a seasonal pattern (McWilliam et al., 1981; Sammarco and

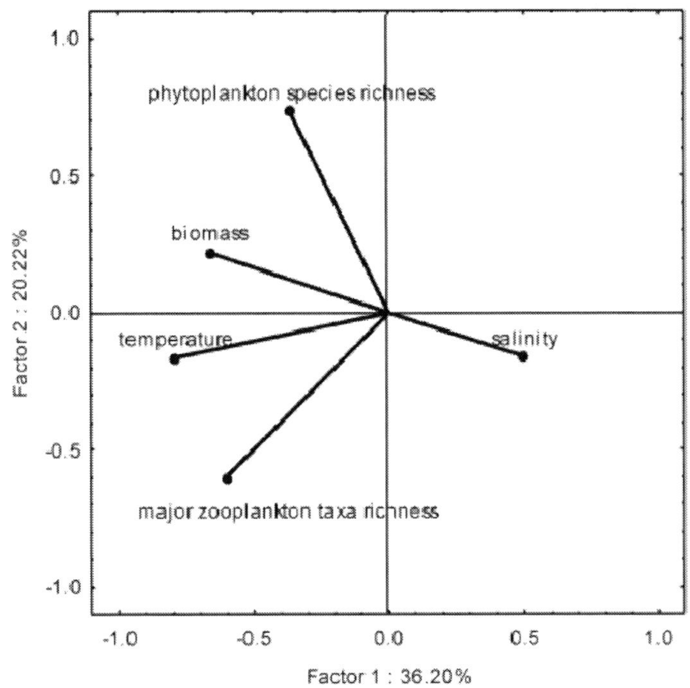

Figure 6. Correlation of variables (wet-weight net-plankton biomass, major zooplankton taxa richness, phytoplankton species richness, temperature and salinity) to factor 1 and 2 resulting from principal component analysis.

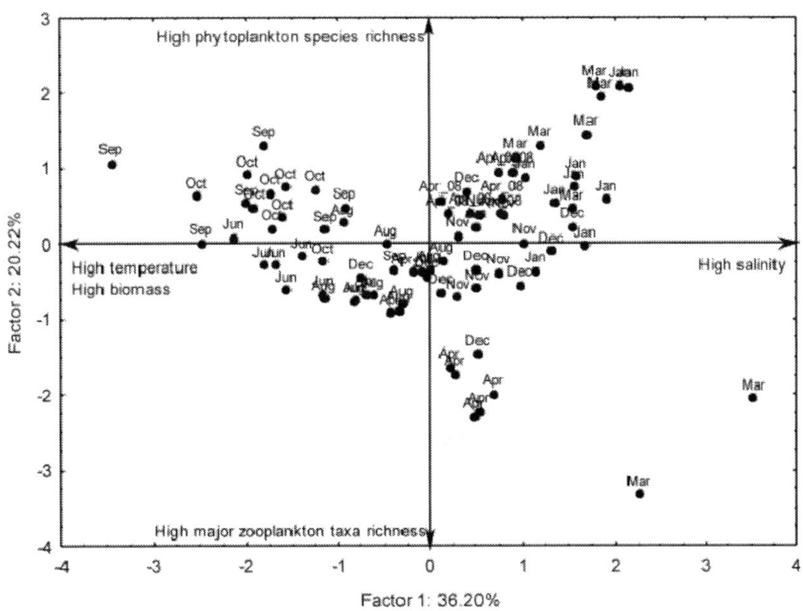

Figure 7. Plot of samples according to factor 1 and 2 resulting from principal component analysis. September and October (rainy months) are associated with higher temperatures and net-plankton biomass.

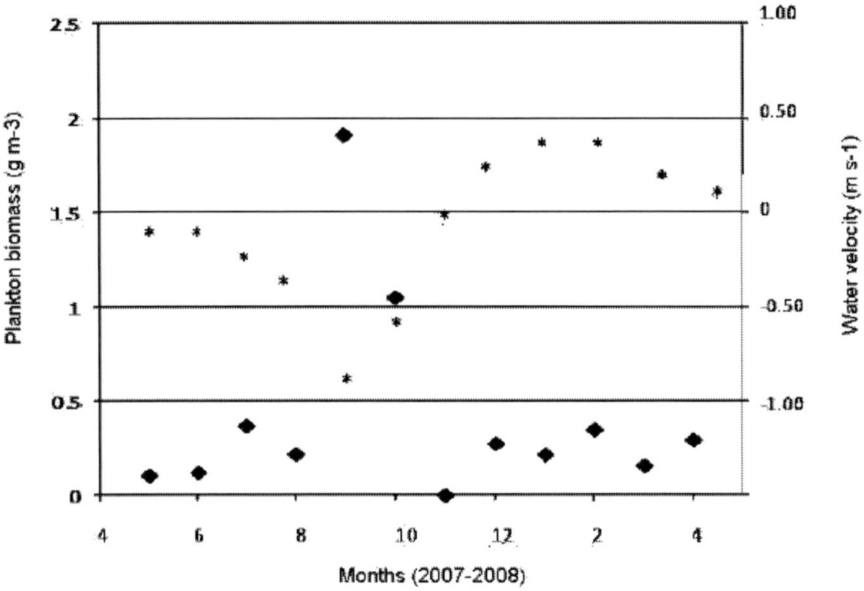

Figure 8. Time series at Stn 8 of the plankton biomass represented with diamonds, and the vertical integrated water velocity three days before sampling represented with asterisks. Negative values indicate southward currents and positive values indicate northward currents, in the National Park Sistema Arrecifal Veracruzano.

Crenshaw, 1984; McKinnon and Thorrold, 1993) or may not (Moore and Sander, 1976; Morales and Murillo, 1996).

In general, the mean values obtained in the present study (Figure 3) lie within the range of values recorded for the Gulf of Mexico by other authors (Table 1). These high biomass values obtained during the present study characterize the NPSAV as a very productive system. In September, an extraordinarily high value of biomass (3637 g m^{-3}) was recorded. During this month, due to the location of the coral reefs and the coastline, a cyclonic eddy is favored in the northwestern area of the park (Salas-Monreal et al., 2009). Cyclonic eddies bring nutrient-rich deep water to the surface, enhancing high-productive areas. Due to the wind conditions (>50 km h^{-1}), the water column mixes in the shallower zone of the NPSAV, an additional reason for high plankton biomass values in the NPSAV.

Our annual cycle analysis showed that no spatial differences in biomass values could be detected within the NPSAV. Plankton standings stocks demonstrated a twofold variation in monthly mean abundance between stations (from 400 to 800 mg m^{-3}), similar to that reported from the Mexican Caribbean (Suarez and Gasca, 1994). Our observations of the highest plankton biomass during the rainy period also correspond to the data obtained from the Mexican Caribbean. In addition, taxa richness did not differ significantly among stations. These results indicate that physical oceanographic processes within the system are able to maintain a constantly high plankton production in different zones within the study area. Although the NPSAV was divided into three subsystems, Veracruz, Jamapa and Antón Lizardo, that contain different bathymetric and hydrological characteristics (Ortíz-Lozano et al., 2009), the production of plankton biomass did not reflect this division. However, variations in taxa composition and abundances must be investigated in more detail to determine the partial contribution of different taxa at generic or specific levels to the plankton biomass in each area of the park.

Although salinity differed significantly in one coastal station (Stn 1; Figure 5D) near the Jamapa River, this variable did not explain as much of the variation as was shown by the PCA. On the other hand, temperature is an important variable in the temporal characterization, and biomass values were positively correlated with it. Also, phytoplankton species richness explained a high proportion of the variation, indicating that it was an important factor in the temporal variation of the plankton biomass evaluated during the sampling period. The major taxa that contributed to zooplankton abundance were mainly calanoid copepods; these dominated over other groups such as cladocerans, which has been recorded as the main group in similar environments (Moore and Sander, 1976; Ferraris, 1982; Vaissiere and Seguin, 1984; Palomares et al., 2002). Among planktonic microalgae, diatoms, primarily centrics, were the main component in the NPSAV, although the cyanophyte *Trichodesmium erythraeum,* a tropical component (Berman-Frank et al., 2007), was also common throughout the entire sampling period. In studies of plankton biomass the comparison of results obtained by different sampling methods, fixation, preservation and measurement remains a major obstacle. The maximal biomass values in the NPSAV are 5 to 8 times higher than in the highly productive coastal zone influenced by the Orinoco River, Venezuela. Increased plankton biomass was previously reported from the area north of the port of Veracruz, with values ranging from 0.3 to 0.4 g m^{-3}; plankton was more abundant in winter, according to Khromov (1965a) and Bogdanov et al. (1968). To our knowledge, the highest biomass values for the Gulf of Mexico, more than 3.0 g m^{-3}, have been reported near the Mississippi River. According to Sorokin (1990), the actual biomass of reef zooplankton, considering demersal species taken from night-time tows, ranges between 1 and 10 g m^{-3}, which is similar to that in the most productive pelagic areas. In the offshore central Gulf of Mexico and in the Caribbean Sea, the plankton biomass is usually less than 0.1 g m^{-3}. However, it may sometimes reach values up to 0.5 g m^{-3} (Khromov, 1965a, 1965b, 1967). In the southern Gulf of Mexico, in the Bank of Campeche, north of the Yucatan Peninsula the plankton biomass may exceed 1.0 g m^{-3} (Khromov, 1967). In the outermost portions of the Campeche shelf, biomass values of up to 1.3 g m^{-3} have been reported (Gasca et al., 1995). In the Gulf of Mexico, upwelling and river runoff result in higher plankton biomass and more pronounced seasonal changes (Bogdanov et al., 1968).

In conclusion, plankton biomass within the NPSAV is unimodal in annual dynamics, with a peak in September-October at the end of the rainy season with the highest precipitation when the currents change direction. Taxa richness contributes to the explanation of seasonal variations, and calanoid copepods and centric diatoms are the most conspicuous plankton components. Spatial analysis of annual plankton biomass values shows that the NPSAV area is rather homogenous.

ACKNOWLEDGMENTS

We are grateful to Cipriano Anaya (ICIMAP) for logistic support during sampling, Eduardo Morteo (ICIMAP) and his team for financial and logistic support, Eduardo Suárez-Morales (ECOSUR, Chetumal, Quintana Roo, Mexico) and Xenia N. Kosobokova (Shirshov

Institute of Oceanology, Moscow, Russia) for their help with literature, and Marcia M. Gowing (University of California at Santa Cruz, California, USA) for improving the writing style. Financial support of Consejo Nacional de Ciencia y Tecnología (CONACYT), Mexico, and Gobierno del Estado de Veracruz to the FOMIX CONACYT-Veracruz project "Fuentes orgánicas de carbono y nitrógeno y su función sobre la estructura trófica en el Sistema Arrecifal Veracruzano" (2007-2010) given to HPE, support of Dirección General de Investigaciones de la Universidad Veracruzana to the project "Algas de la zona arrecifal Veracruzana, Golfo de México, con énfasis en las algas rojas, diatomeas y dinoflagelados" given to YBO (2007-2009), support of Programa de Mejoramiento del Profesorado (PROMEP) to the project "Patrones de distribución de la diversidad y biomasa de grupos funcionales clave para el Sistema Arrecifal Veracruzano" (2011-2012), and support of CONACYT to the project "Dinámica estacional fitoplanctónica y detección de especies nocivas y potencialmente tóxicas en la zona costera de Veracruz" (CONACYT-UV MOD-ORD-70-08 PCI-1054-11-08; 2008-2009) given to JAAC are also appreciated. Finally, the ADCP were acquired with financial support from grant 78773 given to DSM by CONACYT-Ciencia Básica (2007-2010).

References

Aguayo-Saviñón, M. A. (1965). *Contribución al conocimiento de los copépodos de la zona arrecifal de Veracruz, Ver. (sistemática y distribución).* Tesis Profesional, Facultad de Ciencias, Universidad Nacional Autónoma de México, México, D. F.

Alldredge, A. and King J. M. (1977). Distribution, abundance and substrate preferences of demersal reef zooplankton at Lizard Island Lagoon, Great Barrier Reef. *Mar. Biol., 41(4)*, 317-333.

Alldredge, A. and King J. M. (1980). Effects of moonlight on the vertical migration patterns of demersal zooplankton. *Exp. Mar. Biol. Ecol., 44*, 133-156.

Alldredge, A. and King J. M. (1985). The distance demersal zooplankton migrate above the benthos: implications for predation. *Mar. Biol., 84*, 253-260.

Alvarez-Cadena, J. N., Suárez-Morales, E. and Gasca R. (1998). Copepod assemblages from a reef-related environment in the Mexican Caribbean Sea. *Crustacena, 71(4)*, 411-433.

Berman-Frank, I., Quigg A., Finkel Z. V., Irwin, A. J. and Haramaty L. (2007). Nitrogen-fixation strategies and Fe requirements in cyanobacteria. *Limnol. Oceanogr., 52*, 2260-2269.

Bogdanov, D. V., Sokolov, V. A. and Khromov N. S. (1968). Regions of high biological and commercial productivity in the Gulf of Mexico and the Caribbean Sea. *Oceanology (USSR), 8(3)*, 466-478. (In Russian; with English summary).

Campos, A. (1980). Distribución y abundancia relativa de los copépodos planctónicos en el Golfo de México y el Mar Caribe. *Cienc. Biol., 5*, 57-74.

Campos-Hernández, A. and Suárez-Morales, E. (1994). *Copépodos pelágicos del Golfo de México y Mar Caribe. I. Biología y sistemática.* Regina de los Ángeles, Quintana Roo, México: Centro de Investigaciones de Quintana Roo.

Carrillo, L., Horta-Puga, G. and Carricart-Ganivet, J. P. (2007). Climate and oceanography. In: J. W. Tunnell Jr., E.A. Chávez and K. Withers (Eds.), *Coral reefs of the southern Gulf of Mexico,* pp. 34-40. Corpus Christie, Texas, USA: Texas A&M University Press.

Castellanos-Osorio, I. A. and Suárez-Morales, E. (1997). Observaciones sobre el zooplankton de la zona arrecifal de Mahahual, Quintana Roo (Mar Caribe mexicano). *Anales Inst. Biol. Univ. Nac. Autón. México, Ser. Zool., 68(2),* 237-252.

Cebrián, J. and Valiela, I. (1999). Seasonal patterns in phytoplankton biomass in coastal ecosystems. *J. Plankton Res., 21*, 429-444.

Clarke, K. R. and Warwick R. M. (2001). *Change in marine communities: an approach to statistical analysis and interpretation.* Plymouth , UK: Primer Ltd.

CONANP (2007). *Programa de conservación y manejo Parque Nacional Sistema Arrecifal Veracruzano (anteproyecto).* Veracruz, México: Comisión Nacional de Áreas Naturales Protegidas.

De la Cruz, A. (1972). Zooplancton de la región sureste del Golfo de México. *Ciencias, Serie 4. Cienc. Biol., Universidad de la Habana, 24*, 1-55.

Domínguez-Castanedo, N., Granados-Barba, A. and Solís-Weiss, V. (2007). Estudio preliminar de la macrofauna bentónica presente en la laguna del arrecife Sacrificios. In: A. Granados-Barba, L. G. Abarca-Arenas and J. M. Vargas-Hernández (Eds.). *Investigaciones científicas en el Sistema Arrecifal Veracruzano,* pp. 113-126. Campeche, Campeche, México: Universidad Autónoma de Campeche.

Emery, A. R. (1968). Preliminary observations on coral reef plankton. *Limnol. Oceanogr., 13*, 293-303.

Ferraris, J. D. (1982). Surface zooplankton at Carrie Bow Cay, Belize. *Smith. Contrib. Mar. Sci., 12*, 143-151.

Flores-Coto, C. (1965). Notas preliminares sobre la identificación de las apendicularias de las aguas veracruzanas. *Anales Inst. Biol. Univ. Nac. Autón. México, 36*, 293-296.

Flores-Coto, C. (1974). Contribución al conocimiento de las apendicularias del Arrecife "La Blanquilla" Veracruz, México con descripción de una nueva especie. *Anales Centro Cienc. Mar Limnol. Univ. Nac. Autón. México, 1(1)*, 41-60.

Flores-Coto, C., Sanvicente, L., Pineda-López, R. and Rodríguez, M. (1988). Composición, distribución y abundancia ictioplanctónica del sur del Golfo de México. *Universidad y Ciencia (México), 5(9)*, 63-84.

García-Abdeslem, J. (1987). *Atlas/memoria del levantamiento geofísico de la Zona Económica Exclusiva y margen continental oeste de México.* México, D. F.: Secretaría de Marina SMPO8710.

Gasca, R, E. Suárez, E. and Castellanos, I. (1995). Biomasas zooplánctícas en aguas superficiales del Golfo de México durante verano e invierno de 1991. *Caribb. J. Sci., 31(1-2)*, 128-140.

Gómez, P. (2007). Inventario de las esponjas del Parque Nacional Sistema Arrecifal Veracruzano, con nuevos registros de especies (Porifera: Demospongiae). In: A. Granados-Barba, L. G. Abarca-Arenas and J. M. Vargas-Hernández (Eds.), *Investigaciones científicas en el Sistema Arrecifal Veracruzano,* pp. 51-72. Campeche, Campeche, México: Universidad Autónoma de Campeche.

González, J. I., Soto, R. and Ochoa, J. (2009). *Predicción de mareas en México.* Ensenada, Baja California, México: Oceanografía Física, Centro de Investigación Científica y de

Educación Superior de Ensenada (CICESE). [on-line] <http://oceanografia.cicese.mx/predmar/>

Gutiérrez de Velasco, G. and Winant, C. D. (1996). Seasonal patterns of wind stress and wind stress curl over the Gulf of Mexico. *J. Geophys. Res., 101(C8)*, 18127-18140.

Heidelberg, K. B., Sebens, K. P. and Purcell J. E. (2004). Composition and sources of near reef zooplankton on a Jamaican forereef along with implications for coral feeding. *Coral Reefs, 23*, 263-276.

Heidelberg, K. B., O'Neil, K. L., Bythell, J. C. and Sebens, K. P. (2010). Vertical distribution and diel patterns of zooplankton abundance and biomass at Coch Reef, Florida Keys (USA). *J. Plankton Res., 32(1),* 75-91.

Hobson, E. S. and Chess, J. R. (1979). Zooplankton that emerge from the lagoon floor at night at Midway Atoll, Hawaii. *Fish. Bull. U.S., 77*, 275-280.

Khromov, N. S. (1965a). Distribution of plankton in the Gulf of Mexico and some features of the seasonal dynamics. In: *Soviet-Cuban Fishery Research (VNIRO-CIP), 1*, pp. 47-69. Moscow. (In Russian; with Spanish summary).

Khromov, N. S. (1965b). About the quantitative distribution of plankton in the northeastern Caribbean Sea and the Gulf of Mexico. *Proceed. VNIRO, 57*, 381-391. (In Russian).

Khromov, N. S. (1967). Studies on plankton in the Gulf of Mexico and the Caribbean Sea. In: *Soviet-Cuban Fishery Research (VNIRO-CIP), 2*, pp. 39-57. Moscow. (In Russian; with Spanish summary).

Lara, M., Padilla, C., García, C. and Espejel, J. J. (1992). Coral reefs of Veracruz, Mexico, I. Zonation and community. In: R.H. Richmond (Ed.), *Proceedings of the 7th International Coral Reef Symposium (22-27 June 1992), vol. 1,* pp. 535-544. Guam, Micronesia: University of Guam Press, UOG Station.

Leal-Rodríguez, D. G. (1965). Distribución de pterópodos en Veracruz, Ver. *Anales Inst. Biol. Univ. Nac. Autón. México, 36*, 249-251.

Lewis, J. B. and Boers J. J. (1991). Patchiness and composition of coral reef demersal plankton. *J. Plankton Res., 13(6)*, 1273-1289.

Martínez-Hernández, J. A. and Camacho-Olivares, B. (2007). Ictiofauna del arrecife artificial "Ex-Cañonero C-50, General Vicente Riva Palacio" en el Sistema Arrecifal Veracruzano, México. In: A. Granados-Barba, L. G. Abarca-Arenas and J.M. Vargas-Hernández (Eds.), *Investigaciones científicas en el Sistema Arrecifal Veracruzano,* pp. 209-220. Campeche, Campeche, México: Universidad Autónoma de Campeche.

McKinnon, A. D. and Thorrold, S. R. (1993). Zooplankton community structure and copepod egg production in coastal waters of the central Great Barrier Reef lagoon. *J. Plankton Res., 15(12)*, 1387-1411.

McWilliam, P. S., Sale, P. F. and Anderson, D. T. (1981). Seasonal changes in resident zooplankton sampled by emergence traps in One Tree Lagoon, Great Barrier Reef. *J. Exp. Mar. Biol. Ecol., 52*, 185-203.

Moore, E. and Sander F. (1976). Quantitative and qualitative aspects on the zooplankton at two Caribbean coral reef stations. *Estuarine Coastal Mar. Sci.,* 4, 589-607.

Morales, A. and Murillo, M. M. (1996). Distribution, abundance and composition of coral reef zooplankton, Cahuita National Park, Limon, Costa Rica. *Rev. Biol. Trop., 44(2)*, 619-630.

Okolodkov, Y. B. (2003). A review of Russian plankton research in the Gulf of Mexico and the Caribbean in the 1960-1980s. *Hidrobiológica, 13(3)*, 207-221.

Okolodkov, Y. B. (2008). *Protoperidinium* Bergh (Dinophyceae) of the National Park Sistema Arrecifal Veracruzano, Gulf of Mexico, with a key for identification. *Acta Bot. Mex., 84*, 93-149.

Okolodkov, Y. B., Campos-Bautista, G., Gárate-Lizárraga, I., González-González, J. A. G., Hoppenrath, M. and Arenas, V. (2007). Seasonal changes of benthic and epiphytic dinoflagellates in the Veracruz reef zone, Gulf of Mexico. *Aquat. Microb. Ecol., 47(3)*, 223-237.

Ortiz-Lozano, L., Granados-Barba, A. and Espejel, I. (2009). Ecosystemic zonification as a management tool for marine protected areas in the coastal zone: Applications for the Sistema Arrecifal Veracruzano National Park, Mexico. *Ocean and Coast. Manage., 52*, 317–323.

Palomares-García, R., Martínez-López, A. and Gárate-Lizárraga, I. (2002). Plankton community changes in Bahía Concepción, México. *CICIMAR Oceánides, 17(2)*, 113-128.

Porter, J. W. and Porter, K. J. (1977). Quantitative sampling of demersal plankton migrating from coral reef substrates. *Limnol. Oceanogr., 22*, 553-556.

Porter, J. W., Porter, K. G. and Batac Calan, Z. (1977). Quantitative sampling of Indo-Pacific demersal reef plankton. In: *Proc. 3rd Int. Coral Reef Symp., Miami,* pp. 105-112.

Salas-de-León, D. A., Monreal-Gómez M. A., Signoret, M. and Aldeco, J. (2004a). Anticyclonic-cyclonic eddies and their impact on near-surface chlorophyll stocks and oxygen supersaturation over the Campeche Canyon, Gulf of Mexico. *J. Geophys. Res., 109*, C05012. <doi:10.1029/2002JC001614>.

Salas-de-León, D. A., Díaz-Flores, M. A. and Monreal-Gómez, M. A. (2004b). Circulation and vorticity in the southern Gulf of Mexico. In: W. Schroeder (Ed.), *Hans Ertel Memorial Book,* pp. 229-243. Mexico, D. F.: German Commission of History of Geophysics and Cosmical Physics.

Salas-de-León, D. A., Monreal-Gómez, M. A., Díaz-Flores, M. A., Salas-Monreal, D., Velasco-Mendoza H., Riverón-Enzástiga, M. L. and Ortiz-Zamora, G. (2008). Role of near-bottom currents in the distribution of sediments within the Southern Bay of Campeche, Gulf of Mexico. *J. Coast. Res., 24*, 1487-1494.

Salas-Monreal, D., Salas-de-León, D. A., Monreal-Gómez, M. A. and M. L. Riverón-Enzástiga. (2009). Current rectification in a tropical coral reef system. *Coral Reefs, 28*, 871-879.

Salas-Pérez, J. J., Salas-Monreal, D., Arenas-Fuentes, V. E., Salas-de-Leon, D. A. and Riveron-Enzastiga, M. L. (2008). Tidal characteristics in a coral reef system from the western Gulf of Mexico. *Cienc. Mar., 34*, 467-478.

Sale, M. C., McWilliam, P. S. and Anderson, D. T. (1976). Composition of the near reef zooplankton at Heron reef. *Mar. Biol., 34*, 596.

Sammarco, P. W. and Crenshaw, H. (1984). Plankton community dynamics of the central Great Barrier Reef lagoon. *Mar. Biol., 82*, 167-180.

Sorokin, Y. I. (1990). Plankton of the reef ecosystems. In: Z. Dubinsky (Ed.), Coral reefs. *Ecosystems of the world, 25,* pp. 291-324. Amsterdam etc.: Elsevier.

Suárez, E. (1991). *Composición, distribución, abundancia y zoogeografía de los copépodos pelágicos (Crustacea) del Golfo de México y Mar Caribe mexicanos.* Tesis Doctoral, Facultad de Ciencias, Universidad Nacional Autónoma de México, México, D. F.

Suárez, E. (1992). Lista faunística de los copépodos calanoideos (Copepods: Calanoida) del Golfo de México: consideraciones zoogeográficas. *Cienc. Mar., 18(2)*, 119-151.

Suárez, E. and Gasca, R. (1990). Variación dial del zooplancton asociado a praderas de *Thalassia testudinum* en una laguna arrecifal del Caribe mexicano. *Universidad y Ciencia (México), 7(13)*, 57-64.

Suárez, E. and Gasca, R. (1994). Zooplankton biomass fluctuations in a Mexican Caribbean bay (Bahía de la Ascensión) during a year cycle. *Caribb. J. Sci., 30(1/2)*, 166-123.

Suárez-Morales, E. and Gasca, R. (1996). Planktonic copepods of Bahía de la Ascención Caribbean coast of Mexico: a seasonal survey. *Crustaceana, 69*, 162-174.

Suárez-Morales, E. and Gasca, R. (2000). The planktonic copepod community at Mahahual reef, western Caribbean. *Bull. Mar. Sci., 66(1)*, 255-267.

Suárez-Morales, E. and Rivera-Arriaga, E. (1998). Zooplancton e hidrodinámica en zonas litorales y arrecifales de Quintana Roo, México. *Hidrobiológica, 8(1)*, 19-32.

Suárez-Morales, E., Segura-Puertas, L. and Gasca, R. (1999). A survey of the reef-related medusa (Cnidaria) community in the western Caribbean Sea. *Gulf Res. Rep., 11*, 23-31.

Tovar, E., Suáres-Morales, E. and Castillo, L. (2009). Multiscale variability of the Chaetognatha along a Caribbean reef lagoon system. *Mar. Ecol. Prog. Ser., 375*, 151-160.

Tranter, D. J. and George, J. (1972). Zooplankton abundance at Kavaratti and Kalpeni Atolls in the Laccadives. In: *Proc. Symp. on Coral Reefs, Mar. Biol. Assoc. India,* pp. 239-256.

Vaissiere, R. and Seguin, G. (1984). Initial observations on the zooplankton microdistribution of the fringing coral reef at Aqaba (Jordan). *Mar. Biol., 90,* 1-11.

Vega-Rodríguez, F. (1965). Distribución de Chaetognatha en Veracruz, Ver. *Anales Inst. Biol. Univ. Nac. Autón. México, 36*, 229-247.

Vinogradov, M. E. and Shushkina, E. A. (1987). *Functioning of the plankton communities of the epipelagial zone of the ocean.* Moscow: Nauka. (In Russian).

Zavala-García, F. and Flores-Coto, C. (1989). Medición de biomasa zooplanctónica. *Anales Centro Cienc. Mar Limnol. Univ. Nac. Autón. México, 16(2)*, 273-278.

In: Zooplankton and Phytoplankton
Editor: Giri Kattel, pp. 89-105

ISBN 978-1-61324-508-8

Chapter 4

THE UPSIDE OF GRAZER-PERIPHYTON INTERACTIONS: A REVIEW

Surjya Kumar Saikia, Santanu Ray and Joyita Mukherjee
Ecological Modelling Laboratory, Department of Zoology,
Visva Bharati University, Santiniketan,
India 731235

ABSTRACT

Periphyton ecology is affected by abiotic factors like light limitation, waterflow, nutrients and physical and chemical properties of substrates. However, the significant influence of biotic factors is contributed through grazing, for algal dominant periphytic communities, the herbivory. Grazing usually causes a reduction to periphytic algal biomass, but individual algal populations can exhibit a positive response to increased grazing intensity. Such positive responses can overweight negative responses in certain nutrient environments. Differential grazing on overstory periphyton canopy stimulate competitive interaction among algae for enhanced understory nutrient. Grazing resistant algae, in addition, contributes to periphytic standing crop with higher nutrient retaining algae. Further, selective grazing on competitively dominant algal species clears the substrates for colonization by slow-growing and deterrent species increasing periphytic micro-algal diversity. Thereby, a heterogeneic periphytic community could be initiated on the grazed surface. These producer-consumer interactions vary specifically on spatial gradients. Stoichiometrically, periphytic C:P and N:P ratios are decreased under grazing pressure, whereas, C:N and C:Chl ratios remain unaffected. Low levels of grazing, in turn, stimulate increase in nutrient cycling and hence algal productivity. Extensive grazing has negative impact since such grazing rates could exceed reproductive rates of periphytic algae. Grazer mouth morphology and feeding preferences may further determine the direction of ecological growth of periphyton.

Keywords: periphyton, algal species, C:nutrient ratio, herbivory

INTRODUCTION

Food webs are a central idea in ecology and represent graphically how nutrients and energy are cycled through species interactions within ecosystems (Wilbur, 1997). In aquatic ecosystems, periphyton being the primary producer plays a key role in determining the structure and function of the aquatic community through periphyton based food web. Periphyton, literally, the layer of microorganisms that forms on many submerged surfaces in water bodies, usually a complex mixture of bacteria, algae, fungi, microscopic animals and non-living organic matter—is a primary food source for grazing invertebrates and some vertebrates (Lock, 1981; Feminella and Hawkins, 1995). In periphyton dominated system, producers and consumers live in tighter spatial connection (Kahlert and Baunsgaard, 1999). Differential taxa and morphological forms of periphyton layer, such as algae and cyanobacteria vary in the extent to which particular grazers are able to consume and assimilate them (Dudley et al., 1986; Peterson, 1987; Hill et al., 1992; Steinman et al., 1992).

In addition, periphytic algal communities are very suitable model systems to test general ecological mechanisms, and especially food-web interactions, since algae have short generation times and trophic cascades are strongest among the benthic communities (Shurin et al., 2002). Looking to this advantage, most studies were performed on effect of periphytic algal diversity and grazer interactions. Some studies reported significant decreases of algal biomass due to grazer activity (Feminella and Hawkins, 1995; Steinman, 1996). Several herbivore types (especially gastropods and trichopteran larvae) can dramatically reduce periphytic biomass to only a few percent (Lamberti et al., 1987; Hill et al., 1992; Hillebrand et al., 2000). However, grazing may result in increased nutrient content or increased biomass-specific productivity of grazed compared to ungrazed periphyton (McCormick and Stevenson, 1991; Rosemond et al., 1993). This positive effect of consumers on their prey can be through excretion of nutrients, removal of senescent cells, or increased uptake of nutrients by the remaining cells (Lamberti et al., 1987; McCormick and Stevenson, 1991; Rosemond et al., 1993; Kahlert and Baunsgaard, 1999). This heterogenous effect of grazers in terms of biomass and nutrient on periphytic algal community is generally overlooked and still to be understood as advantageous mechanism for benthic ecosystem. Present review is an effort to focus a light on this subject.

GRAZING INCREASES ALGAL HETEROGENEITY OF PERIPHYTON

One of the long run, fundamental issues in ecology is to understand what determines species richness. As determining factor on periphyton availability, grazing could increase species richness via grazer mediated co-existence if preferred resource species as food are competitively dominated but reduce richness via extinction of rare species (Lubchenco, 1978; Liess and Hillebrand, 2004). Conceptual models predict a reversal of grazing impacts on floral resources through species richness in nutrient-poor ecosystems (Proulx and Mazumder, 1998; Worm et al., 2002) resulting enhanced nutrient status by increasing the ability of plants to re-grow after grazing, thereby reducing the probability of extinction. That is why extinction or nich collapse could be a rare show for periphytic community. Graphically, a shift in periphytic species richness can be expected from low to high nutrient availability depending on whether grazer pressure is low or high, since grazers decrease richness at low nutrient condition but increase it at high nutrient condition (Liess and Kahlert, 2007). But, the strength of the grazing effect i.e. reduction in species richness in nutrient-poor habitats may not be reached, rather, such effect can be an offset, at least to some degree, by grazer-mediated nutrient excretion (Liess and Kahlert, 2007). In other words, grazing and competition can modify the species composition of algal assemblages (Duffy and Hay, 2000; Nielsen, 2001) generating heterogeneity through temporal or spatial (patch selection) resilience, but not extinction, in littoral plant communities (Lubchenco, 1978).

Patch formation is a phenomenon very often seen in periphyton-grazer relationship. Macroalgal consumption by grazers (Malm et al., 1999; Duffy and Hay, 2000; Engkvist et al., 2000; Nielsen, 2001) can create empty patches on substrates and reduce algal competition by thinning out the community (Lubchenco, 1978; Connell and Keough, 1985; Gaines, 1985). An opposite effect of intra specific competition among periphytic members could be expected if grazer resistant algae succeed following differential patch module grazing (Mc Cormik and Stevenson, 1991). However, selective grazers may prefer competitively dominant algal species thereby clearing substrates for colonization by slow-growing and deterrent species, thus increasing macroalgal diversity (Paine, 2002; Worm et al., 2002). Such effects of grazing may, further, differ along the depth gradient, as grazers, and also algae, are limited by several abiotic factors such as wave energy in open waters, light intensity, light quality, light penetration, sedimentation, etc. (Lubchenco and Menge, 1978; Menge and Sutherland, 1987; Hurd, 2000). As an outcome, grazers may reduce macroalgal species in both the colonization (Lotze et al., 1999, 2001) as well as in the adult phases (Lubchenco, 1978; Duffy and Hay, 2000) through variable random and/or selective ways, and such a regulation is often effective in maintaining variation in both species composition and total community biomass (Lotze et al.,1999, 2001; Duffy and Hay 2000; Nielsen, 2001; Belliveau and Paul, 2002; Worm et al., 2002; Korpinen et al., 2007) leading to periphytic heterogeneity. On time scale, too, grazing algal colony experience most intense grazing during early colonization phase (Lotze et al., 1999, 2001; Lotze and Worm, 2000; Korpinen et al., 2007). Any variation in grazer densities at this time could have a strong impact on the subsequent macro algal assemblages generating heterogeneity. In some cases, such grazing may facilitate emergence of new species (Korpinen and Jormalainen, 2008).

Alvarez and Peckarsky (2005) further observed increase in algal heterogeneity with increasing consumer mobility. Using simple two patch models, Abrams (2000) predicted that the pattern of resource heterogeneity along an axis of grazer mobility should fit a hump-shaped curve. Assuming that grazers have homogeneous distributions and equal access to all patches, they predicted that immobile grazers should homogenize resources, variability among resource patches should be greatest for intermediate rates of movement, and highly mobile consumers should decrease spatial variation in food abundance by overexploitation.

Grazing Effect of Algal Biomass and Productivity

Grazing is considered as most important determinant of periphyton biomass. It, in general, decreases benthic algal accumulation rates on substrates and thereby causes a reduction in total algal biomass (Lamberti and Resh, 1983; McAuliffe, 1984; Hart, 1985; Steinman et al., 1987, 1989). Exclusion and removal experiments on coral reefs showed that algal standing crop could be 1.5 to 15 fold when grazers were excluded (Hatcher and Larkum, 1983). In presence of multiple competitive grazers, the result could be reversed. McCollum et al., (1998) observed that the fish, redear sunfish (*Lepomis microlophus*) had a positive effect on periphytic biomass by reducing the grazing of snail on periphyton. Presence of predators of grazers may also exert positive results on periphytic biomass. It was observed in streams of Panama where bands of high periphyton biomass occurred in the top water layer which is avoided by the fish for fear of predation (Power et al., 1989). Though grazing results reduction in periphytic biomass, algal productivity increases due to reduced competition among algal members (Carpenter, 1986; Mc Cormick and Stevenson, 1989). Coral reef algae grazed by sea urchins had a lower biomass than ungrazed algae but productivity was not reduced (Carpenter, 1986). Similarly, Norberg (1999), using transparent incubation chambers, measured a 4-fold increase in periphyton specific productivity in grazed periphyton compared to ungrazed controls. Moreover, the grazer presence increased the Chl: biovolume ratio, as reported from streams (Hill and Knight, 1987). Although other explanations are possible, the decrease in filamentous algae and the increase in Chl: biovolume ratio presumably reflects a preferential removal of upright, large species by grazing herbivores.

Though reduction of algal biomass due to interaction of grazers and grazer preferred algal individuals occurs, individual non or less preferred algal populations can exhibit a positive response to increases in grazing intensity (McCormick and Stevenson, 1989; McCormick, 1990; Steinman et al., 1989) and effect on Chl: biovolume ratio. This positive affect on individual or understory periphytic communities may come through exposure to resources (e.g., nutrients) or other abiotic facilities (e.g. light) (McCormick and Stevenson, 1989) due to grazing.

Grazer Effect on Understory Periphyton

Intake or understory periphyton communities have low nutrient uptake rates due to boundary effects and diffusion inhibition (Ribber and Wetzel, 1987). Cuker (1983), studying in an arctic lake with snail as grazer observed that grazing activity caused a shift in relative algal composition. Larger diatoms ($\geq$20 μm) dominated the epilithic algae at lower grazer densities but were relatively less important at the higher grazer loads, while coccoid forms and small diatoms ($\leq$20 μm) increased in relative importance. Filamentous forms showed little response to snail load. Absolute algal abundance calculated from the relative counts and chlorophyll measurements showed significant decreases for large diatoms, small diatoms, and filaments; absolute levels of coccoids were independent of snail density.

Studies of McCormick and Stevenson (1991) on periphytic diatom and filamentous algal species by snail *Goniobasis* showed interactive response for understory as well as overstory periphytic algae with grazer presence through nutrient enrichment. The pattern of response of *Stigeoclonium,* an understory and grazing resistant filamentous alga, to nutrient enrichment and grazing indicates that grazers have a strong effect on the nutrient response of understory species in benthic algal assemblages. In the absence of snail grazing, *Stigeoclonium* exhibited no significant response to nutrient enrichment. In the presence of grazers, the response of

Stigeoclonium to nutrient enrichment was proportional to the density of snails. Two mechanisms could be hypothesized for such positive effects of grazers on the grazer-resistant algae, first, grazers themselves may excrete nutrients and directly increase nutrient supply (Friedl, 1974; Grimm, 1988); and second, grazers may indirectly increase nutrient supply to understory algae by removing overstory algal community. Evidence supporting such hypothesis shows that transport rates of nutrients into understory periphyton can be slowed down by even relatively low densities of overstory periphyton (Riber and Wetzel, 1987; Stevenson and Glover, 1990). It is possible, in their experiment that removal of the overstory diatom community by *Goniobasis* resulted in increased penetration of nutrients to *Stigeoclonium* cells underneath. Although grazing may also increase the availability of other resources (e.g., light) in a similar manner, the primary effect could be on the availability of the amended nutrients. On excretory nutrient supplementation through grazer, McCormick and Stevenson's (1991) result on increase in *Stigeoclonium* density at higher grazer densities is important. It suggests that direct effects of nutrient excretion by *Goniobasis* may be possible. McCormick and Stevenson (1991) further concluded that grazers apparently affect competitive interactions among benthic algae by differentially reducing overstory abundances, thereby increasing the supply of nutrients, and possibly other resources, to the understory. Grazers also may increase nutrient supply directly by excretion, and results of McCormick and Stevenson (1991) suggest that those nutrients may stimulate growth of grazer resistant species. Indeed, if nutrient supply is adequate and a species is present that is highly resistant to grazing, as is *Stigeoclonium* in their experiment, grazing can increase benthic algal standing crop if reproduction exceeds grazing rates. However, Hillebrand et al (2001) considered such routes of nutrient enrichment are significant for periphyton nutrient stoichiometry.

Grazing - Periphyton Interaction and Nutrient Stoichiometry

Although, most of the times nutrient and grazing effects have been separately studied, few investigations have considered them simultaneously (McCormick and Stevenson, 1991; Hill et al., 1992; Rosemond, 1993; Rosemond et al., 1993; Pan and Lowe, 1994; Neckles et al., 1993). In lakes, Marks and Lowe (1989) showed an increase of grazing effects on periphyton composition with nutrient enrichment. Hillebrand et al. (2000) reported significant and counteracting effects of both factors in a spring grazing experiment at the Baltic coast.

Periphyton assemblages are embedded in a mucilage matrix and surrounded by a boundary layer which slows down the rate of diffusion of nutrients into the community and limits the access to water column nutrients by algal members (Bothwell, 1989; Burkholder et al., 1990; Hill et al., 1992). Therefore, it is evident that the structure and productivity of periphyton are highly regulated by the interplay of grazers (Steinman, 1996) and abiotic factors such as nutrients (Hill et al., 1992, Borchardt, 1996, Hillebrand, 2002) and light (Hill, 1996). Grazers though reduce periphyton biomass; the nutrients they recycle can be an important source of nutrients to periphyton (Grimm, 1988; Mulholland et al., 1991). The ratio at which nutrients are recycled depends on the nutrient composition of both consumers and their food, i.e. periphytic assemblage (Elser et al., 1988; Urabe, 1993). Furthermore, as described earlier, mobile grazers may actively select high-quality or nutrient rich food colonies of periphyton and increase homogeneity through stabilizing local grazer density (Hill et al., 1992; Nisbet et al., 1997; Cruz-Rivera and Hay, 2000). Thus, the relationship of nutrient availability and grazer response holds the core of the principle of theory of ecological stoichiometry in periphytic system.

Light, as abiotic component interacts with grazing and nutrients (Rosemond, 1993; Rosemond et al., 2000), in shaping and maintaining autotroph community composition. As with increased nutrients, higher irradiance may help plants counteract grazing pressure through resuming community biomass and productivity. Thus, the interacting effects of consumers and nutrients will be stronger when light intensities are higher. In nutrient-poor environments, increased light intensities make nutrient limitation more intense (Frost and Elser, 2002) via a general increase in autotroph C: nutrient ratios (Hessen et al., 2002; Urabe et al., 2002). In nutrient-rich environments, high light intensities increase autotroph growth rates and counteract extinction probabilities.

The C: N: P ratios are widely used to indicate nutritive status of phytoplankton (Hecky et al., 1993) and this concept has been established also for benthic microalgae (Kahlert, 1998; Hillebrand and Sommer, 1999). Based on such measurements, algal communities like diatoms and cryptophytes are recorded as higher nutrient rich algal members than green algae (Chlorophytes) which are in turn are higher nutrient rich members than cyanophytes and therefore, availability of heterogenous periphytic taxa on substrate could determine nutrient sources (in terms of availability of C and N measured as C: N) (Ahlgren *et al.*, 1990; Lundtstedt and Brett, 1991) for grazers. Regarding such taxonomic nutrient availability, filamentous algae (mostly chlorophytes, but also rhodophytes) are reduced most by grazing, whereas the relative importance of filamentous cyanobacteria is often enhanced by grazer presence (Hillebrand and Kahlert, 2001; Lowe and Hunter, 1988; Rosemond et al., 1993; Sommer, 1997). The detritus or heterotrophic components in the periphyton could also influence C: N: P ratios. Nutrients may increase either N or P content or change the algal composition to species with a different C: N: P ratio, whereas, grazers may reduce the detritus content or non algal component, change the composition of algae either by preferential grazing or avoidance, or increase the supply of nutrients to algae through excretion or removal of overstory community (Hillebrand and Kahlert, 1991). This suggests the algal taxonomic composition in periphytic systems as partial nutrient management tool or an unlikely explanation with supposedly high P demand such as cyanobacteria or low P demand such as chlorophytes (Sommer, 1996).

However, these positive effects of grazing on nutrient content may not outweigh consumptive losses, i.e., algal biomass is not increased even though grazer presence (Steinman, 1996; De Mazancourt et al., 1998). Hillebrand and Kahlert (1991) found strongest effects of grazer presence on C:P and N:P ratios, whereas C:N and C:Chl were less or not affected. An important role of changed nutrient supply by grazing can be mediated by removal of the canopy layer leading to a more efficient nutrient uptake or by nutrient excretion (McCormick and Stevenson, 1991). Hillebrand et al. (2008) described three pathways for grazer mediated periphytic interactions affecting nutrient stoichiometry (Figure 1). First, the non algal component, which could be a dominant part of the organic material of periphyton assemblage (Frost et al., 2002a) is reduced by unselective grazing. Benthic invertebrates graze upon both detritus and algal component of periphyton but only algae regenerate. Therefore, grazing not only reduces non algal component of periphyton, but also facilitates the growth of live component within it.

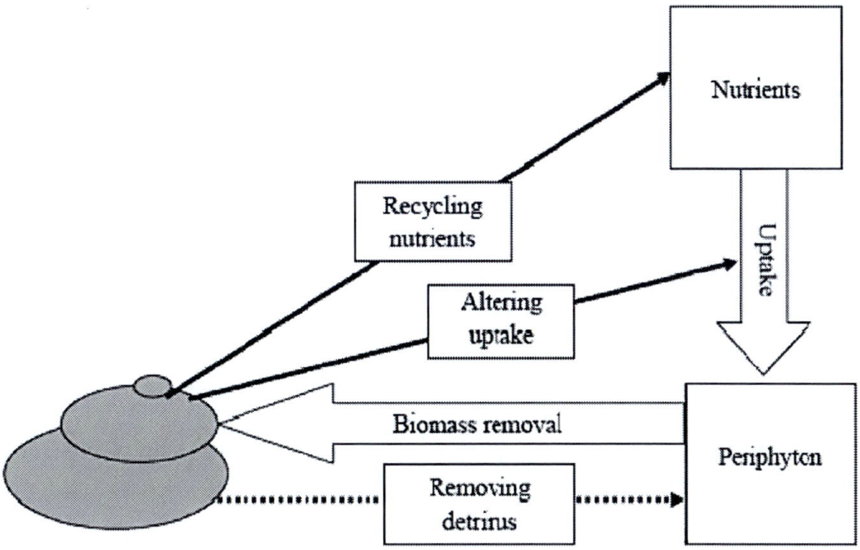

Figure 1. Direct and indirect effects of grazer on nutrient stoichiometry (Hillebrand et al., 2008). Three different pathways affecting nutrient stoichiometry are described. They are 1. Altering uptrake, recycling uptake and removing detritus.

Thus grazing has an indirect effect on reducing C:nutrient stochiometry on periphytic substrate. Secondly, nutrient uptake of intact periphyton mats is often slower than cell specific uptake rates as boundary effects reduce the uptake ability of the benthic algae (Riber and Wetzel, 1987; Bothwell, 1989; Burkholder et al., 1990). By altering the architecture as well as increasing heterogeneity of the periphyton community and by increasing the relative availability of nutrients (through reducing C:nutrient ratio) per unit biomass, grazers may enhance periphytic nutrient uptake and as a consequence increase its nutrient content. Nutrient uptake rates could further change periphyton stoichiometry if grazers alter species composition or simply enhance biomass turnover. Thirdly, the excretion or egestion of nutrients or both by grazers also increase the supply of nutrients to the periphytic assemblages. Grazers may spatially recycle nutrients that increase the availability and uptake of nutrients by the periphyton. However, in streams, grazers may increase the export of nutrients (Mulholland et al., 1991).

Grazed communities are often dominated by either prostrate species, which adhere tightly to the substrate, or by small understory species which can avoid being grazed by virtue of their size (Hunter, 1980; Hunter and Russell-Hunter, 1983; Sumner and McIntire, 1982). Studies have shown that highly grazed periphyton has a low C:N ratio which is also characteristic of periphyton experiencing rapid turnover during early succession (Hunter, 1980; McMahon et al., 1974). In addition, it has been confirmed through oxygen evolution experiments that grazed periphyton has accelerated turnover rates (Lamberti and Resh, 1983).

Thus, it is clear that grazing can maintain periphyton communities in physiologically earlier successional stages.

Sterner and Elser (2002) predicts that the amount of available C, N and P in primary producer biomass can limit grazer growth. Grazers should sequester the nutrient limiting growth and excrete excess nutrients. Here, grazers may excrete nitrogen (Friedl, 1974; Grimm, 1988) and increase nutrient supply directly. This process of differential excretion of nutrients can alter periphyton stoichiometry by altering nutrient availability. Frost et al. (2002b) suggested that benthic grazers can affect the elemental composition of their food in a manner similar to their planktonic counterparts. In some experiments, nutrient supply and the presence of herbivores have increased periphyton P content (decreased C:P and N:P ratios), suggesting that grazer excretion of P may be responsible for an increase in algal P (Hillebrand and Kahlert, 2001). Higher food quality (lower C:P and N:P ratios) can lead to higher growth rates of benthic grazers, especially when algal biomass is low (Stelzer and Lamberti, 2002). Liess and Hillebrand (2006) showed that decreasing C:P ratio with P enrichment can be maintained with introduction of grazers. It, further, depends upon the mouthpart morphology of grazer. Liess and Hillebrand (2006) revealed that greater C:P ratio in periphyton C:P ratio is greater when grazers of scraping, gathering and shredding natures are introduced. However, in rasping and scraping types of grazers, such C:P ratio does not increase. N:P ratios also followed the same trend.

MODELLING PERIPHYTON INTERACTIONS IN ECOLOGY

Mathematical modeling is a useful tool to investigate the qualitative and quantitative aspects of ecological systems. Now a day it is one of the major tools in ecological research. Periphyton is an important topic in ecological system, many models both theoretical and simulation models have been proposed and analysed by many researchers in different countries.

Buzzelli et al. (2000) developed a periphyton growth model and use it to identify critical factors in ecosystem phosphorus dynamics in pristine freshwater wetlands of Everglades National Park, USA. With the simulation of their proposed model they showed the changes of periphyton biomass under increased of total phosphorus. Buzzelli et al. (1999) in another model described the spatial structure and growth of periphyton community. From morphological differences in the growth patterns of filamentous and non-filamentous algae, the model simulated the vertical structure of a periphyton mat. The effects of shear stress on the colonization and the detachment process, and the overall response of the community to water flow were formulated. They evaluated the thickness of the periphyton mat, which determines dynamically the light attenuation in the periphyton mat. The growth rate and vertical light distribution inside the periphyton mat were simulated with three values of the biotic light attenuation coefficient. A new model for biofilm growth is based on a logistic approach but accounts for detachment of the biofilm by the running water and the existence of a threshold point for growth was proposed by Momo (2000). Above a critical current velocity, development of periphyton is not stable. A stability analysis shows the existence of a fold catastrophe in the model and the amplification of water velocity fluctuations by the algal biomass. Stream ecosystems include both surface and subsurface components that are connected by the flow of water. Processes occurring in subsurface sediments affect those in the surface, and vice versa. A model was developed by Cattaneo et al. (1993) to investigate the effects of nutrient transformations occurring in subsurface sediments on the growth of surface-dwelling periphyton. In the model the stream is divided into three zones: free-flowing surface water, a surface storage zone where flow is minimal and where periphyton growth occurs, and a subsurface zone. The behavior of the model was examined both at steady state and as it approached steady state after periphyton biomass was reduced to low values, simulating the effects of a scouring flood. The model also shows that processes in the subsurface zone can have a large effect on surface organisms, and illustrates how surface periphyton communities interact with subsurface microorganisms through the recycling of nutrients. Uehlinger et al. (1996) evaluated the the dominant processes that control periphyton biomass, measured data were compared with a set of simulations from an empirical dynamic periphyton model. Different combinations of process hypotheses were systematically activated and deactivated in order to assess their importance. Kim et al. (1992) used a convection-dispersion hydrologic transport model by coupling a transient storage submodel and a biotic uptake submodel based on Michaelis-Menten kinetics. In this model they simulate nitrate retention in response to change in load in a third-order stream, differentiate biotic versus hydrologie factors in nitrate retention, and produce a research tool whose properties are consistent with laboratory and field observations. Hydrodynamic parameters were fitted from chloride concentration during a 20-day chloride-nitrate coinjection. Nitrate concentration from the 20-day coinjection experiment served as a base for model validation. The complete transport retention model reasonably predicted the observed nitrate concentration. Using Artificial Neural Network in the model study is widely accepted and practiced by many ecologists, Saravia et al. (1998) made an attempt for modelling the periphyton dynamics with this tool exemplified by the oxbow lake reopening study (The Słupia River, Northern Poland). They designed Artificial Neural Network architectures and trained in order to create models of interactions between 18 macrozooperiphyton, microzooperiphyton and phytoperiphyton taxa in the changing ecosystem..Neural network models allowed a quantitative insight into periphyton dynamics and indicated trophic relationships, both predator-yprey and competitive. Son and Fujino (2003) develped a model of the development of a periphyton community in a stream under the influence of nutrients and light. Coupling the model with a nutrient transport model clarified the longitudinal distribution of both periphyton community and nutrients. The thickness of the periphyton mat, an important factor regulating nutrient exchange between the mat and overflowing water, was determined from water velocity and the periphyton biomass. They validated the simulated results with with three observational data sets, first of all a comparison of electrical conductance in two channels with different periphyton biomasses validated the model in the mass exchange between stationary and flowing water zones; secondly a comparison of the temporal variation in periphyton biomass and nutrient concentration in a once-through and a re-circulated water channel, validated the relationships among the periphyton biomass, the nutrient uptake rate, and the nutrient concentrations in the stationary water zone; and finally a longitudinal distribution of the algal species composition of *Stigeoclonium* and *Chamaesiphon*, and the nutrient concentrations of a 140 m reach was reproduced and compared with measured data. The light intensity indirectly controlled the nutrient gradients along the stream by the periphyton biomass in the third application. Toda and Tsujimoto (2010) used numerical modeling of interspecific competition between filamentous and non filamentous periphyton on a flat channel bed. In their study they described the processes of the interspecific competition which are an integral part of the modeling of growth strategies. The temporal growth of each kind of periphyton is calculated by balancing primary production, immigration, respiration, and detachment. For the modeling of primary production and respiration, transportation of nutrients and other substances inside the biofilm is estimated using diffusion equations including the effect of turbulent diffusivity near the channel bed surface. In estimating the immigration rate, different immigration functions for filamentous and nonfilamentous periphyton are used in order to represent the different effect of turbulent flow on the immigration strategy of each type of periphyton. A series of numerical computations under different hydraulic conditions shows that the degree of dominance of filamentous periphyton increases with time, and the net primary production of periphyton increases with increasing friction velocity.

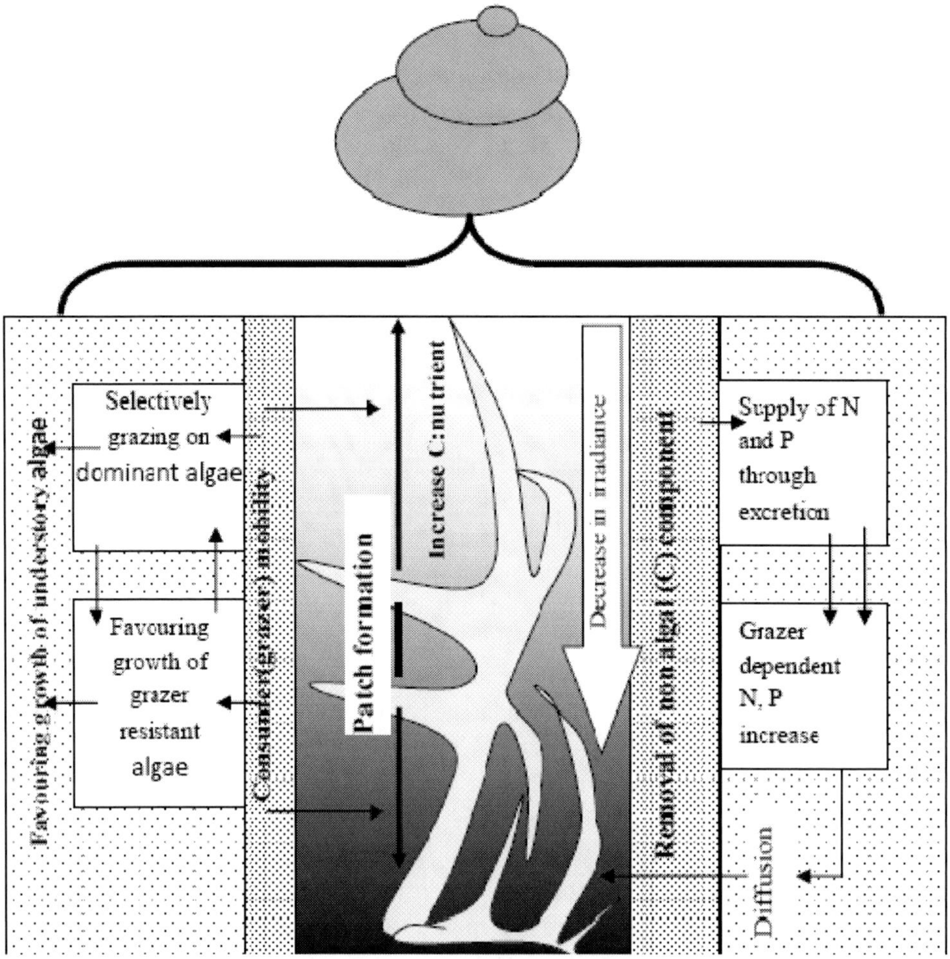

Figure 2. Positive effects of grazer on periphyton. The main driving interactions are 1. Favoring growth of understory algae, 2. Consumer mobility and 3. Removal of non algal component from periphyton layer.

Concluding Remarks

As conceptualized for many ecosystems, periphyton-grazer regulated systems are less vulnerable to environmental disturbances for its multifold positive effects of grazing (Figure 2). Although modeling is very important tool in ecology, the works on periphyton modelling are not enriched. New tools on modelling have been developed, among these tools are structural dynamic modelling, network modelling, use of Self Organizing Map in spatial modelling etc. With the applications of these new modelling tools on periphyton ecology, more new aspects will be highlighted in future in determining the community structure and trophic transfer of nutrients or energy, periphytic-grazer relationship and these could represent an ideal biotic tool for ecologists.

References

Abrams, P.A. (2000). The impact of habitat selection on the spatial heterogeneity of resources in varying environments. *Ecology, 81*, 2902-2913.

Ahlgren G., Lundstedt L., Brett M. T. and Forsberg C. (1990). Lipid composition and food quality of some freshwater phytoplankton for cladoceran zooplankters. *Journal of Plankton Research, 12*, 809-818.

Alvarez, M. and Peckarsky B. L. (2005). How do grazers affect periphyton heterogeneity in streams? *Oecologia, 142*, 576-587.

Asaeda, T., and Son, D. H. (2000). Spatial structure and populations of a periphyton community: a model and verification. Ecological Modelling, 133, 195-207.

Belliveau, S. A. and Paul, V. J. (2002). Effects of herbivory and nutrients on the early colonization of crustose coralline and fleshy algae. *Marine Ecology Progress Series, 232*, 105-114.

Biggs, B. J. E., Stevenson, R. J. and Lowe R. L. (1998). A habitat matrix conceptual model for stream periphyton. *Archiv fuer Hydrobiologie*, 143, 21-56.

Borchardt, M. A. (1996). Nutrients. In R. J. Stevenson, M. L. Bothwell and R. L. Lowe (Eds.), *Algal ecology* (pp.183-227). Academic Press, London, UK.

Bothwell, M. L. (1989). Phosphorus-limited growth dynamics of lotic periphyton diatom communities: Areal biomass and cellular growth rate response. *Canadian Journal of Fisheries and Aquatic Sciences, 46*, 1293-1301.

Bronmark, C. (1985). Interactions between macrophytes, epiphytes and herbivores: An experimental approach. *Oikos,* 45, 26-30.

Burkholder, J. M., Wetzel, R. G. and Klomparensm, K. L. (1990). Direct comparison of phosphate uptake by adnate and loosely attached microalgae within an intact biofilm matrix. *Applied and Environmental Microbiology, 56*, 2882-2890.

Buzzelli, C. P., Wetzel, R. L., and Meyers, M. B. (1999). A Linked Physical and biological Framework to Assess Biogeochemical Dynamics in a Shallow Estuarine Ecosystem. *Estuarine, Coastal and Shelf Science, 49,* 829–851.

Buzzelli, C. P., Childers, D. L., Dong, Q, Jones, R. D. (2000).Simulation of periphyton phosphorus dynamics in Everglades National Park. *Ecological Modelling, 134,* 103-115.

Carpenter, R.C. (1986) Partitioning hervibory and its effects on coral reef algal communities. *Ecological Monographs.* 56, 345-363.

Cattaneo, A., Legendre, P., Niyonsenga, T. (1993). Exploring Periphyton Unpredictability. *Journal of North Americal Benthological Society, 12,* 418-430.

Connell, J. H. and Keough, M. J. (1985). Disturbance and patch dynamics of subtidal marine animals on hard substrata. In S. T. A. Picket and P. S. White (Eds.), *The ecology of natural disturbance and patch dynamics* (pp. 125-151). New York : Academic Press.

Cruz-rivera, E. and Hay, M. E. (2000). Can quantity replace quality? Food choice, compensatory feeding, and fitness of marine mesograzers. *Ecology, 81*, 201-219.

Cuker, B. E. (1983). Grazing and nutrient interactions in controlling the activity and composition of the epilithic algal community of an arctic lake. *Limnology and Oceanography, 28*, 133-141.

De Mazancourt, C., Loreau, M. and Abbadie, L. (1998). Grazing optimization and nutrient cycling: When do herbivores enhance plant production? *Ecology*, *79*, 2242-2252.

Denicola, D. M., Mcintire, C. D., Lamberti, G. A., Gregory, S. V., and Shkenas, L. R. A. (1990). Temporal patterns of grazer periphyton interactions in laboratory streams. *Freshwater Biology, 23*, 475-489.

Dent C. L., Henry, J. C. (1999). Modelling nutrient-periphyton dynamics in streams with surface-subsurface exchange. *Ecological Modelling, 122,* 97-116.

Dudley, T. L., Cooper, S. D., and Hemphill, N. (1986). Effects of macroalgae on a stream invertebrate community. *Journal of the North American Benthological Society, 5*, 93-106.

Duffy, J. E., and Hay, M. E., (2000). Strong impacts of grazing amphipods on the organization of a benthic community. *Ecological Monographs*, *70*, 237-263.

Duffy, J. E., Richardson, J. P., and France, K. E., (2005). Ecosystem consequences of diversity depend on food chain length in estuarine vegetation. *Ecology Letters, 8*, 301-309.

Elser, J. J., Elser, M. M., Mackay, N. A., and Carpenter, S. R. (1988). Zooplankton-mediated transitions between N- and P-limited algal growth. *Limnology and Oceanography, 33*, 1-14.

Engkvist, R., Malm, T., and Tobiasson, S. (2000). Density dependent grazing effects of the isopod Idotea baltica Pallas on *Fucus vesiculosus* L in the Baltic Sea. *Aquatic Ecology, 34*, 253-260.

Feminella, J. W., and Hawkins, C. P. (1995). Interactions between stream herbivores and periphyton: a quantitative analysis of past experiments. *Journal of the North American Benthological Society*, *14,* 465-509.

Friedl, F. E. (1974). Nitrogen excretion by the freshwater snail *Lymnaea stagnalis jugularis* Say. *Comparative Biochemistry and Physiology*, *49,* 617-622.

Frost, P. C., Stelzer, R. S., Lamberti, G. A., and Elser, J. J. (2002a). Ecological stoichiometry of trophic interactions in the benthos: understanding the role of C:N:P ratios in lentic and lotic habitats. *Journal of the North American Benthological Society*, *21,* 515-528.

Frost, P. C., Elser, J. J., and Turner, M. A. (2002b). Effects of caddisfly grazers on the elemental composition of epilithon in a boreal lake. *Journal of the North American Benthological Society*, *21*, 54-63.

Frost, P.C., and Elser, J.J. (2002). Effects of light and nutrients on the net accumulation and elemental composition of epilithon in boreal lakes. *Freshwater Biology, 47,* 173-183.

Gaines, S. D. (1985). Herbivory and between-habitat diversity: The differential effectiveness of defences in a marine plant. *Ecology*, *66*, 473-485.

Goecker, M. E., and Kåll, S. E. (2003). Grazing preferences of marine isopods and amphipods on three prominent algal species of the Baltic Sea. *Jounal of Sea Research, 50*, 309-314.

Grimm, N. B. (1988). Role of macroinvertebrates in nitrogen dynamics of a desert stream. *Ecology, 69*, 1884-1893.

Hatcher, B.G. and Larkum, A.W.D. (1983). An experimental analysis of factors controlling the standing crop of the epilithic algal community on a coral reef. *Journal of Experimental and Marine Biology and Ecology,* 69: 61-84.

Hart, D. D. (1985). Grazing insects mediate algal interactions in a stream benthic community. *Oikos, 44*, 40-46.

Hecky, R. E., Campbell, P., and Hendzel, L. L. (1993). The stoichiometry of carbon, nitrogen, and phosphorus in particulate matter of lakes and oceans. *Limnology and Oceanography, 38*, 709-724.

Hessen, D. O., Faerovig, P. J., and Andersen, T. (2002). Light, nutrients, and P:C ratios in algae: grazer performance related to food quality and quantity. *Ecology*, *83*, 1886-1898.

Hill, W. R. (1996). Effects of light. In R. J. Stevenson, M. L. Bothwell, and R. L. Lowe (Eds.) *Algal ecology* (pp.121-148). London, UK: Academic Press.

Hill, W.R., and Knight, A. W. (1987). Experimental analysis of the grazing interaction between a mayfly and stream algae. *Ecology*, *68*, 1955-1965.

Hill, W. R., Boston, H. L., and Steinman, A. D. (1992). Grazers and nutrients simultaneously limit Lotic primary productivity. *Canadian Journal of Fisheries and Aquatic Sciences, 49*, 504-512.

Hill, W. R., Weber, S. C. and Stewart, A. J. (1992). Food limitation of two lotic grazers: quantity, quality, and size-specificity. *Journal of the North American Benthological Society, 11*, 420-432.

Hillebrand, H. (2002). Top-down versus bottom-up control of autotrophic biomass—a meta-analysis on experiments with periphyton. *Journal of the North American Benthological Society, 21*, 349-369.

Hillebrand, H. and Kahlert, M. (2001) Effect of grazing and nutrient supply on periphyton biomass and nutrient stoichiometry in habitats of different productivity. *Limnology and Oceanography*, 46(8), 1881-1898

Hillebrand, H. and Sommer, U. (1997). Response of epilithic microphytobenthos of the Western Baltic Sea to in situ experiments with nutrient enrichment. *Marine Ecology Progress Series*, *160*, 35-46.

Hillebrand, H., Worm, B. and Lotze, H. K. (2000). Marine microbenthic community structure regulated by nitrogen loading and grazing pressure. *Marine Ecology Progress Series, 204*, 27-38.

Hillebrand, H., and Kahlert, M. (2001). Effect of grazing and nutrient supply on periphyton biomass and nutrient stoichiometry in habitats of different productivity. *Limnology and Oceanography*, *46,* 1881-1898.

Hillebrand, H., Bennett, D. M., and Cadotte, M. W. (2008). Consequences of dominance: a review of evenness effects on local and regional ecosystem processes. *Ecology, 89,* 1510-1520.

Hunter, R. D. (1980). Effects of grazing on the quantity and quality of freshwater aufwuchs. *Hydrobiologia, 69*, 251-259.

Hunter, R. D. and Russell-Hunter, W. D. (1983). Bioenergetic and community changes in intertidal aufwuchs grazed by Littorina Zittorea. *Ecology, 64,* 761-769.

Hurd, C. L. (2000). Water motion, marine macroalgal physiology, and production. *Journal of Phycology*, *36*, 453-472.

Kahlert, M. (1998). C:N: P ratios of freshwater benthic algae. Archiv fuer Hydrobiologie, *Special Issue Advances in Limnology, 51*, 105-114.

Kahlert, M. and Baunsgaard, M. T. (1999). Nutrient recycling-a strategy of a grazer community to overcome nutrient limitation. *Journal of the North American Benthological Society, 18,* 363-369.

Kim, B. K. A., Jackman, A. P., and Triska, F. J. (1992). Modeling biotic uptake by periphyton and transient hyporrheic storage of nitrate in a natural stream. *Water Resources Research,* 28, 2743–2752.

Korpinen, S., Jormalainen, V. and Honkanen, T. (2007). Effects of nutrients, herbivory, and depth on the macroalgal community in the rocky sublittoral. *Ecology, 88,* 839-852.

Korpinen, S. and Joralainen, V. (2008). Grazing effects in macroalgal communities depend on timing of patch colonization. *Journal of experimental Marine biology and Ecology*, *360*, 39-46.

Lamberti, G. A., Ashkenas, L. R., Gregory, S. V. and Steinman, A. D. (1987). Effect of three herbivores on periphyton communities in laboratory streams. *Journal of experimental Marine biology and Ecology*, *6,* 92-104.

Lamberti, G.A. and Resh, V.H. (1983). Stream periphyton and insect herbivores: an experimental study of grazing by a caddisfly population. *Ecology, 64*, 1124-1135.

Liess, A. and Hillebrand, H. (2004). Direct and indirect effects in herbivore periphyton interactions. *Archiv für Hydrobiologie, 159*, 433-453.

Liess, A. and Hillebrand, H. (2006). Role of nutrient supply in grazer-periphyton interactions: reciprocal influences of periphyton and grazer nutrient stoichiometry. *Journal of the North American Benthological Society*, *25*, 632-642.

Leiss, A. and Kahlert, M. (2007). Gastropod grazers and nutrients, but not light, interact in determining periphytic algal diversity. *Oecologia*, 152, 101-111.

Lock, M. A. (1981). River epilithon – alight andorganic energy transducer. In M. A. Lock, and D. D. Williams (Eds), *Perspectives in Running Water Ecology* (pp. 3-40). New York: Plenum.

Lotze, H. K., Schramm, W., Schories, D. and Worm, B. (1999). Control of macroalgal blooms at early developmental stages: *Pilayella littoralis* versus *Enteromorpha* spp. *Oecologia, 119*, 46-54.

Lotze, H. K. and Worm, B. (2000). Variable and complementary effects of herbivores on different life stages of bloom-forming macroalgae. *Marine Ecology Progress Series, 200,* 167-175.

Lotze, H. K., Worm, B. and Sommer, U. (2001). Strong bottom-up and top-down control of early life stages of macroalgae. *Limnology and Oceanography*, *46*, 749-757.

Lowe, R. L., and Hunter, R. D. (1988). Effect of grazing by *Physa integra* on periphyton community structure. *Journal of the North American Benthological Society, 7*, 29-36.

Lubchenco, J. (1978). Plant species diversity in a marine intertidal community: importance of herbivore food preference and algal competitive abilities. *American Naturalist, 112*, 23–39.

Lubchenco, J., Menge, B. A. (1978). Community development and persistence in a low rocky intertidal zone. *Ecological Monographs, 48*, 67-94.

Lundstedt L. and Brett M. T. (1991). Differential growth rates of 3 cladoceran species in response to mono-algal and mixed-algal cultures. *Limnology and Oceanography, 36,* 159-165.

Malm, T., Engkvist, R., Kautsky, L., (1999). Grazing effects of two freshwater snails on juvenile Fucus vesiculosus in the Baltic Sea. *Marine Ecology Progress Series, 188*, 63-71.

Marks, J. C., and Lowe, R. L. (1989). The independent and interactive effects of snail grazing and nutrient enrichment onstructuring periphyton communities. *Hydrobiologia*, 85, 9-17.

McAuliffe J. R. (1984). Resource depression by a stream herbivore: effects on distributions and abundances of other grazers. *Oikos, 42,* 327-333.

McCormick P. V. (1990). Direct and indirect effects of consumers on benthic algae in an ephemeral stream food web. *Canadian Journal of Fisheries and Aquatic Sciences , 47,* 2057-2065.

McCormick, P. V. and Stevenson, R. J. (1991). Grazer control of nutrient availability in the periphyton. *Oecologia, 86,* 287-291.

McCormick P. V., Stevenson R. J. (1989). Effects of snail grazing on benthic algal community structure in different nutrient environments. *Journal of the North American Benthological Society, 8*, 162-172.

McMahon, R. F., R. D. Hunter and Russell-Hunter, W. D. (1974). Variation in aufwuchs at six freshwater habitats in terms of carbon biomass and of carbon: nitrogen ratio. *Hydrobioogia,. 45,* 391-404.

McCollum, E. W., Crowder, L. B. and McCollum, S. A. (1998). Complex interactions of fish, snails, and littoral zone periphyton. *Ecology,* 79, 1980-1994.

Menge, B. A., Sutherland, J. P. (1987). Community regulation: Variance in disturbance, competition, and predation in relation to environmental stress and recruitment. *American Naturalist, 130,* 730-757.

Momo, F. R. (2000). A new model for periphyton growth in running waters, *Hydrobiologia, 299,* 215-218.

Mulholland, P. J., Steinman, A. D., Palumbo, A. V., Elwood, J. W., and Kirschtel, D. B. (1991). Role of nutrient cycling and herbivory in regulating periphyton communities in laboratory streams. *Ecology, 72,* 966-982.

Nicotri, M. E. (1977). Grazing effects of four marine intertidal herbivores on the microflora. *Ecology, 58,* 1020-1032.

Nielsen, K. J. (2001). Bottom-up and top-down forces in tide pools: Test of a food chain model in an intertidal community. *Ecological Monographs, 71*, 187-217.

Nisbet, R. M., Diehl S., Wilson, W. G., Cooper, S. D., Donaldson, D. D., and Kratz, K. (1997). Primary-productivity gradients and short-term population dynamics in open systems. *Ecological Monographs, 67,* 535-553.

Neckles, H. A., Wetzel, R. L., and Orth R. J. (1993). Relative effects of nutrient enrichment and grazing on epiphyte-macrophyte (Zostera marina L.) dynamics. *Oecologia, 93,* 285-295.

Norberg, J. (1999). Periphyton fouling as a marginal energy source in tropical tilapia cage farming. *Aquaculture Research,* 30, 427-430.

Orav-Kotta, H., Kotta, J. (2003). Seasonal variations in the grazing of Gammarus oceanicus, Idotea baltica, and Palaemon adspersus on the benthic macroalgae. *Proceedings of the Estonian Academy of Sciences Biology Ecology, 52,* 141-148.

Paine, R.T. (2002). Trophic control of production in a rocky intertidal community. *Science, 296,* 736-739.

Pan, Y., and Lowe, R. L. (1994). Independent and interactive effects of nutrients and grazers on benthic alga1 community structure. *Hydrobiologia, 291*, 201-209.

Peterson, C. G. (1987). Gut passage and insect grazer selectivity of lotic diatoms. *Freshwater Biology, 18,* 455-460.

Power, M. E., Dudley, T. L., and Cooper, S. D. (1989) Grazing catfish, fishing birds, and attached algae in a Panamanian stream. *Environmental Biology of Fishes*, 26, 285-294.

Proulx, M., and Mazumder, A. (1998) Reversal of grazing impact on plant species richness in nutrient-poor vs. nutrient-rich ecosystems. *Ecology, 79,*2581-2592.

Riber, H. H., and Wetzel, R. (1987). Boundary-layer and internal diffusion effects on phosphorus fluxes in lake periphyton. *Limnology and Oceanography, 32*, 1181-1194.

Rosemond, A. D. (1993). Interactions among irradiance, nutrients, and herbivores constrain a stream algal community. *Oecologia, 94,* 585-594.

Rosemond, A. D., Mulholland, P. J., and Elwood, J. W. (1993). Top-down and bottom-up control of stream periphyton: Effects of nuttients and herbivores. *Ecology, 74,* 1264-1280.

Rosemond A. D., Mulholland P. J., Brawley, S, H, (2000). Seasonally shifting limitation of stream periphyton: response of algal populations and assemblage biomass and productivity to variation in light, nutrients, and herbivores. *Canadian Journal of Fisheries and Aquatic Sciences, 57*, 66-75.

Saravia, L. A., Momo F., Boffi Lissin, L.D. (1998). Modelling periphyton dynamics in running water. *Ecological Modelling, 114,* 35-47.

Shurin J. B., Borer E. T., Seabloom E. W., Anderson K., Blanchette C. A., Broitman B., Cooper S. D., Halpern B. S. (2002). A crossecosystem comparison of the strength of trophic cascades. *Ecology Letters, 5,* 785-791.

Sommer, U. (1996). Nutrient competition experiments with periphyton from the Baltic Sea. *Marine Ecology Progress Series, 140,* 161-167.

Sommer, U. (1997). Selectivity of Idothea chelipes (Crustacea, Isopoda) grazing on benthic microalgae. *Limnology and Oceanography, 42,* 1622-1628.

Sommer, U. (1999). The susceptibility of benthic microalgae to periwinkle (Littorina littorea, Gastropoda) grazing in laboratory experiments. *Aquatic Botany, 63,* 11-21.

Son, D. H. and Fujino, T. (2003). Modeling Approach to Periphyton and Nutrient Interaction in a Stream. *Journal of Environmental Engineering, 129,* 834-843.

Steinman, A. D, Gregory, S. V., Lamberti, G. A., and Ashkenas, L. R. (1987). Effects of herbivore type and density on the taxonomic structure and physiognomy of algal assemblages in laboratory streams. *Journal of the North American Benthological Society, 6,* 175-188.

Steinman, A. D., McIntire, C. D., Gregory, S. V., Lamberti, G. A. (1989). Effects of irradiance and grazing on lotic algal assemblages. *Journal of Phycology, 25,* 478-485.

Steinman, A. D., P. J. Mulholland and W. R. Hill, 1992. Functional responses associated with growth form in stream algae. *Journal of the North American Benthological Society, 11,* 229-243.

Steinman, A. D. (1996). Effects of grazers on benthic freshwater algae. In R. J. Stevenson, M. L. Bothwell, and R. L. Lowe (Eds.), *Algal ecology* (pp.341-373). London, UK.: Academic Press.

Stelzer, R. S., and Lamberti, G. A. (2002). Ecological stoichiometry in running waters: periphyton chemical composition and snail growth. *Ecology, 83,* 1039-1051.

Sterner, W. R., and Elser J. J. (2002). *Ecological stoichiometry*. Princeton, New Jersey and Oxford, UK: Princeton University Press.

Stevenson, R. J., and Glover, R. M. (1990). Interactive effects of algal density and current on nutrient diffusion into periphyton mats. *Bulletin of the North American Benthological Societ, 7,* 74.

Sumner, W. T., McIntire, C. D. (1982). Grazer-periphyton interactions in laboratory streams. *Archiv fuer Hydrobiologie, 93,* 135-157.

Toda, Y., and Tsujimoto, T. (2010). Numerical modeling of interspecific competition between filamentous and nonfilamentous periphyton on a flat channel bed. *Landscape and Ecological Engineering, 6,* 81-88.

Uehlinger, U., Buhrer, H., Reichert, P. (1996). Periphyton dynamics in a floodprone prealpine river: evaluation of significant processes by modelling. *Freshwater Biology, 36,* 249-263.

Urabe, J. (1993). N and P cycling coupled by grazers' activities: food quality and nutrient release by zooplankton. *Ecology, 74,* 2337-2350.

Urabe, J., Kyle, M., Makino, W., Yoshida, T., Andersen, T., and Elser J. J. (2002). Reduced light increases herbivore production due to stoichiometric effects of light/nutrient balance. *Ecology, 83,* 619-627.

Wilbur, H. M. (1997). Experimental ecology of food webs: complex systems in temporary ponds. *Ecology, 78,* 2279-2302.

Worm, B., Lotze, H. K., Hillebrand, H., and Sommer, U. (2002) Consumer versus resource control of species diversity and ecosystem functioning. *Nature, 417,* 848-851.

In: Zooplankton and Phytoplankton
Editor: Giri Kattel, pp. 107-122

ISBN 978-1-61324-508-8

Chapter 5

USING EPILITHIC FILAMENTOUS GREEN ALGAE COMMUNITIES AS INDICATORS OF WATER QUALITY IN THE HEADWATERS OF THREE SOUTH AFRICAN RIVER SYSTEMS DURING HIGH AND MEDIUM FLOW PERIODS

Paul J. Oberholster *
CSIR Natural Resources and the Environment, Stellenbosch,
Department of Paraclinical Sciences, Faculty of Veterinary Science,
University of Pretoria,
Onderstepoort, South Africa

ABSTRACT

Physical and chemical characteristics of the head waters of the Olifant's, Lephalala and Mokolo rivers and their tributaries were assessed over a period of one year to compare effects of seasonal, spatial and human environmental factors on epilithic filamentous green algae communities as indicators of different sources of pollution. The epilithic filamentous green algae assemblage differed among the three rivers and their tributaries as a function of drainage basin characteristics, but exhibited common seasonal changes related to temperature and electrical conductivity (EC). Human impacts were assumed to cause higher epilithic algal densities in stream basins with increase in TP due to progressive eutrophication. In the study frequent floods in the raining season truncate filamentous green algal successional sequence and thereby favouring either tightly attached, resistant forms or those recolonization and growth rates high enough to accumulate biomass between flood periods. It was evident from the study that certain filamentous green algal species was indicators of land use activities and can therefore be used as indicators of pollution during medium and high flow regimes.

---

* Address: CSIR Natural Resources and the Environment, P.O. Box 320, Stellenbosch 7599, Department of Paraclinical Sciences, Faculty of Veterinary Science, University of Pretoria, P/Bag X04, Onderstepoort 0110, South Africa.

INTRODUCTION

Water quality parameters alone often do not indicate all human effects to rivers, whereas the resident biota is thought to respond to an integration of all human impacts (Karr 1991). According to an earlier study conducted by Yoder (1991) the assessment of Ohio streams in the USA using biota correctly identify the presence of human influences 49.8% of the time when it was not identified by water quality variables. A range of biota which includes diatoms, macroinvertebrate and fish have been employed to monitor fresh water quality (Oberholster et al. 2005, 2007) Periphyton algae in general and diatoms in particular possess many attributes that make them ideal organisms to employ in water quality studies as indicators of river health. Phytoplankton's are ubiquitous, ecologically important, and sensitive to a broad range of environmental stressors and respond rapidly to changes in water chemistry due to anthropogenic pollution (McCormick and Stevenson 1998). Periphyton algae in general play several roles of fundamental importance to river ecosystems. As organisms at the base of the food web, they are often the dominant producers and principal energy source to higher trophic levels in rivers and streams (Salmaso and Braioni 2008).

Different methods based on epilithic diatom community composition have proven especially effective in monitoring nutrient enrichment, pH and organic pollution (Silva-Benavides 1995; Oberholster et al. 2005, 2010a). According to Dodds et al. (1998) the short fall of previous trophic indices developed by e.g. Kelly and Whitton (1995), is that they are specific to a limited region, requires sophisticated taxonomic skills and excludes filamentous green algae which are most commonly associated with eutrophic conditions. However, although non diatom algal community based approaches are less commonly used in stream or river monitoring, they have proven to be effective in classifying New Zealand and Austrian rivers (Biggs 1985, 1990).

On the other hand the use of plantonic phytoplankton as indicators of pollution impacts in rivers are constrained by the advective losses that characterize flowing waters (rate of turbulent mixing) causing the drift of populations and the decrease of the population growth rates (Reynolds and Descy 1996). Alternatively epilithic filamentous green algae communities in rivers and streams may be more sensitive to a given impact than plantonic phytoplankton, since they are stationary and their community characteristics develop entirely within the environmental conditions of a specific site (Stevenson and White 1995). The objective of this study was to compare the variation in epilithic filamentous green algae communities that was related to seasonal and spatial flow regimes and human-related environmental factors in the head waters of the Olifant's, Mokolo and Lephalala rivers and their tributaries. Finally, the question of whether or not characterization of epilithic filamentous green algae was practical and an informative tool in the assessment of environmental impacts (i.e., mining, industry, agriculture sector) as well as indicators of organic pollution from wastewater treating plants in the headwaters of the studied river ecosystems during medium and high flows was addressed.

Methods and Materials

Three river systems were selected to verify if filamentous benthic green algae can be used as indicators of progressive eutrophication and mining activities. Permanent sampling stations were chosen with the following criteria: (1) potential nutrient limitation sites as control (determined by TN/TP ratios); (2) sampling sites with a wide range of TP (± 0.025-0.075 mg l-1); sulphate concentrations and electric conductivity (EC); (3) substrate velocities of ≤ 2 m2.s-1 and ≥ 4 m2.s-1; and (4) sampling sites with stony substrate in unshaded riffle areas.

Study Area

In the analyses, epilithic filamentous green algae communities and physical/chemical data from the headwaters of the Olifant's (Mpumalanga Province), Mokolo and Lephalala (Limpopo Province) rivers was used. The rivers surveyed were chosen with reference to their pollution status in order to cover a wide range of conditions. In the headwaters of the Olifant's River, acid mine drainage, untreated sewage and industrial pollution and extensive agriculture are the main sources of anthropogenic stressors on the aquatic environment (Oberholster et al. 2010a). The situation is exacerbated by a historic lack of policy implementation to address acid mine drainage (AMD), especially at mine closure. All rivers and tributaries and drainage basin characteristics are listed in Table 1. Land use in the upper Mokolo River catchment is dry land agriculture activities with urban settlements while the land use in the Lephalala River is agriculture and predominant extensively game farming (Oberholster et al. 2010b).

Table 1. Land use and sampling locations in the head waters of the upper catchment of the Olifant's, Mokolo and Lephalala rivers

Site Number	Description	Geographical Co-ordinates	Stressors Causing Main Impacts
1	Wilge River tributary of the main stem Olifants River	S25° 37.292' E28° 59.941'	Commercial agriculture and Sewage effluent from the town Bronkhorstspruit
2	Klip River upstream of confluence with Olifants River	S25 ° 37.290' E29 ° 12.752'	Mining (acid mine drainage from abundant coal mines), Industrial and Sewage effluent from the city of Witbank
3	Olifants River main stem upstream of confluence with Klip River	S25 ° 37.435' E29 ° 13.011'	Mining (Coal), Industry, Sewage effluent and Agriculture
4	Small Olifants River, upstream of confluence with main stem of the Olifants	S25 ° 41.723' E29 ° 21.130'	Mining (Coal), Industry, Sewage effluent from the town of Middelburg and Agriculture

Table 1. (Continued)

Site Number	Description	Geographical Co-ordinates	Stressors Causing Main Impacts
5	Olifants River main stem, upstream of confluence with the small Olifants	S25 ° 42.088' E29 ° 17.930'	Mining (Coal), Industry, Sewage effluent and Agriculture
6	Olifants – downstream of Riverview sewage works	S25 ° 37.236' E29 ° 12.850'	Mining, Industry, Sewage effluent and Agriculture
7	Tributary of Small Olifants – downstream of industry and mining	S25 ° 46.256' E29 ° 28.957'	Mining (Coal), Industry and Sewage effluent
8	Koffiespruit – upper Wilge River catchment	S25 ° 59.683' E28 ° 39.767'	Agriculture (irrigation, dryland and commercial farming) and Sewage effluent
9	Upper main stem Olifants River	S26 ° 13.320' E29 ° 27.753'	Agriculture(dryland commercial farming)
10	Steenkoolspruit tributary of the main stem of the Olifants River	S26 ° 8.243' E29 ° 16.189'	Mining (Coal) and Agriculture
11	Rhenosterspruit, near-pristine headwater tributary of the Mokolo River, south of the town of Vaalwater	S24 ° 30.4' E27 ° 52.00'	Agriculture (game farming)
12	Mokolo River, north of the town of Vaalwater	S23 ° 58.09' E27 ° 43.35'	Agriculture (game and irrigation farming)
13	Palala River, near-pristine headwater tributary of the Lephalala River	S24 °1 8.27' E28 ° 31.19'	Agriculture (game, cattle farming)
14	Lephalala River, in the Lapalala Wilderness Reserve	S23 ° 52.46' E28 ° 20.09'	Agriculture (game farming)

Epilithic Filamentous Green Algae Sampling

Epilithic green filamentous algae samples were collected from natural substrate over a period of 1 year during high and low flow regimes. A known area (5 cm in diameter) of a rock's surface was isolated with a cylindrical open ended Plexiglas tube with basal neoprene seal, scraped with brushes after squirted 100 ml stream water into the jar. The algal slurry was pulled from the chamber with a syringe extended with a tygon tube (Douglas 1958; Hauer and

Lamberti 2006). Five discrete epilithic algae samples were collected at each site on six sampling occasions and combined in a composite sample. Epilithic samples was preserved in the field by addition of acidic Lugol's solution to a final concentration of 0.7%, followed after one hour by the addition of buffered formaldehyde to a final concentration of 2.5%. These samples were transported to the laboratory in cooler boxes on ice and processed for analyses of chlorophyll *a* (chl-*a*) according to Porra et al. (1989).

All algal identifications were made with a compound microscope at 1250 x magnification (Van Vuuren et al. 2006). Aliquots (10-100 ml) were sedimented depending on the abundance of the organisms in the samples. Strip counts were made until at least 100 individuals of each of the dominant phytoplankton species had been counted (American Public Health Association 1992). Epilithic algae abundance in the samples was evaluated by counting the presence of each species (as cells in a filament or equal number of individual cells). Filamentous algae biovolume was calculated by measuring the corresponding dimensions using the geometric formulae given by Willen (1976). Filamentous benthic algae species composition throughout the whole text refers to biomass, measured as cell volume.

The Berger-Parker dominance index (Berger and Parker 1970) was used to measure the evenness or dominance of epilithic filamentous green algae at each sampling site:

$$D = N_{max}/N \qquad (1)$$

where N_{max} = the number of individuals of the most abundant species present in each sample, and N = the total number of individuals collected at each site.

Epilithic filamentous green algae diversity was calculated using the Shannon's diversity index (Shannon and Weaver 1949):

$$H = -\left(\sum (\rho i \ln \rho i)\right) \qquad (2)$$

where ρi is the proportion of the total number of specimens, I, expressed as a proportion of the total number of species for all species present in the ecosystem. The product of ρi ln ρi for each species in the aquatic ecosystem is summed, and multiplied by –1 to give *H*.

Equilibrial epilithic filamentous green algae species (*sensu* Naselli-Flores 2003) at all sampling sites were determined over the study period of one year. The following criteria of Naselli-Flores (2003) were used: (1) 1, 2 or 3 species of filamentous algae contribute more than 80% of the total biomass, (2) their existence or coexistence persists for more than 1-2 weeks, and (3) during this period, filamentous algae biomass does not increase significantly. The current directly above the sampled rocks' upper surface was measured using a current meter in order to determine the current velocity actually experienced by the filamentous periphyton.

PHYSICAL AND CHEMICAL CHARACTERISTICS

Dissolved oxygen, temperature, pH and conductivity of the sampling sites were measured in situ with a Hach sensionTM 156 portable multiparameter (Loveland, USA). Water samples for chemical analysis were collected and filtered through GF/filters. The GF/F filtered water was preserved in prerinsed polyethylene bottles with sulfuric acid (to pH 2.0) for analysis of dissolved nutrients. The polyethylene bottles were placed in a coolbox while in transit to the laboratory. All analyses were done according to standard methods (American Public Health Association (APHA), 1992). Concentrations of TP were determined as reactive phosphorus after digestion of the sample with alkalis (Grasshoff et al. 1983). The nitrate concentration was determined with an ion chromatograph Metrohm Compact IC. Ammonia was analyzed on an autoanalyzer by the cadmium reduction method, whereas soluble reactive phosphorus was determined by the ascorbic acid method. All sampling sites were classified according to the total cationic charge (ΣZ+) (Stallard and Edmond 1983). The ΣZ+ was determined as the milliequivalent sum of potassium (K+), sodium (Na+), magnesium (Mg2+) and calcium (Ca2+). The ΣZ+ of the sampling sites in the headwaters of the Olifant's (Mpumalanga Province), Mokolo and Lephalala (Limpopo Province) rivers are summarised and grouped in Figure 1 with corresponding Stiff diagrams representative of each group. The summer near-surface average nutrient concentrations by Dodds et al. (1997) were used for classification of the trophic state of studied rivers.

Data Treatment

Subsequent multiple comparisons were done with the Tukey posthoc test at $P < 0.05$ to compare benthic chlorophyll a concentrations as surrogate for biomass with the physical and chemical variables of the surface water (i.e., water flow, water temperature, conductivity, nutrient and sulphate concentrations), as well as with filamentous algae cells between sampling sites. All statistical calculations were done with the statistical package SPSS (SPSS Inc., Chicago, IL., USA).

RESULTS

Physical and Chemical Characteristics

The surface water temperature of all sampling stations range from 11 (± 3) °C in winter to a maximum of 27 (± 4) °C in the summer months at sampling sites 11 and 14 in the Mokolo and Lephalala rivers and their tributaries. The electrical conductivity (EC) values of these rivers indicated an average poor electrolyte range of 11 (± 5) μScm-1 during winter to 45 (± 3) μScm-1 in the summer months. The average lowest EC values measured during high flows (≥ 4 m2.s-1) in the headwaters of the Olifant's River and its tributaries was at sites 1 and 8 with an EC of 26 (± 4) μScm-1 and 32 (± 7) μScm-1. During the winter months (May, June and July) with low average water flows (≤ 2 m2.s-1) EC values increased to 753 (± 18) μScm-1 at site 1 and 324 (± 9) μScm-1 at site 8. DO values were high during both high (≥ 4 m2.s-1) and low flows (≤ 2 m2.s-1) ranging from 7.17 (± 1.2) to 8.13 (± 0.9) mg l1 in the Mokolo and Lephalala rivers which can be attributed to the morphology of the channel, slope and low degree of disturbance in these two rivers. In the head waters of the Olifant's River the average DO values range from 5.13 (± 0.8) to 7.41 (± 1.4). The only sampling site with lower DO values (range from 3.57 to 4.23 mg l1) in comparison with all the other sampling sites was site 6 downstream of the Riverview sewage works in the upper Olifant's River catchment.

The concentrations of TP were not significant different between sites in the Mokolo and Lephalala rivers and also was within the range reported for other trophic rivers in the oligotrophic-mesotrophic range (Ortiz-Zayas et al. 2005). Relationships between average benthic filamentous algae biomass (chl-a mg m-2) and selected environmental parameters in the Upper Olifant's River sampling sites (excluding sampling site 2 impacted by acid mine drainage), during the mid winter season (July) showed a significant correlation between average filamentous algae biomass (chl-a 68 mg m-2) and average TP (0.060 mgl-1) ($p \geq 0.04$); chl-a 68 mg m-2 and average temperature (10 ± 4 °C) ($p \geq 0.02$), and chl-a 68 mg m-2 and average EC 410 ($\pm$ 28) μScm-1 ($p \geq -0.01$). The lowest TP concentrations (0.013 and 0.022 mgl-1) measured in the upper Olifant's River catchment were during high flows ($\geq$ 4 m2.s-1) in midsummer (January 2010) at sampling sites 2 and 8.

Grouping (A to E), of the sampling sites of the Olifant's, Lephalala and Mokolo rivers according to the total cationic charge ($\Sigma Z+$) with a corresponding Stiff diagrams for each group is showed in Figure 1. Group A represents the sampling sites along the Mokolo and Lephalala rivers. This group was characterized by low cationic charge and good charge balance. These sites can be classified as near pristine, and are only slightly influenced by local game farming and agriculture activities. Groups D and E (sites 2 and 7) were the most severely impacted by outside influences e.g. acid mine drainage and industries. Although site 7 has a higher $\Sigma Z+$, the corresponding Stiff diagram indicates a charge balance in the tested water samples indicating that this sampling site was contaminated from various sources. The low alkalinity, high sulphate and low pH of site 2 was a clear indication that the site has been strongly influenced by acid mine drainage from surrounding abundant coal mines in Witbank area within the headwaters of the Olifant's River.

The piper diagram in Figure 2 gives further insight into the groupings and influences on the various sampling sites. Sites 3, 4, 5 and 6 (*group C* in Figure 1) were again closely grouped according to their water chemistry. It can be assumed that these sites on the confluences of the small Olifant's and Olifant's rivers were impacted by common sources e.g. sewage works and agriculture activities. *Group A* is more distributed in Figure 2 due to the higher flow velocities of the sites on the Mokolo River (Site 11 and 12) leading to quicker fresh recharge of the river water.

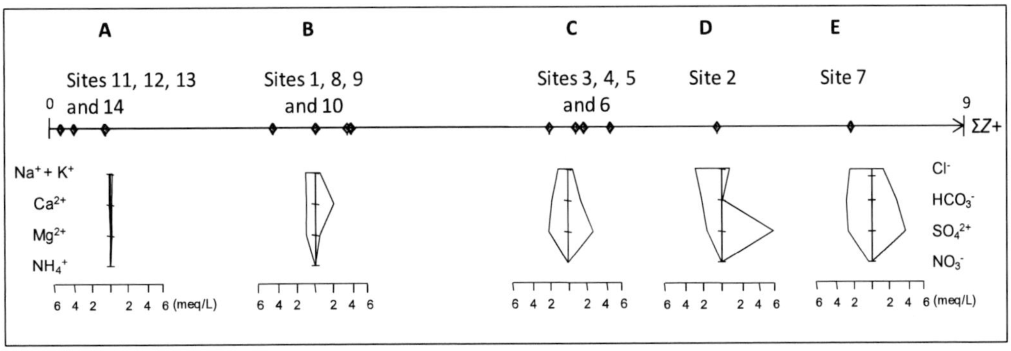

Figure 1. Grouping, A to E, of the Olifant's, Lephalala and Mokolo rivers sampling sites according to the total cationic charge ($\Sigma Z+$) with a corresponding Stiff diagrams for each group, as measured for the high rainfall season, November 2009 to February 2010.

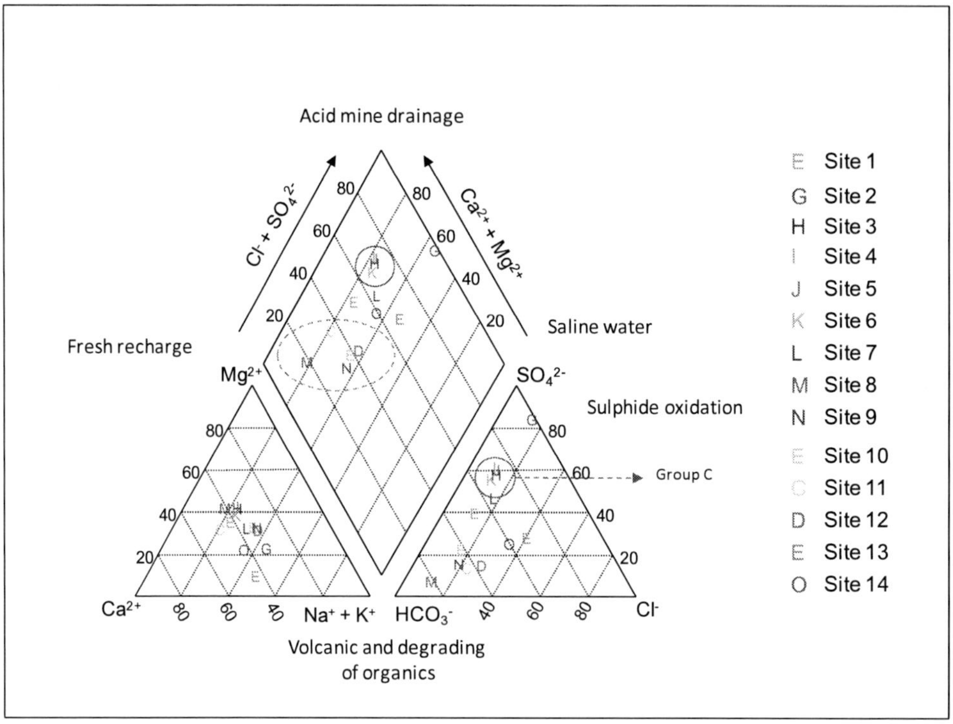

Figure 2. A piper diagram to further classify the water chemistry of the Olifant's, Lephalala and Mokolo rivers sampling sites measured in the high rainfall season (November 2009 to February 2010).

Although sites 9 and 10 (Group B) have higher charge balances as well as sulphate content compared to *Group A,* the high flow rates measured for these sites also assist in the fresh recharge of the sample sites. Site 2 was again singled out to sulphide oxidation in the water supply and a higher likelihood of impact from acid mine drainage. Other conclusions that can be made from the Piper diagram in Figure 2 was that no dominant cations occurred in any of the sites as well as no saline sites were observed due to overall low occurrences of chloride in the summer months during high flow regimes.

Figure 3 shows the sulphate concentration, total phosphate concentration and EC for each of the sampling sites. The electrical conductivity increased with increasing sulphate content, except for sites 9 and 10 which had higher electrical conductivities due to higher alkalinity as CaCO3 compared to the more 'polluted' sites. For sulphate and phosphate concentrations a balance was reached after approximately 100 mg.l-1 of sulphate after which the concentration of phosphate was quenched. For site 2 which is impacted by acid mine drainage, the significant sulphate concentrations and low pH served to further reduce the phosphate concentrations which were indicative of lower filamentous algae biomass.

ALGAE ASSEMBLAGE IN RELATIONSHIP WITH ENVIRONMENTAL DRIVERS

The dominant epilithic green filamentous algae taxa recorded during the rainy season at all 14 sampling sites from November 2009 to July 2010 are listed in Table 2. The main events that induce changes in green epilithic filamentous community composition at all sampling sites were related to the fluctuation in temperature, EC and flow regimes of the different rivers and its tributaries during the winter and summer sampling period. In April to October, discharge decreased ($\leq$ 2 m2.s-1) with more stable conditions during the winter and spring months causing an increase in filamentous algae diversity (average $H' = 3.1$) at all sampling sites excluding site 2. This period was also characterized by lower surface water temperature (average 11˚C ± 3) in the winter months and improved light conditions for green filamentous epilithic communities. The late summer epilithic filamentous algae in the head waters of the Olifant's River had a low average biomass (chl-a 23 mg m-2) and diversity (average $H' = 1.9$) at all sampling sites (excluding site 2) in incomparison with the winter sampling period (filamentous algae average biomass chl-a 58 mg m-2).

Strong relationships were observed between different filamentous epilithic taxa and water column temperature and EC at all sampling sites between the winter and summer sampling period in the Olifant's, Lephalala and Mokolo rivers. The filamentous epilithic community composition in the Mokolo and Lephalala rivers were dominated (Berger and Parker Index, 0.332 and 0.211) by the green filamentous algae Zygnema frigidum during the winter and Draparnaldia glomorus in the summer season (Berger and Parker Index, 0.213 and 0.194). A much lower average chl-a biomass of 28.3 mg m-2 in winter and 6.1 mg m-2 in summer were measured in these two river systems in comparison to the upper Olifant's catchment. These filamentous species was totally absent in the upper Olifant's River and its tributaries during the whole study period and correlated strongly ($p \geq 0.05$) with the low EC values (15-45 µScm-1) in the headwaters of the Mokolo and Lephalala rivers (Table 2.). Another green filamentous epilithic algal that occurred in the Mokolo and Lephalala rivers — but in much lesser abundance — was Spirogyra script (average cells in winter 773 cells cm2 and 375 cells cm2 in summer).

In the upper Olifant's River catchment, the Klip stream a tributary of the Olifant's River (sampling site 2) which was mainly impacted by acid mine drainage from abundant coal mines, was dominated by a single green filamentous algal namely Ulothrix punctata with a low average chl-a biomass of 2.4 mg m-2). This species only occurred during the winter months of April to July and was replaced in summer by the filamentous cyanobacteria Schizothrix. The occurrence of Ulothrix punctata during the mid winter sampling correlated positively with the low average pH of 3.1 (± 1.3); ($p \geq 0.01$) and the high sulphate of 367 (± 29) mg l-1; ($p \geq 0.04$), as well as with high EC average value of 2500 (± 630) µScm-1; ($p \geq 0.02$) and low water column temperature values between 9 and 11 ˚C; ($p \geq 0.05$) (Table 2).

In the upper main stem of the Oliphant's River (sampling sites 5) the filamentous periphyton Cladophora glomorus was the dominant algal species (Berger and Parker Index, 0.581) during the mid winter months from April to July 2010. The decline of Cladophora glomorus cell numbers (from 934 cells cm2 to 364 cells cm2) during the summer months at this site correlated positively ($p \geq 0.03$) with an increase in summer water column temperature from 11 (± 3)˚C in winter to 26 (± 4) ˚ C. The latter site had the highest average chl-a biomass of all sites (152 mg m -2) measured in winter. The average TP and pH measured at this sampling site in winter was 0.081 mg l-1 and 8.34 mg l-1 (Table 2).

The dominance (Berger and Parker Index 0.231) of Rhizoclonium hierglyphicum during the summer months (October to February) at sampling sites 9 in the small Olifant's River was strongly correlated ($p \geq 0.04$) with an increase in water velocity ($\geq$ 4 m2.s-1) at this site. The filamentous algal Stigeoclonium tenue dominated (Berger and Parker Index 0.353) the periphyton throughout the study at sampling 6 downstream of the Riverview sewage works in the head water of the Olifant's River and correlated strongly ($p \geq 0.02$) with the higher TP (0.85 mg l-1) measured at this site in comparison with the other sites in the upper Olifant's catchment. The maximum chl-a biomass of 121.3 mg m -2 was measured during July at this sampling site. Spirogyra reinhardi occurred also at this site but at lesser numerical numbers (314 cells cm2). Sites 1 and 9 in the upper Olifant's River that was mainly impacted by agriculture activities was dominated (Berger and Parker Index 0.214) by Spirogyra reinhardi in the winter months, while Rhizoclonium hierglyphicum dominated (Berger and Parker Index 0.243) during the summer months (Table 2).

Steenkoolspruit, a tributary of the main upper Oliphant's River impacted by active mining and agriculture activities were dominated (Berger and Parker Index 0.271) in the summer months by Rhizoclonium hierglyphicum while Ulothrix punctata dominated the winter months.

Table 2. Qualitative characterization of the most significant environmental variables for each taxonomic site group during the summer rainfall period

Classification		Characterization of each site				
Site number	Dominant periphyton	E.C. µS.cm-1	Flow velocity m2.s-1	Tphos mg.l-1	SO42- mg.l-1	pH
1	Spirogyra reinhardi	low	High	Med	Low	slightly basic
2	Ulothrix punctata	high	High	Low	very high	Acidic
3	Stigeoclonium tenue	med	High	med-high	High	slightly basic
4	Stigeoclonium tenue	high	High	High	High	slightly basic
5	Cladophora glomorus	med-high	High	High	High	slightly basic
6	Stigeoclonium tenue	med-high	very high	High	High	slightly basic
7	Stigeoclonium tenue	high	Medium	med-high	High	slightly basic
8	Spirogyra reinhardi	low	Medium	Low	Low	slightly basic
9	Rhizoclonium hierglyphicum	med	High	low-med	Low	slightly basic
10	Rhizoclonium hierglyphicum	med-high	High	Med	Low	slightly basic
11	Draparnaldia glomorus	low	Medium	Low	Low	Basic
12	Draparnaldia glomorus	low	High	Low	Low	slightly basic
13	Draparnaldia glomorus	low	Medium	Low	Low	slightly basic
14	Draparnaldia glomorus	low	Medium	Low	Low	slightly basic
Criteria for classification (DWAF 1996a,1996b, 1996c)						
Low or acidic		<100	< 2	< 0.025	<50	< 5.5
High or basic		>180	> 4	> 0.075	>100	> 8.5
slightly basic						7.5 - 8.5

The small Olifant's River, a tributary (sites 4 and 7) of the Olifant's River was dominated by Stigeoclonium tenue upstream of its confluence with the main stem of the Olifant's River during the winter and summer sampling period. The average maximum biomass measured at these sites was chl-a 66 mg m-2 and chl-a 71 mg m-2 in winter. Ulothrix punctata at site 2 and Stigeoclonium tenue at site 6 was the only two species accounted for more than 80% of the bio-volume of the benthic community for the study period, indicating that they were 'equilibrial species" (sensu Naselli-Flores et al., 2003). In our study filamentous epilithic algae mats at all selected sampling sites were the most dominant during prolong periods of stable flows (low and medium flow regimes) from mid winter (May to August) to early summer months (October to November). A decrease in filamentous green algae mat biomass was observed during periods of stable flow in early summer at sites 3 and 8. The algae biomass at these sites was lower in comparison to sites 4 and 5. Total Phosphorus measured in the water column of sites 3 and 8 were also lower (0.062 and 0.018 mg l-1).

Autecology of Different Green Filamentous Epilithic Algae

According to Dodds et al. (1998) 0.025 mg l-1 of TP is the limit for oligotrophy in a lotic system; whereas TP values $\geq$ 0.075 mg l-1 are indicative of eutrophication. According to these TP criteria for summer sampling, sites 1, 3, 7, 9 and 10 would be considered as mesotrophic (range of 0.025 to 0.040 mg TP l-1) and sites 2, 8, 11, 12, 13, 14 as oligothrophic. The nutrient values of the upper Olifant's River and its tributaries differ due to land use in each drainage basin. Sites 4, 5 and 6 fall within the eutrophic range ($\geq$ 0.075 mg TP l-1). Sampling sites 9, 6 and 10 was indicative of eutrophication conditions with an average chl-a measurement for both summer and winter. The low values NO3-N could be attributed to immobilization of N by leaf litter decomposition in the Mokolo and Lephalala River channels (McDowell and Asbury 1994). The Benthic chl-a observed at all sampling sites of the Mokolo and Lephalala rivers was low in comparison with other benthic stream biomass measured by Dodds et al. (1998) and Biggs (1996), ranging from 0.38 to 27.6 mg m2. However, the higher chl-a of benthic phytoplankton biomass measured during the winter months in the upper catchment of the Olifants River except site 2 was above the ranges detected by Biggs (1996) and Dodds et al. (1998). In streams with an EC value higher than 1000 μScm-1 or below 45 μScm-1, the benthic chl-a biomass correlated negatively with the stream water EC ($p \geq -0.05$). Similar responses were also observed in other studies that found EC to be a major influence on benthic algae (Newall and Walsh, 2005). In general our observations in the Lephalala and Olifant's rivers are in relationship with previous findings of Biggs (1996). According to Biggs (1996), streams poor in electrolytes and nutrient concentrations have been found to have low benthic biomass levels. Welch et al. (1988) indicated that more than 20 percent coverage of filamentous species could be expected if biomass exceeded 100 mg chl-a m-2.

Both sites 5 and 6 had lower concentrations of iron in comparison with other sites in the head waters of the upper Olifant's catchment. However, Cladophora glomorus dominated site 5 while Stigeoclonium tenue dominated site 6. This phenomenon can possibly be related to nutrient and iron concentrations in the water column of these two sites. In a previous study by

McLean and Benson-Evans (1974) the author indicated that Cladophora glomorus was absent from areas with high levels of iron which is within relationship with our observation and the possible reason for the dominance of Cladophora glomorus at site 5. However, the dominance of Stigeoclonium tenue at site 6 with the same iron concentration range as site 5 can be related to the differences in nutrient concentration at this sampling site in comparison with site 5. McLean and Benson-Evans (1974) indicated that Cladophora glomorus tolerated only slightly narrower ranges of nutrient concentrations than Stigeoclonium tenue. The dominance of Stigeoclonium tenue at sites 3,4,6 and 7 with high sulphate concentrations indicated that this species is resistant to metal pollution from coal acid mine drainage and is an indicator of organic pollution as previously describe by Whitton (1970) and Blum (1957).

A study conducted by Pawlik-Skowronska (2001) showed that alkaline pH values as observed at sites 3,4,6 and 7 in our study is highly significant for Stigeoclonium due to its influence on the chemical metal speciation and availability in mining water, but also due to its effects on the inorganic carbon speciation essential for algal metabolism, including the phytochelatin synthesis. Francke and Cate (1980) indicated from their study that phosphate saturation values for Stigeoclonium strains is around 10 mgl-1 which imposed that Stigeoclonium is tolerant to high nutrient levels. A further possible reason for the dominance of Stigeoclonium tenue at sites 3, 4, 6 and 7 was the detection of higher concentrations of chloride during the winter sampling at these sites in comparison with all other sampling sites except for site 2 with a high chloride value but a low pH value of ± 3. In a study conducted by Francke and Cate (1980) they indicated that different strains of Stigeoclonium were tolerant to high levels of chloride. The dominance of Ulothrix punctata at sites 2 can be related to the low surface water pH value measured at this site, since Niyogi et al. (1999) reported that Ulothrix spp. could tolerate pH values below 4 in a Rocky Mountain headwater stream that received acidic mine drainage. A possible explanation for the decline of algae mats at sampling sites 3 and 8 with lower levels of TP in the overlying surface water during periods of stable flows in the early summer months can be related to a shift from cellular control growth rate limitations of the algae mats to community growth rate as mats developed. This phenomenon was also observed in a previous study by Bothwell (1989) in the Thompsen River system and by Borchart (1996). It was unlikely that in case of our study growth of cells at the surface of the filamentous algae mats was P-limited at sites 3 and 8 with lower measured values of TP in the overlying water. But as the thickness of the mats develops during late winter and the onset of summer with higher water temperatures and longer day lengths, diffusion of P molecules to the base of these mats are restricted possibly due to uptake by cells nearer to the upper surface or through restricted water movement in the mats. This results in P-starvation of the underlying layers and eventual decaying of the mats. At sampling sites 4 and 5 with higher levels of P concentrations in the overlying water, rates of P diffusion to the underlying layers increased resulting in the maintenance of high biomass for much longer periods at these sampling sites. According to Hilton et al. (2006) the filamentous algae Cladophra which dominated site 5 obtains its nutrients direct from the water column and could not grow well under low nutrient conditions, but grows quickly under high P concentrations which possibly explain its domination at site 5 with higher TP values.

Although several previous studies suggested an active role played by zooplankton in controlling phytoplankton dynamics in rivers (Ferrari et al. 2006), it was not the case in our study. We did observe an increase of zooplankton during lower flow regimes with higher algae mat growth. However, the densities of zooplankton were very low (≤ 20 ind. l-1) in the headwaters of all three systems, suggesting a negligible role played by zooplankton grazing. Possible explanations for this phenomenon was the oligotrophic states of the, Lephalala and Mokolo rivers, while the adverse effects of high pollution may have played a role in the Olifant's River system.

Conclusion

From the data generated in the study it is evident that green filamentous epilithic algal taxa can be used as indicators of different physical and chemical variables in the aquatic environment. Physical disturbances such as flooding can cause the detection of much lower numbers or different community assemblages of macroinvertebrates, water column phytoplankton and fish during catchment monitoring programmes. These results can give a false interpretation of the impacts of physical and chemical parameters at a specific sampling site due to drift of these organisms. However, green filamentous epilithic algae are stationary at stony substrate in unshaded riffle areas and their community characteristics develop entirely within the environmental conditions of a specific site. Furthermore, macroinvertebrates and water column phytoplankton communities can also take much longer time periods after a flood disturbances to recover, which can be time constraining and not cost effective for monitoring programmes.

References

American Public Health Association (APHA), American Water Works Association (AWWA), and Water Pollution Control Federation (WPCF), 1992. Standard Methods for the Examination of Water and Wastewater. (19th edition). APHA, AWWA, and WPCF, Washington, D.C., USA.

Berger, W.H., Parker, F.L. (1970). Diversity of planktonic *Foraminifera* in deep sea sediments. *Sci. 168*: 1345-1347.

Biggs, B.J.E, (1985). The use of periphyton in the monitoring of water quality. Biological Monitoring in Freshwaters: proceedings of a seminar In: Pridmore, R.D., Cooper, A.B., (eds), Biological monitoring in freshwater. Ministry of works and Development, Wellington, New Zealand, pp. 117-142.

Biggs, B.J.F. (1996). Patterns in benthic algae of streams In: Stevenson R.J., Bothwell M.L., Lowe, R.L. (eds), Algal ecology: freshwater benthic ecosystems. Academic Press, San Diego, pp. 31-56.

Biggs, B.J.F. (1990). Periphyton communities and their environmentsin New Zealand rivers. *N ZJ Mar Freshwater Res. 24*: 367-386.

Blum, J.L. (1957). An ecological study of the algae of the Saline Rivers, Michigan. *Hydrobiol. 9*: 361-408.

Borchard, M.A. (1996). Nutrients. In: Stevenson R.J., Bothwell M.L., Lowe, R.L. (eds), Algal ecology: freshwater benthic ecosystems. Academic Press, San Diego, pp. 184-227.

Bothwell, M.L. (1985). Phosphorus limitation of lotic periphyton growth rates: an intersite comparison using continuous-flow throughs (Thompsen River system: British Columbia). *Limnol. and Oceanogr. 30*: 526-542.

DWAF (Department of Water Affairs and Forestry) (1996a). *South African Water Quality Guidelines, Volume 1: Domestic Use*. Pretoria, Department of Water Affairs and Forestry.

DWAF (Department of Water Affairs and Forestry) (1996b). *South African Water Quality Guidelines, Volume 4: Agricultural Use: Irrigation*. Pretoria, Department of Water Affairs and Forestry.

DWAF (Department of Water Affairs and Forestry) (1996c). *South African Water Quality Guidelines, Volume 7: Aquatic Ecosystems*. Pretoria, Department of Water Affairs and Forestry. Meybeck, M. Global occurrence of major elements in rivers. In: J.I. Drever, Editor, *Treatise on Geochemistry, Surface and Ground Water, Weathering, and Soils*, Elsevier (2003), pp. 207–223.

Dodds, W.K., Smith, V.H., Zander, B. (1997) Developing nutrient targets to control benthic chlorophyll levels in streams: A case study of the Clark Fork River. *Wat. Res. 31*: 1738-1750.

Dodds, W.K., Jones, J.R., Welch, E.B. (1998). Suggested classification of stream trophic state: distribution of temperate streamthopes by chlorophyll, total nitrogen and phosphorus. *Wat. Res. 32*: 1455-1462.

Douglas, B. (1958). The ecology of the attached diatoms and other algae in a small stony stream. *J. of Ecol. 46*: 295-322.

Ferrari, I, Viglioli S, Viaroli P, Rossetti, G. (2006). The impact of the drought event of summer 2003 on the zooplankton of the Po River (Italy). *Verh. Verein. Limnol. 29*: 2143-2149.

Francke, J.A., Cate, H.J. (1980). Ecotypic differentiation in reponse to nutritional factors in the algal genus *Stigeoclonium tenue* Kütz (Chlorophyceae). *Br. Phycol. J. 15*: 343-355.

Grasshoff, K., Ehrhardt, M., Kremling, K.(1983). Methods of seawater analyses. Second, Revised and extended edition. Verslag chemie: Weinheim.

Hauer, F.R., Lamberti, G.A. (2006). Methods in stream ecology.Elsevier, Academic Press, Amsterdam, pp.357-377.

Hilton, J., O'Hare, M., Bowes, M.J., Jones, J.I. (2006). How green is my river? A new paradigm of eutrophication in rivers. *Sci. Tot. Environ. 365*: 66-83.

Karr, J. R. (1991). Biological integrity: a long neglected aspect of water resource management. *Ecological Applications 1*: 66-84.

Kelly M.G., Whitton B.A., (1995). The trophic Diatom Index: a new index for monitoring eutrophication in rivers. *J. Appl. Phycol. 7*: 433-444.

McCormick, P.V., Stevenson, R.J. (1998). Periphyton as a tool for ecological assessment and management in the Florida Everglades. *J. Phycol. 4*: 726-733.

McDowell, W.H., Asbury C.E. (1994). Export of carbon, nitrogen and major ions from three tropical montane watersheds. *Limnol. Oceanogr. 39*: 111-113.

Mclean, R.O., Benson-Evans, K. (1974). The distribution of *Stigeoclonium tenue* Kütz in South Wales in relation to its use as an indicator of organic pollution. *Br. Phycol. J.* 9: 83-89.

Naselli-Flores, L., Padisak, J., Dokulil, M.T., Chorus, I., (2003) Equilibrium/steady-state concept in phytoplankton ecology. *Hydrobiol. 502*: 395-403.

Newall, P., Walsh, C.J. (2005). Response of epilithic diatom assemblage to urbanization influences. *Hydrobiol. 532*: 53-67.

Niyogi, D.K., McKnight, D.M., Lewis, W.M. (1999). Influences of water and substrate quality for periphyton in a Montana stream affected by acid mine drainage. *Limnol. Oceanogr. 44*: 804-809.

Oberholster, P.J, Botha, A-M, Cloete, E.T. (2007). Biological and chemical evaluation of sewage water pollution in the Rietvlei Nature Reserve wetland area, South Africa. *Environ. Pol.* 156: 184-192.

Oberholster, P.J., Botha, A-M, Cloete, T.E. (2005). Using a battery of bioassays, benthic phytoplankton and the AUSRIVAS method to monitor long-term coal tar contaminated sediment in the Cache la Poudre River, Colorado. *Wat. Res. 39*: 4913-4924.

Oberholster, P.J., Myburgh, J.G., Ashton, P.J., Botha, A-M., (2010a). Responses of phytoplankton upon exposure to a mixture of acid mine drainage and high levels of nutrient pollution in Lake Loskop, South Africa. *Eco. Environ. Saf. 73*: 326-335.

Oberholster P.J., Ashton P.J., A-M. Botha, G. Fritz (2010b). First report of the freshwater ciliate *Oprydium versatile* in an African river. *Water SA 36*: 315-321.

Ortiz-Zayas, J.R., Lewis, W.M., Saunders J.F., McCutchan J.H., Scatena, F.N. (2005). Metabolism of a tropical rainforest stream. *J. N. Am. Benthol. Soc24*: 769-783.

Pawlik-Skowronska, B. (2001). Phytochelatin production in freshwater algae in response to heavy metals contained in mining water, effects of some environmental factors. *Aquat. Tox. 52*: 241-249.

Porra, R.J., Thompson, W.A., Kriedemann, P.E., 1989. Determination of accurate extinction coefficient and simultaneous equations for assaying chlorophylls *a* and *b* extracted with four different solvents: verification of the concentration of chlorophyll standards by atomic absorption spectrometry. *Biochim. Biophys. Acta 975*: 384-394.

Reynolds, C.S., Descy J.P. (1996). The production, biomass and structure of phytoplankton in large rivers. *Arch. Hydrobiol. Suppl. 113*: 161-187.

Salmaso, N., Braioni, M.G.(2008). Factors controlling the seasonal development and distribution of the phytoplankton community in the lowland course of a large river in Northern Italy (River Adige). *Aquat. Ecol. 42*: 533-545.

Shannon, C.E., Weaver, W. (1949). The Mathematical Theory of Communications. University of Illinois Press, Urbana, Illinois, USA.

Silva-Benavides, A.M. (1995). The use of water chemistry and benthic diatom communities for qualification of a polluted tropical river in Costa Rica. *Rev. Biol. Trop. 44*: 395-416.

Stallard, R.F., J.M. Edmond. (1983) Geochemistry of the Amazon 2. The influence of geology and weathering environment on the dissolved load. *J. Geophys. Res. 88*: 9671-9688.

Stevenson, R.J., White, K.D. (1995). A comparison of natural and human determinants of phytoplankton communities in the Kentucky River Basin, USA. *Hydrobiologia 297*: 201-216.

Van Vuuren, S., Taylor, J.C., Gerber, A., Van Ginkel, C. (2006). Easy identification of the most common freshwater algae. – North-West University and Department of Water Affairs and Forestry, Pretoria, South Africa, pp. 1-200.

Welch, E.B., Jacoby J.M., Horner, R.R., Seely, M.A. (1988). Nuisance biomass levels of periphytic algae in streams. *Hydrobiol. 157*: 161-168.

Whitton, B.A. (1970). Toxicity of heavy metals to Chlorophyta from flowing waters. *Arch. Microbiol. 72*: 353-360.

Willen, E. (1976). A simplified method of phytoplankton counting. *Br. J. Phycol. 11*: 265-278.

Yoder, C.O. (1991). Answering some concerns about biological criteria based on experiences in Ohio. In: Water quality standards for the twenty-first century. US. Environmental Protection Agency, Washington, D.C., USA, pp. 95-104.

In: Zooplankton and Phytoplankton
Editor: Giri Kattel, pp. 123-140

ISBN 978-1-61324-508-8

Chapter 6

SPATIAL AND TEMPORAL DISTRIBUTION PATTERNS OF ZOOPLANKTON IN A SHALLOW LOWLAND COASTAL LAKE, LAKE WAIHOLA IN NEW ZEALAND

Giri R. Kattel* and Gerard P. Closs

Department of Zoology, University of Otago
Dunedin, New Zealand

ABSTRACT

Communities in shallow coastal lakes are inclined to change with environmental variations quite frequently due to their proximity to marine environments. The pelagic zooplankton community of shallow lowland coastal lake, Lake Waihola in South Island, New Zealand was examined by monthly day and night sampling by hand-operated bilge pump. Over 15 sampling trips, each consisted of diel (D/N) measurement at 50 m apart with five replicate samples at four positions, shallow-inshore (Sh/In), shallow-offshore (Sh/Of), deep-inshore (Dp/In) and deep-offshore(Dp/Of) corresponded a total of 600 samples being collected. Zooplankton individuals collected in D/N samples in four positions were pooled and averaged for each season. Distribution of zooplankton community over temporal and spatial scales were visualized in ordination using non-metric multi-dimensional scaling (MDS) followed by permutational multivariate ANOVA in PRIMER. Four-factor interaction (Date*D/N*Sh/Dp*In*Of) reveals that the effect of sampling dates was significant on distribution of zooplankton, but the distribution across the diel timeframe over spatial scales (D/N*In/Of*Sh*Dp) was insignificant. Some degree of distribution was observed along vertical (Sh/Dp) gradient, but no significant patterns were evident across the horizontal (In/Of) gradient. Amongst zooplankton, cladocerans such as *Daphnia*, *Ceridaphnia* and *Bosmina* showed a poor and patchy distribution pattern where *Bosmina* being dominant in January 1998. Except the Sh/Of day sample of February 1998, when calanoid copepods were absent, the distribution of calanoid copepods, cyclopoid copepods and amphipods was relatively common throughout the period of study. The overall distribution patterns of zooplankton

* Present address: Corresponding author, Centre for Environmental Management School of Science and Engineering, University of Ballarat (Mt Helen), PO Box 663, Ballarat, Vic, 3353, AUSTRALIA, E-mail: giri.kattel@uclmail.net.

in Lake Waihola indicate that there may be significant differences amongst zooplankton community (e.g., cladocerans vs non-cladocerans) to respond to local environmental and seasonal changes, consequently our understanding of shallow lentic coastal ecosystems in South Island, New Zealand is becoming increasingly complex.

Keywords: Lake Waihola, seasonal variation; diel variation; pelagic zooplankton; planktonic community; DVM; DHM; New Zealand; shallow coastal lake.

INTRODUCTION

The spatial and temporal variation in distribution patterns of zooplankton community across the littoral and open water habitats of many lakes around the world is a commonly observed feature (Burks et al. 2002). Ontogenetic shift in habitat use for food and reproductive behaviour have been widely regarded as the causes of the variation in abundance and composition of zooplankton over space and time (Romare et al. 2003). Diel variation in abundance of zooplankton has been observed frequently at various locations of the littoral and pelagic environments (Lauridsen et al. 1999). Diel vertical and diel horizontal migration (DVM and DHM) behaviours in many species of zooplankton are extensively documented (Gliwicz 1986; Smiley and Tessier 1998). Predation by zooplanktivorous fish has been described as one of main causes of DVM and DHM mechanisms in zooplankton (Jeppesen et al. 2000). However, fish predation may not be the only reason that triggers DVM and DHM behaviours in zooplankton. For example, juvenile common bullies in Lake Waihola, New Zealand selectively preferred amphipod zooplankton irrespective of their diel migration patterns (Wilhelm et al. 2007).

Variation in composition and abundance of zooplankton community over seasonal time frame is strongly influenced by changes in physical and chemical variables (Romare et al. 2003). Seasonal variation in temperature, salinity, turbidity, pH and nutrients (TP and TN) of water have been described as causal variables for significant changes in spatial and temporal distribution patterns of phytoplankton and zooplankton community in many shallow lakes (e.g., Mackenzie and Gillespie 1986; Hall and Burns 2002 a and b; Downs et al. 2008). In shallow coastal lakes, wind-induced waves and regular tidal cycles can result in strong spatial and temporal variability in salinity gradients and phytoplankton dynamics (Hamilton and Mitchell 1996; Schallenberg and Burns 2004; Schallenberg et al. 2010). As a result of differences in primary productivity, and optima and tolerances of individual zooplankton to salinity and temperature gradients in shallow lowland coastal lakes, the distribution patterns of zooplankton community in space and time can vary significantly (Schallenberg et al. 2003; Hall and Burns 2002c).

Presence of macrophytes in shallow lakes also plays a crucial role for zooplankton dynamics through direct and indirect mechanisms. Aquatic macrophytes enhance water quality through nutrient cycling and provide refuge for migratory zooplankton against vertebrate predators such as fish (Meerhoff et al. 2007). Macrophytes support food habitats of water birds (McKinnon and Mitchell 1994), which can indirectly influence the food web structure and dynamics of shallow coastal lakes and estuaries by regulating fish populations (Erwin 1996).

Recently Jensen et al. (2010) have reported that temporal variability in salinity may be more powerful factor than dynamics of macrophyte density for structuring the composition and abundance of zooplankton in shallow coastal brackish lakes. Consequently the top-down or bottom-up controls of lentic food webs of shallow lowland lakes can vary significantly over space and time. Lake Waihola (Figure 1), a shallow, tidally-influenced lowland coastal lake experiences diel and seasonal fluctuations in salinity, and seasonal changes in temperature (Hall and Burns 2002c; Kattel and Closs 2007). Until recently, the study of diel and seasonal dynamics of zooplankton community in Lake Waihola over one year cycle has not been undertaken comprehensively. In this chapter, we have made an attempt to understand temporal and spatial variability in distribution patterns of composition and abundance of six common zooplankton (*Daphnia*, *Ceriodaphnia*, *Bosmina*, calanoid copepods, cyclopoid copepods and amphipods) in Lake Waihola between January 1998 and January 1999.

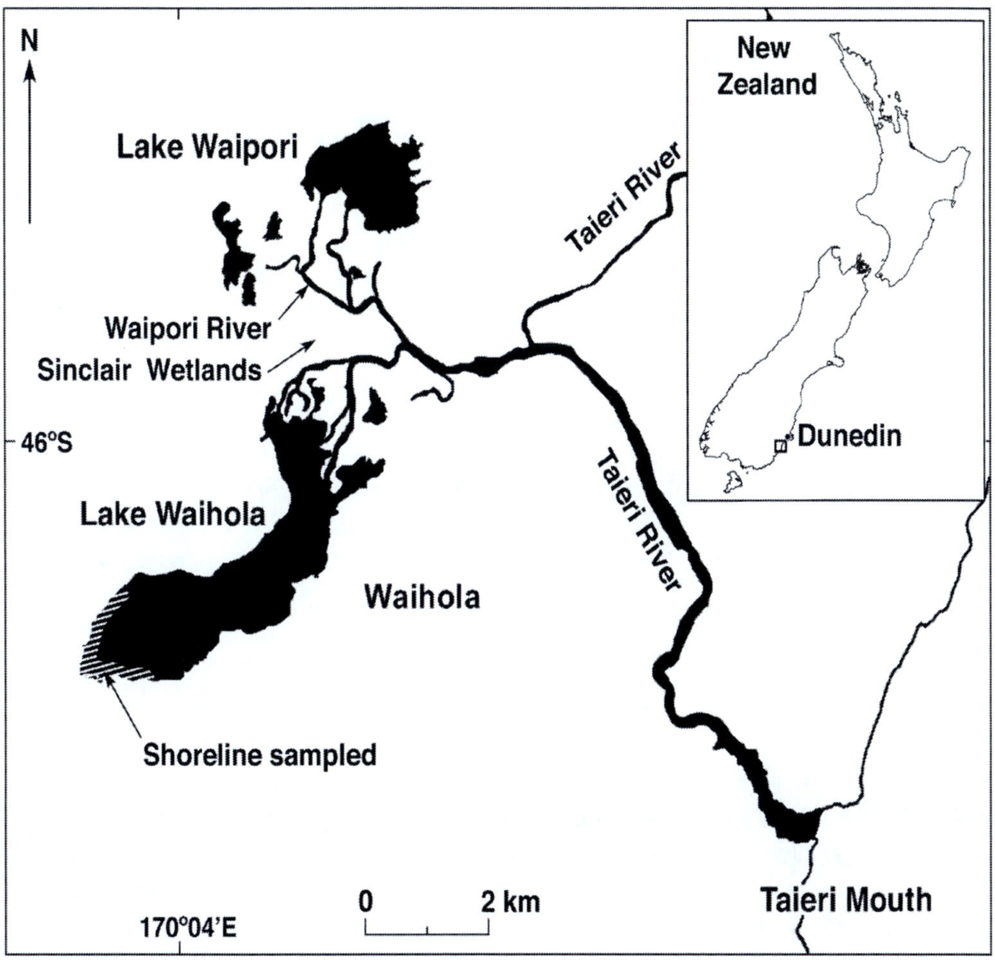

Figure 1. Lake Waihola south Island, New Zealand and associated rivers and wetlands. Barred areas represent sampling areas for zooplankton.

Study Site

Lake Waihola is a small (6.35 sq. km), shallow, eutrophic and phosphorous-limited tidal coastal lake. It is located in 46° S, 170° E of South Island, New Zealand (Figure 1, also see Schallenberg and Burns 2003 for details). The mean depth of the lake is <2 m. The lake is fed by the Waipori River. However, the water flow along the Waipori River, and inflows from the local catchments are relatively small (Robertson 1987; Hamilton 1990; Ryder 1995). A number of freshwater fish and zooplankton species have been recorded in Lake Waihola over time (Kattel 1999; Levine et al. 2005). Calanoid copepods (*Bockella hamata*), *Bosmina meridionalis* and *Daphnia carinata* have been recorded as some of common species of zooplankton in the lake (Hall and Burns 2001). The lake has excessive growth of shoreline and emergent macrophytes (Hamilton and Mitchell 1996) and the increased dynamics of pelagic fish, zooplankton and phytoplankton communities during summer (Kattel 1999; Jeppeson et al. 2000; Sutherland and Closs 2001; Hall and Burns 2001; Flöder and Burns 2004; Kattel and Closs 2007). However, at a period of low macrophyte density, the lake experiences a prolonged turbidity due to wind-induced currents and sediment resuspension (Hamilton and Mitchell 1996, 1997).

Method

Zooplankton was sampled using a hand-operated bilge-pump to pump water through a 150 μm mesh net. A total of 17 litres of water was filtered for each sample. Zooplankton samples were either collected from inshore locations along the edge of macrophyte (predominantly *Typha* sp.) beds, or 50 metres offshore. In addition, samples were collected from either the surface (20 cm below the surface) or bottom (20 cm above the substrate) at an inshore or offshore location. Inshore/offshore (In/Of), shallow/deep (Sh/Dp) samples were collected in random order at 50 metre intervals along the southern shore of Lake Waihola (Figure 1) with monthly basis between January 1998 and January 1999. In every visit, there were at least five replications for each sample in different location (for details please see Kattel 1999).

Statistical Analysis

In the first step, changes in distribution of the composition of the zooplankton community comprising *Daphnia*, *Ceriodaphnia*, *Bosmina*, calanoid copepods, cyclopoid copepods and amphipods (N = 600 samples) across the temporal and spatial timeframes (Date, D/N, In/Of and Sh/Dp) in Lake Waihola were examined using a permutational multivariate ANOVA (PERMANOVA) in PRIMER (Table 1). In the second step, visual distribution of zooplantkton samples was examined in non-metric multi-dimensional (MDS) ordination. In order for better understanding through visualization of the zooplankton community dynamics across temporal and spatial scales, zooplankton count from similar replicate samples were pooled and averaged before ordination.

Table 1. Permutational ANAOVA (PERMANOVA) of 600 samples over diel and seasonal timeframes in Lake Waihola in 1998/99

Source	df	SS	MS	F	P
Date	14	2.1	15298	90.89	0.001
D/N	1	6370.1	6370.1	37.84	0.001
In/Off	1	5753.6	5753.6	34.18	0.001
Sh/Dp	1	4900.5	4900.5	29.11	0.001
Date × D/N	14	16523.0	1180.2	7.01	0.001
Date × In/Of	14	10758.0	768.44	4.56	0.001
Date × Sh/Dp	14	9615.3	686.80	4.08	0.001
Date × In/Of	1	176.4	176.43	1.04	0.416
Date × Sh/Dp	1	516.8	516.83	3.07	0.053
In/Off × Sh/Dp	1	5969.2	5969.2	35.46	0.001
Date × D/N × In/Of	14	4353.7	310.98	1.84	0.008
Date × D/N × Sh/Dp	14	4984.4	356.03	2.11	0.003
Date × In/Of × Sh/Dp	14	2843.0	203.07	1.20	0.256
D/N × In/Of × Sh/Dp	1	58.2	58.256	0.34	0.697
Date × D/N × In/Of × Sh/Dp	14	4973.8	355.27	2.11	0.002
Res	480	80785	168.3		
Total	599	3.727			

Replicates of the spatial (shallow-offshore, deep-inshore, deep-offshore and shallow-inshore) in each seasonal (Date) and diel (D/N) timeframe were averaged. Each sample was then coded with respect to sampling date, time (D/N) and location (In/Of and Sh/Dp) sample. For example, the sample collected from shallow-inshore location during the day of January 1998 was coded as D1DaShIn. Altogether there were 120 pooled samples (Appendix 1). In MDS ordination, distribution of samples was also examined across the four seasons, Summer (December, January, February), Autumn (March, April, May), Winter (June, July, August) and Spring (September, October, November) followed by a trajectory curve was drawn for the course of study.

RESULTS

Composition and Abundance of Zooplankton Community

Marked seasonal variation in the abundance of zooplankton in Lake Waihola over the period of study was significant (Table 1). Densities of zooplankton were generally greater at night compared with the day from January to June 1998, although the difference was not always great (Figure 2). In contrast, densities of zooplankton were higher by day compared with the night in November and December 1998, with the pattern again reversing in January 1999 (Figure 2). The mean night abundance (23.59±4.49 SE/L) for one year was slightly higher than the mean day abundance (22.59±4.78 SE/L) (Figure 2).

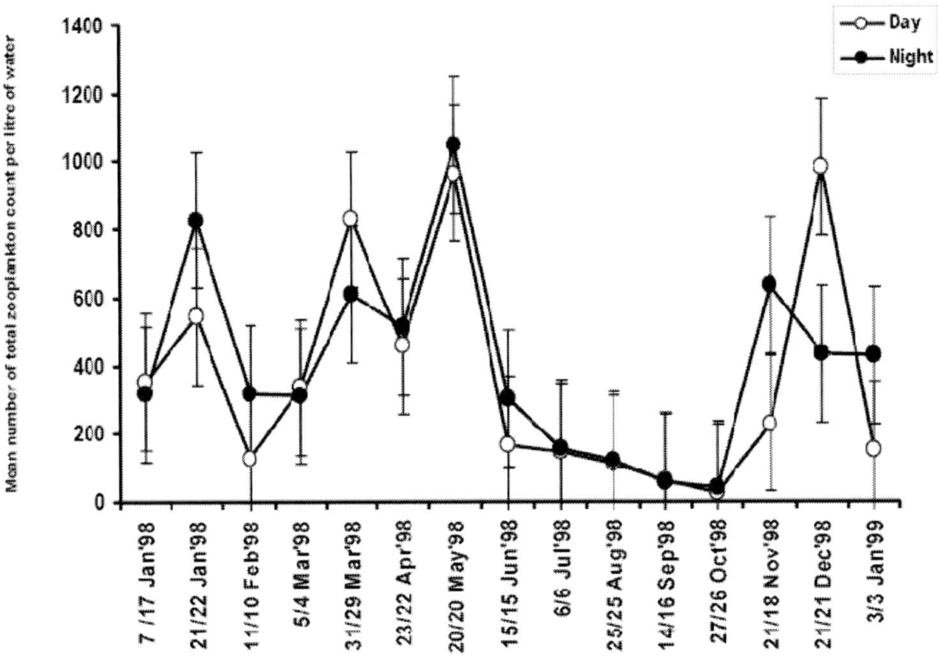

Figure 2. Variation in the abundance of total zooplankton taxa in Lake Waihola collected between Jan'98 and Jan'99. Open circle represents day and solid circles represent night. X-axis represents sampling dates in which day and night are given before and after the slash respectively.

The four-factor interaction (Date*D/N*Sh/Dp*In/Of) suggests that seasonal variation in zooplankton abundance is highly consistent through space and time, however, diel variation in abundance of zooplankton is inconsistent. Three-factor interaction (D/N*In/Of*Sh/Dp) indicates inconsistencies in diel (DVM and DHM) distribution pattern in zooplankton over vertical and horizontal gradients (Table 1).

Densities of zooplankton were relatively high from January to May 1998, although a marked drop in density was evident in February and early March 1998 (Figure 2). Densities declined after May 1998, and remained low until at least late October 1998 (Figure 2). A peak in zooplankton abundance occurred in November and December 1998 (Figure 2). However, the highest (61.57±15.34 SE/L) and lowest (1.5±0.55 SE/L) abundance per litre of water sample were recorded in 20 May and 27 October 1998 respectively (Figure 2).

The diel and spatial distribution patterns of zooplankton was examined in MDS ordination if they were distinctly clustered. However, the clustering of the pooled 120 zooplankton samples was not overwhelming (Fig 3). At 20% similarly level, three groups of zooplankton are formed, in which sample 1 and sample 17 collected during the day (Appendix 1) are formed as separate groups (Figure 3). At 80% similarity level, six groups of zooplankton are formed (Appendix 1, Figure 3).

Since the influence of seasonal variation in distribution patterns of zooplankton composition and abundance across the vertical and horizontal gradients was significant, zooplankton collected in each sampling occasion was grouped into four seasons (summer, autumn, winter and spring) and examined the patterns in MDS ordination (Figure 4).

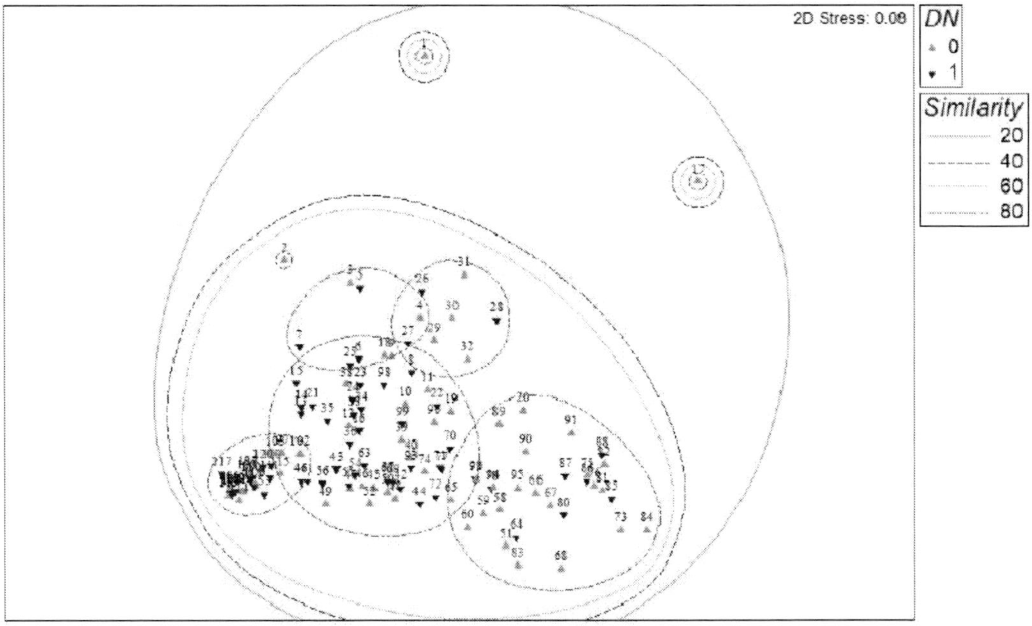

Figure 3. Two dimensional MDS configuration of 120 samples clustering at 20%-80% similarity levels. DN represents Day (0) and Night (1).

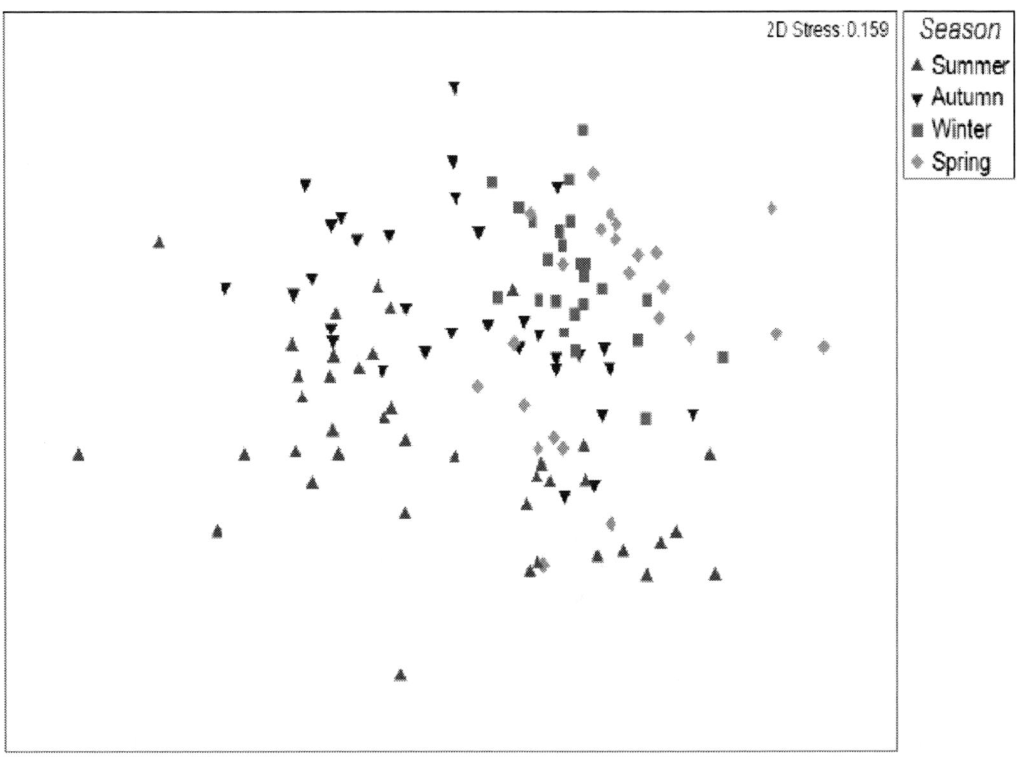

Figure 4. Bray-Curtis Similary Matrix of zooplankton abundance in Lake Waihola in four seasons.

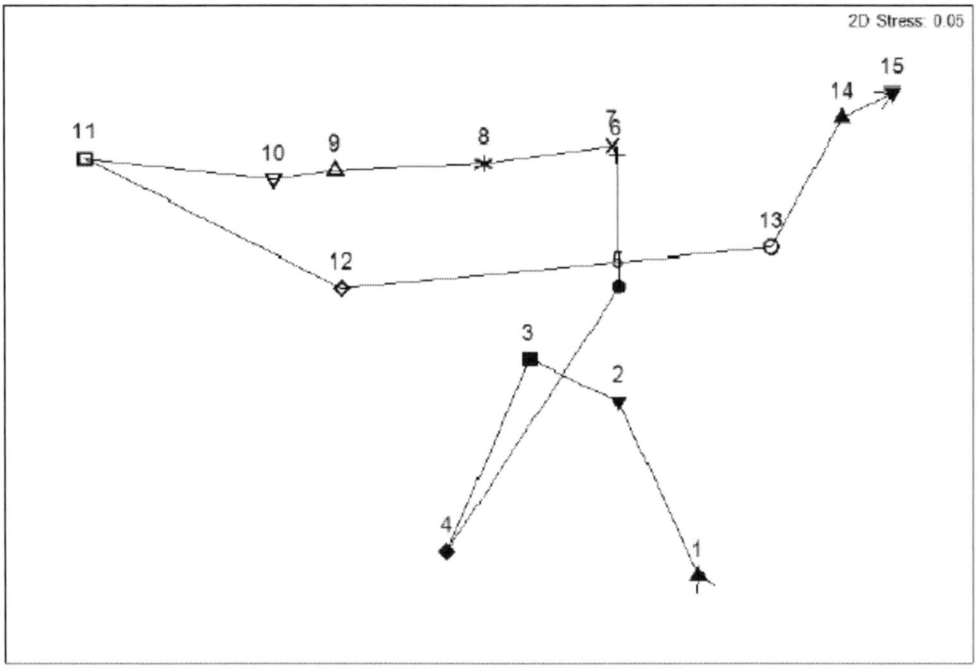

Figure 5. Trajectory curve of the zooplankton community over 15 sampling occasions between January 1998 and January 1999 in MDS ordination. Please use Figure 2 (X-axis, left-right) for each sampling date from 1-15.

Seasonal distribution pattern in ordination are clearly visualised. Summer samples are distributed further left, autumn samples are in the middle and winter and spring samples are distributed towards upper-right corner of the ordination space (Figure 4). Zooplankton abundance over 15 sampling trips in Lake Waihola is presented by a trajectory curve in MDS. The trajectory curve shows that except April and May 1998 there was a gradual shift in zooplankton abundance with changes in seasonal timeframes (Figure 5).

Variation in Population Structure of Zooplankton

The composition of the zooplankton population during the course of study was variable, although at any one time, it was generally dominated by calanoid copepods (*Boeckella hamata*), cyclopoid copepods or the amphipods (*Paracalliope fluviatilis*). Calanoid copepods contributed 77% of total plankton followed by amphipods, which comprised 14% of total plankton populations. Other taxa recorded include cyclopoid copepods, *Bosmina*, *Daphnia* and *Ceriodaphnia* contributing 3%, 2%, 2% and 0.03% of total plankton populations respectively (Table 2).

The bubble plot diagram of three cladocerans, *Daphnia*, *Ceriodaphnia* and *Bosmina* show a poor and patchy distribution pattern over the course of study. Densities of *Bosmina* were recorded high in January 1998 samples while the population of *Daphnia* increased in March 1998 (Figure 6). High density of *Bosmina* was recorded in only one sampling occasion (e.g., D1DaShOf) during the day of 7 January 1998 (Appendix 1).

Table 2. Percentage composition of zooplankton in Lake Waihola between January 1998 and January 1999

Zooplankton	% composition
Daphnia	1.73
Ceriodaphnia	0.03
Bosmina	1.90
Calanoid copepods	77.22
Cyclopoid copepods	3.20
Amphipods	13.6

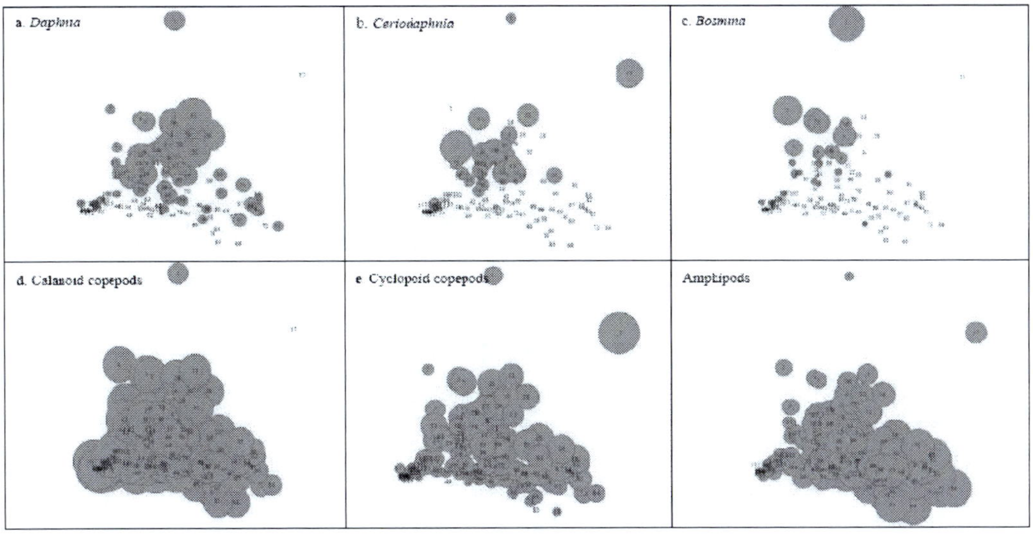

Figure 6. Bray-Curtis Similarity Matrix of each zooplankton (a. Daphnia, b. Ceriodaphnia, c. Bosmina, d. calanoid copepods, e. cyclopoid copepods, f. amphipods) in 120 samples presented as bubble plots in MDS (Stress = 0.08). See Appendix 1 for labelling of each sample (1 to 120).

The bubble plot diagrams for density of zooplankton in Lake Waihola show that calanoid copepods, cyclopoid copepods and amphipods were commonly present over the period of study. However, calanoid copepods were not recorded in D3DaShOf sample (Figure 6, also see Appendix 1).

DISCUSSION

The general distribution patterns of composition and abundance of zooplankton community across the temporal and spatial scales of Lake Waihola in 1998/1999 indicate that seasonal variations such as changes in temperature and life history behaviour of zooplankton could have played a significant role (e.g., Forsyth and James 1991 a and b). Ontogenetic shift in habitat use for food and reproduction could be the main causes of the variation in distribution of the zooplankton community over space and time (Romare et al. 2003). Majority of shallow New Zealand lakes have shown a significant variability in biotic and abiotic compositions through time resulting in complex dynamics of structure and function of ecosystem in these lakes. Phytoplankton and macrophyte densities of some New Zealand lakes such as Tomahawk Lagoon have shown irregular and inverse cycles of dynamics over time and such dynamics have been regulated by vertebrate populations including the fish and water birds (McKinnon and Mitchell 1994). A rapid resuspension of sediments has also been reported to Tomahawk Lagoon due to a presence of vertebrate populations (McKinnon and Mitchell 1994). Shallow coastal lakes, in particular, are sensitive to temporal variability of weather patterns including temperature, rainfalls and winds. Significant effects of wind-induced resuspension and turbulent diffusion on shallow coastal lake systems have been reported in New Zealand extensively (Hamilton and Mitchell 1997; Thomas and Schallenberg 2008). Mechanisms including the mixing of sediment minerals and increased phytoplankton assemblages can reduce light penetration (Schallenberg et al. 2010). Changes in temperature, light and rainfalls along with seasonal variations can directly influence rapid nutrient dynamics and resuspension of sediments in shallow lakes. MacKenzie and Gilespie (1986) reported that occurrence of winter/spring diatom blooms in Tasman Bay was caused by nano-planktonic species bloom following the rainfall and large flood event.

Evidence of various studies suggests that survival and reproduction of zooplankton community in shallow lakes are greatly dependent on temperature and mixture of nutrients and chemicals in the water column. Coastal shallow lakes receive salts during regular tidal intrusions in bi-weekly or monthly basis. The concentration of salts and temperature of water determine the reproduction of fish and zooplankton. For example, survival and egg production of calanoid copepods was highest when salinity ranged from 20-2700 mgL^{-1} Cl and temperature ranged between10^{o} C and 20° C in Lake Waihola. However, in reduced temperature, the adaptive capacity of calanoid copepods also declined with increasing salinity (Hall and Burns 200c). Seasonal changes in weather patterns and temperature as well as occasional stochastic events (e.g. stormy conditions) can have significant implications in competition for food and ultimate species success of zooplankton community in many New Zealand lakes (Chapman and Green 1999; Greenwood et al. 1999). Sometimes other factors can also influence the growth and productivity of zooplankton over temporal time frames. For example, in Lake Mangakaware (North Island), competition amongst cladocerans such as *Bosmina*, *Daphnia* and *Ceriodaphnia* resulted in significant decline in lipid contents and fecundity (Greenwood et al. 1999). Zooplankton succession revealed by a trajectory curve in MDS ordination (Figure 5) indicates a strong seasonal effect on the dynamics of the zooplankton community in Lake Waihola.

The patchy and poor results, such as an increased abundance of *Bosmina* only for one sampling occasion in the summer 1998 (D1DaShOf) and low abundance of *Daphnia* in early autumn 1998 in Lake Waihola were unexpected (Figure 6). However, both *Bosmina* and *Daphnia* were recorded when the lake was dominated by macrophytes (Kattel and Closs 2007). This indicates an increasing role of macrophytes on micro-habitat structure of cladocerans in Lake Waihola. At a time when the loss of micro-habitat structures and deterioration of water quality occurs, *Bosmina* and *Daphnia* can display a narrow range of resistance to thrive in brackish water environments (Hall and Burns 2002 a, b). However, unlike other zooplankton, there was surprisingly a high density of calanoid copepods particularly during winter and autumn 1998 in Lake Waihola. One investigation suggests that calanoid copepods have higher ability to colonise the coastal brackish waters (Jamieson 1998). Relatively equal proportion of composition of calanoid copepods and cyclopoid copepods have been reported in many New Zealand lakes throughout the year (e.g., Forsyth and MacCallum1980; Hall and Burns 2001). Differences in proportion of calanoid copepods and cyclopoid copepods in total plankton community recorded during the course of this study are not known, however some difficulties in taxonomic identification were experienced between calanoid copepds and cycloppoid copepods particularly in their early stage of development (Table 2).

The high occurrence in winter but a low abundance of amphipods (*Paracalliope fluviatilis*) in Lake Waihola in summer is not yet clear. Wilhelm et al. (2007) reported a significantly high selective predation by juvenile common bullies (<32 mm) on small amphipods for various occasions of sampling in Lake Waihola in 2000/2001. Even though no significant effects of fish predation on composition of amphipods in Lake Waihola were reported, amphipods were found abundantly in the gut of common bullies (Kattel 1999; Kattel and Closs 1999). It is indicative that predation by juvenile common bullies could have reduced amphipod populations during summer 1998/99. A low predation pressure by co-occurring mysids (e.g., *Tenagomysis chiltoni*) can also be anticipated during summer. We did not include the records of *T. chiltoni* as they were not a part of our study. However, during the course of this study, we noticed an unaccounted number of *T. chiltoni* in the water column particularly during the summer. In summer, macrophyte density was abundant in Lake Waihola. However, density of macrophytes reduced significantly with relatively high level of turbidity during winter 1998 (Kattel and Closs pers. observation). Both laboratory and field based studies suggest that *T. chiltoni* can be passive in predation at turbid environment with limited or no macrophyte density and often switch their feeding habits from amphipods to cladocerans (e.g., Wilhelm et al. 2002). At low macrophyte density and increased turbidity, fish can also switch their feeding habit from cladocerans to mysids (Hays and Rutledge 1991) indicating that there may be a presence of a complex trophic structure and function in most shallow lowland coastal lentic ecosystems in New Zealand.

The general pattern of the diel variations in zooplankton community in Lake Waihola is inconsistent through time indicating that the role of fish predation on zooplankton is relatively weak over the course of study. Our result does not support the diel migrations behaviour in zooplankton of Lake Waihola which might have triggered by fish predation (Jeppesen et al. 1997; also see Lieschke and Closs 1999 for Australian shallow lakes). Significant effect of Sh/Dp gradient on zooplankton community (Table 1), however, suggests that presence of some degree of vertical movements in zooplankton in Lake Waihola. In summer 1998, Kattel and Closs (2007) reported a relatively high density of juvenile perch and inanga in the water column of Lake Waihola at night. Presence of juvenile perch may have resulted in irregular (possibly reverse) migration behaviour amongst zooplankton community along vertical (Sh/Dp) gradient. However, it is difficult to justify if the tendency of such migrations behaviour in zooplankton along vertical gradient was mediated by fish predation.

Unlike vertical migration behaviour, the DHM in zooplankton is a commonly observed feature in many shallow lakes. In order to avoid predation from planktivorous fish, zooplankton migrate towards shoreline littoral habitats during day (Lauridsen and Buenk 1996; Lauridsen et al. 1996; Burks et al. 2002), and utilize the open water habitat at night when the visibility of fish on zooplankton is poor (Burks et al. 2002). Physico-chemical variation in water quality (e.g., temperature, turbidity, pH and salinity) can also cause DHM in zooplankton in shallow coastal lakes (e.g., Hall and Burns 2002). However, no significant DHM behaviour in zooplankton community was observed in Lake Waihola during the course of this study. Initially we thought that cladocerans such as *Bosmina* and *Daphnia* may have performed DHM in Lake Waihola. However, the contribution of cladocearns to total zooplankton community in the lake was too small and our results (Table 1) also show that the dynamics of cladocerans over space and time in Lake Waihola was not significantly mediated by fish predation.

The zooplankton community of Lake Waihola during the course of this study was dominated by mainly calanoid copepods, some cyclopoid copepods and amphipods. The response of these non-cladocerans to abiotic and biotic environments is different from that of cladocerans (e.g., Meerhoff et al. 2007). Differences in distribution of cladocerans and non-cladocerans in Lake Waihola have been evident in bubble plot diagrams (Figure 6). Although cladocerans (e.g., *Daphnia*) are typically the most important components of the fish diet due to their large body size and rich food quality (Perrow and Irvine 1992), the rare abundance of cladocerans during the period of this study in Lake Waihola may be due to their poor ability to resist degraded water quality caused by periodic changes in maritime environments. Consequently the implications of such changes may have been influencing the dynamics of macrophytes over temporal and spatial scales followed by variability in levels of turbidity and salinity (Hall and Burns 2002 a, b).

Conclusion

The pelagic community structure of zooplankton in Lake Waihola across spatial and temporal scales suggests that ontogenetic shift in diet and life history behaviour plays a significant role. With regard to migration patterns over temporal timeframes, no such strong diel migrations (DVM and DHM) behaviour in zooplankton was observed. However, the effect of seasonal changes over spatial gradients (shallow v deep and inshore v offshore) was highly significant for shaping community structure and dynamics of zooplankton in Lake Waihola. Shallow in nature, frequent changes in the density of emergent and submerged macrophytes, sediment resuspension and the location of the lake from a close proximity to marine environment may be the significant factors for changes in dynamics of biological communities. Seasonal changes in weather patterns and variation in tidal cycles have triggered a rapid movements of saline water followed by changes in physico-chemical properties of water and response of zooplankton community to the natural variability in Lake Waihola.

Although some cladocerans recorded (*Daphnia*, *Ceriodaphnia* and *Bosmina*) in Lake Waihola were significant diet of fish, their contribution to total zooplankton community was relatively poor over the course of study. However, non-cladocerans, calanoid copepods, cyclopoid and amphipods were predominantly occurred in the lake during the period of study. Composition, abundance and distribution of zooplankton community in Lake Waihola over space and time during the course of study indicates that there may be significant differences amongst zooplankton groups (e.g., cladocerans vs non-cladocerans) and their response to local environmental and seasonal changes, consequently obscuring our understanding in their complex nature of migrations, ecosystem structure and functions in shallow lowland coastal lakes in South Island, New Zealand. An in-depth investigation of zooplankton community in Lake Waihola at species level is useful to understand their migrations behaviour and variations in their distribution patterns over space and time.

Appendix 1. Pooled zooplankton samples (1-120), date of sampling and the codes given for indirect gradient analysis in ordination

Sample	Code	Date	Sample description	Sample	Date	Sample code	Description
1	D1DaShOf	7/1/98	Date1-Day-Shallow-Offshore	25	4/3/98	D4NiShOf	Date4-Night-Shallow-Offshore
2	D1DaDpIn	7/1/98	Date1-Day-Deep-Inshore	26	4/3/98	D4NiDpIn	Date4- Night -Deep-Inshore
3	D1DaShIn	7/1/98	Date1-Day-Shallow-Inshore	27	4/3/98	D4NiShIn	Date4- Night -Shallow-Inshore
4	D1DaDpOf	7/1/98	Date1-Day-Deep-Offshore	28	4/3/98	D4NiDpOf	Date4- Night-Deep-Offshore
5	D1NiShOf	17/7/98	Date1-Night-Shallow-Offshore	29	5/3/98	D4DaShOf	Date4- Day -Shallow-Offshore
6	D1NiDpIn	17/7/98	Date1-Night-Deep-Inshore	30	5/3/98	D4DaDpIn	Date4-Day-Deep-Inshore
7	D1NiShIn	17/7/98	Date1-Night-Shallow-Inshore	31	5/3/98	D4DaShIn	Date4-Day -Shallow-Inshore
8	D1NiDpOf	17/7/98	Date1-Night-Deep-Offshore	32	5/3/98	D4DaDpOf	Date4-Day -Deep-Offshore
9	D2DaShOf	21/1/98	Date2-Day-Shallow-Offshore	33	29/3/98	D5NiShOf	Date5- Night-Shallow-Offshore
10	D2DaDpIn	21/1/98	Date2-Day-Deep-Inshore	34	29/3/98	D5NiDpIn	Date5- Night-Deep-Inshore
11	D2DaShIn	21/1/98	Date2-Day-Shallow-Inshore	35	29/3/98	D5NiShIn	Date5- Night-Shallow-Inshore
12	D2DaDpOf	21/1/98	Date2-Day-Deep-Offshore	36	29/3/98	D5NiDpOf	Date5- Night-Deep-Offshore
13	D2NiShOf	22/1/98	Date2-Night-Shallow-Offshore	37	31/3/98	D5DaShOf	Date5- Day -Shallow-Offshore
14	D2NiDpIn	22/1/98	Date2-Night-Deep-Inshore	38	31/3/98	D5DaDpIn	Date5- Day-Deep-Inshore
15	D2NiShIn	22/1/98	Date2-Night-Shallow-Inshore	39	31/3/98	D5DaShIn	Date5- Day-Shallow-Inshore
16	D2NiDpOf	22/1/98	Date2-Night-Deep-Offshore	40	31/3/98	D5DaDpOf	Date5- Day-Deep-Offshore
17	D3DaShOf	11/2/98	Date3-Day-Shallow-Offshore	41	22/4/98	D6NiShOf	Date6- Night -Shallow-Offshore
18	D3DaDpIn	11/2/98	Date3-Day-Deep-Inshore	42	22/4/98	D6NiDpIn	Date6- Night -Deep-Inshore
19	D3DaShIn	11/2/98	Date3-Day-Shallow-Inshore	43	22/4/98	D6NiShIn	Date6- Night -Shallow-Inshore
20	D3DaDpOf	11/2/98	Date3-Day-Deep-Offshore	44	22/4/98	D6NiDpOf	Date6- Night -Deep-Offshore
21	D3NiShOf	10/2/98	Date3-Night-Shallow-Offshore	45	23/4/98	D6DaShOf	Date6- Day t-Shallow-Offshore

Appendix 1. (Continued)

Sample	Code	Date	Sample description	Sample	Date	Sample code	Description
22	D3NiDpIn	10/2/98	Date3-Night-Deep-Inshore	46	23/4/98	D6DaDpIn	Date6- Day -Deep-Inshore
23	D3NiShIn	10/2/98	Date3-Night-Shallow-Inshore	47	23/4/98	D6DaShIn	Date6- Day -Shallow-Inshore
24	D3NiDpOf	10/2/98	Date3-Night-Deep-Offshore	48	23/4/98	D6DaDpOf	Date6- Day -Deep-Offshore
Sample	Code	Date	Description	Sample	Date	Sample code	Description
49	D7DaShOf	20/5/98	Date7-Day-Shallow-Offshore	77	25/8/98	D10NiShOf	Date10-Night-Shallow-Offshore
50	D7DaDpIn	20/5/98	Date7-Day-Deep-Inshore	78	25/8/98	D10NiDpIn	Date10-Night-Deep-Inshore
51	D7DaShIn	20/5/98	Date7-Day-Shallow-Inshore	79	25/8/98	D10NiShIn	Date10-Night-Shallow-Inshore
52	D7DaDpOf	20/5/98	Date7-Day-Deep-Offshore	80	25/8/98	D10NiDpOf	Date10-Night-Deep-Offshore
53	D7NiShOf	20/5/98	Date7-Night-Shallow-Offshore	81	14/9/98	D11DaShOf	Date11-Day-Shallow-Offshore
54	D7NiDpIn	20/5/98	Date7-Night-Deep-Inshore	82	14/9/98	D11DaDpIn	Date11-Day-Deep-Inshore
55	D7NiShIn	20/5/98	Date7-Night-Shallow-Inshore	83	14/9/98	D11DaShIn	Date11-Day-Shallow-Inshore
56	D7NiDpOf	20/5/98	Date7-Night-Deep-Offshore	84	14/9/98	D11DaDpOf	Date11-Day-Deep-Offshore
57	D8DaShOf	15/6/98	Date8-Day-Shallow-Offshore	85	16/9/98	D11NiShOf	Date11-Night-Shallow-Offshore
58	D8DaDpIn	15/6/98	Date8-Day-Deep-Inshore	86	16/9/98	D11NiDpIn	Date11-Night-Deep-Inshore
59	D8DaShIn	15/6/98	Date8-Day-Shallow-Inshore	87	16/9/98	D11NiShIn	Date11-Night-Shallow-Inshore
60	D8DaDpOf	15/6/98	Date8-Day-Deep-Offshore	88	16/9/98	D11NiDpOf	Date11-Night-Deep-Offshore
61	D8NiShOf	15/6/98	Date8-Night-Shallow-Offshore	89	27/10/98	D12DaShOf	Date12-Day-Shallow-Offshore
62	D8NiDpIn	15/6/98	Date8-Night-Deep-Inshore	90	27/10/98	D12DaDpIn	Date12-Day-Deep-Inshore
63	D8NiShIn	15/6/98	Date8-Night-Shallow-Inshore	91	27/10/98	D12DaShIn	Date12-Day-Shallow-Inshore
64	D8NiDpOf	15/6/98	Date8-Night-Deep-Offshore	92	27/10/98	D12DaDpOf	Date12-Day-Deep-Offshore
65	D9DaShOf	6/7/98	Date9-Day-Shallow_Offshore	93	26/10/98	D12NiShOf	Date12-Night-Shallow-Offshore
66	D9DaDpIn	6/7/98	Date9-Day-Deep-Inshore	94	26/10/98	D12NiDpIn	Date12-Night-Deep-Inshore

Sample	Code	Date	Sample description	Sample	Date	Sample code	Description
67	D9DaShIn	6/7/98	Date9-Day-Shallow-Inshore	95	26/10/98	D12NiShIn	Date12-Night-Shallow-Inshore
68	D9DaDpOf	6/7/98	Date9-Day-Deep-Offshore	96	26/10/98	D12NiDpOf	Date12-Night-Deep-Offshore
69	D9NiShOf	6/7/98	Date9-Night-Shallow-Offshore	97	18/11/98	D13NiShOf	Date13- Night -Shallow-Offshore
70	D9NiDpIn	6/7/98	Date9-Night-Deep-Inshore	98	18/11/98	D13NiDpIn	Date13- Night -Deep-Inshore
71	D9NiShIn	6/7/98	Date9-Night-Shallow-Inshore	99	18/11/98	D13NiShIn	Date13- Night -Shallow-Inshore
72	D9NiDpOf	6/7/98	Date9-Night-Deep-Offshore	100	18/11/98	D13NiDpOf	Date13- Night -Deep-Offshore
73	D10DaShOf	25/8/98	Date10-Day-Shallow-Offshore	101	21/11/98	D13DaShOf	Date13- Day -Shallow-Offshore
74	D10DaDpIn	25/8/98	Date10-Day-Deep-Inshore	102	21/11/98	D13DaDpIn	Date13- Day -Deep-Inshore
Sample	Code	Date	Description	Sample	Date	Sample code	Description
75	D10DaShIn	25/8/98	Date10-Day-Shallow-Inshore	103	21/11/98	D13DaShIn	Date13- Day -Shallow-Inshore
76	D10DaDpOf	25/8/98	Date10-Day-Deep-Offshore	104	21/11/98	D13DaDpOf	Date13- Day -Deep-Offshore
105	D14DaShOf	21/12/98	Date14-Day-Shallow-Offshore	113	3/1/99	D15DaShOf	Date15-Day-Shallow-Offshore
106	D14DaDpIn	21/12/98	Date14-Day-Deep-Inshore	114	3/1/99	D15DaDpIn	Date15-Day-Deep-Inshore
107	D14DaShIn	21/12/98	Date14-Day-Shallow-Inshore	115	3/1/99	D15DaShIn	Date15-Day-Shallow-Inshore
108	D14DaDpOf	21/12/98	Date14-Day-Deep-Offshore	116	3/1/99	D15DaDpOf	Date15-Day-Deep-Offshore
109	D14NiShOf	21/12/98	Date14-Night-Shallow-Offshore	117	3/1/99	D15NiShOf	Date15-Night-Shallow-Offshore
110	D14NiDpIn	21/12/98	Date14-Night-Deep-Inshore	118	3/1/99	D15NiDpIn	Date15-Night-Deep-Inshore
111	D14NiShIn	21/12/98	Date14-Night-Shallow-Inshore	119	3/1/99	D15NiShIn	Date15-Night-Shallow-Inshore
112	D14NiDpOf	21/12/98	Date14-Night-Deep-Offshore	120	3/1/99	D15NiDpOf	Date15-Night-Deep-Offshore

Acknowledgment

Field and technical assistance were provided by John Hollows, Ken Miller, Chris Arbuckle, Murray McKenzie, and Ian Dickson. The research was conducted under the New Zealand Ministry of Foreign Affairs and Trade (MFAT) scholarship and the University of Otago, Dunedin for GRK. The research was completed with the approval of the University of Otago Animal Ethics Committee.

References

Burks RL, Lodge DM, Jeppesen E, Lauridsen TL (2002) Diel horizontal migration of zooplankton: costs and benefits of inhabiting the littoral. Freshwater Biology 47: 43-365.

Chapman MA, Green JD (1999) Crustacean zooplankton community in a New Zealand lake during four decades of trophic change. *New Zealand Journal of Geology and Geophysics* 33: 361-373.

Downs TM, Schallenberg M, Burns CW (2008) Responses of lake phytoplankton to micronutrient enrichment: a study in two New Zealand lakes and an analysis of published data. *Aquatic Sciences* 70 (2008) 347 – 360, 1015-1621, doi: 10.1007/s00027-008-8065-6.

Erwin RM (1996) Dependence of water birds and shorebirds in shallow-water habitats in mid-Atlantic coastal regions: an ecological profile and management recommendations. *Estuaries* 19: 213-219.

Flöder S, Burns CW (2004) Phytoplankton diversity of shallow tidal lakes: influence of periodic salinity changes on diversity and species number of a natural assemblage. *Journal of Phycology* 40: 54-61.

Forsyth DJ, MacCallum ID (1980) Zooplankton of Lake Taupo. *New Zealand Journal of Marine and Freshwater Research* 14: 65-69.

Forsyth DJ, James MR (1991a) Seasonal abundance and diel vertical dynamics of zooplankton in Lake Okaro. *Archiv fur Hydrobiologie* 121, 410-429.

Forsyth DJ, James MR (1991b) Population dynamics and production of zooplankton in eutrophic Lake Okaro, North Island, New Zealand. *Archiv fur Hydrobiologie* 120: 287-314.

Gliwicz ZM (1986) Predation and the evolution of vertical migration in zooplankton. Nature 320: 746-748.

Greenwood T, Green JD, Hicks BJ, Chapman MA (1999) Seasonal abundance of small cladocerans in Lake Mangakaware, Waikato, New Zealand. *New Zealand Journal of Marine and Freshwater Research* 33: 399-415.

Hall CJ, Burns CW (2001) Effects of salinity and temperature on survival and reproduction of *Boeckella hamata* (Copepoda: Calanoida) from a periodically brackish lake. *Journal of Plankton Research* 23: 977-103.

Hall CJ, Burns CW (2002a) Mortality and growth responses of *D. carinata* to increases in temperature and salinity. *Freshwater Biology* 47: 451-456.

Hall CJ, Burns CW (2002b) Environmental gradients and zooplankton distribution in a shallow, tidal lake. *Archiv für Hydrobiologie* 154: 485-497.

Hall CJ, Burns CW (2002c) Effects of temperature and salinity on the survival and egg production of *Gladioferens pectinatus* Brady (Copepodas: Calanoida). *Estuarine, Coastal and Shelf Science* 55: 557-564, doi:10.1006/ecss.2001.0923.

Hall CJ, Burns CW (2003) Responses of crustacean zooplankton to seasonal tidal salinity changes in the coastal Lake Waihola, New Zealand. *New Zealand Journal of Marine and Freshwater Research* 37: 31-43.

Hamilton DP (1990) Sediment resuspension by wind in shallow lakes, unpublished PhD thesis, University of Otago, Dunedin, New Zealand.

Hamilton DP, Mitchell SF (1996) An empirical model for sediment resuspension in shallow lakes. *Hydrobiologia* 317: 209-220.

Hamilton DP, Mitchell SF (1997) Wave induced sheer stresses, plant nutrients and chlorophyll in seven shallow lakes. *Freshwater Biology* 38: 159-168.

Hays JW, Rutledge MJ (1991) Relationship between turbidity and fish diets in Lakes Waahi and Whangape, New Zealand. *New Zealand Journal of Marine and Freshwater Research* 25: 297-304.

Jamieson CD (1998) Calanoid copepod biogeography in New Zealand. *Hydrobiologia* 367: 189-197.

Jensen E, Brucet S, Meerhoff M, Nathansen L, Jeppesen E (2010) Community structure and diel migration of zooplankton in shallow brackish lakes: role of salinity and predators. *Hydrobiologia* 646: 215-229, doi: DOI 10.1007/s10750-010-0172-4.

Jeppesen E, Lauridsen T, Mitchell SF, Burns CW (1997) Do planktivorous fish structure the zooplankton communities in New Zealand lake? *New Zealand Journal of Marine and Freshwater Research* 31: 163-173.

Jeppesen E, Lauridsen T, Mitchell SF, Christoffersen K, Burns CW (2000) Trophic structure in the pelagial of 25 shallow New Zealand lakes: changes along nutrient and fish gradients. *Journal of Plankton Research* 22: 951-968.

Kattel GR (1999) Seasonal and diel dynamics of pelagic fish and zooplankton populations in a shallow lowland South Island Lake in New Zealand. Unpublished M. Sc. Thesis, The University of Otago, New Zealand.

Kattel GR, Closs GP (2007) Spatial and temporal variation in the fish community of a south Island, New Zealand coastal lake. *New Zealand Journal of Marine and Freshwater Research* 41 1-11.

Lauridsen TL, Buenk I (1996) Diel changes in the horizontal distribution of zooplankton in the littoral zone of two shallow eutrophic lakes. *Archiv fur Hydrobiologie* 137: 161-176.

Lauridsen TL, Jeppesen E, Mitchell SF, Lodge DM, Burks RL (1999) Diel variation in horizontal distribution of *Daphnia* and *Ceriodaphnia* in oligotrophic and mesotrophic lakes with contrasting fish densities. *Hydrobiologia* 408/409: 241-250.

Lauridsen TL, Pedersen LJ, Jeppesen E, Søndergaard M (1996) The importance of macrophyte bed size for cladoceran composition and horizontal migration in a shallow lake. *Journal of Plankton Research* 18: 2283-2294.

Levine SN, Zehrer RF, Burns CW (2005) Impact of resuspended sediment on zooplankton feeding in Lake Waihola, New Zealand. Freshwater Biology 50: 1515-1536.

Lieschke JA, Closs GP (1999) Regulation of zooplankton composition and distribution by a zooplanktivorous fish in a shallow, eutrophic floodplain lake in southeast Australia. *Archiv fur Hydrobiologie* 146: 397-412.

Mackenzie AL, Gillespie PA (1986) Plankton ecology and productivity, nutrient chemistry, and hydrography of Tasman Bay, New Zealand. *New Zealand Journal of Marine and Freshwater Research* 20:365–395.

McKinnon SLC, Mitchell SF (1994) Eutrophication and black swan (*Cygnus atratus* Latham) populations: tests of two sample relationships. Hydrobiologia 279/280 (Developments in *Hydrobiology* 96: 163-170.

Meerhoff M, Carlos I, De Mello FT, Clemente JM, Jensen E, Lauridsen TL, Jeppesen E (2007) Effects of habitat complexity on community structure and predator avoidance behavior of littoral zooplankton in temperate verus subtropical shallow lakes. *Freshwater Biology* 52: 1009-1021.

Perrow MR, Irvine K (1992) The relationship between cladoceran body size and the growth of underyearling roach (*Rutilus rutilus*) (L.) in two shallow lowland lakes: a mechanism for density-dependent reductions in growth. *Hydrobiologia* 241, 155-161, doi: 10.1007/BF00028638.

Robertson BM (1987) Water quality- Lake Waihola-Waipori investigation. Otago Catchment and Regional Water Board (OCB), Dunedin, New Zealand

Romare P, Berg S, Lauridsen T, Jeppesen E (2003) Spatial and temporal distribution of fish and zooplankton in shallow lake. *Freshwater Biology* 48: 1353-1362.

Ryder G (1995) Factors affecting water quality in the lower Taieri catchment. Otago Regional Council, Dunedin, New Zealand.

Schallenberg M, Burns CW (2003) A temperate, tidal lake-wetland complex – II. Water quality and implications for zooplankton community structure. *New Zealand Journal of Marine and Freshwater Research* 37: 429-447.

Schallenberg M, Burns CW (2004) Effects of sediment resuspension on phytoplankton production: Teasing apart the influences of light, nutrients and algal entrainment. *Freshwater Biology* 49: 143-159.

Schallenberg M, Larned ST, Hayward S, Arbuckle C (2010) Contrasting effects of managed opening regimes on water quality in two intermittently closed and open coastal lakes. *Estuarine, Coastal and Shelf Science,* 86, 587-597, doi: 10.1016/j.ecss.2009.11.001

Smiley EA, Tessier A.J (1998) Environmental gradients and horizontal distribution of microcrustaceans in lakes. *Freshwater Biology* 39: 397-409.

Sutherland DL, Closs GP (2001) Spatial and temporal variations in the abundance and composition of ichthyoplankton in a large South Island New Zealand river estuary. *New Zealand Journal of Marine and Freshwater Research* 35: 1061-1069.

Thomas DB, Schallenberg M. (2008) Benthic shear stress gradient defines three mutually exclusive modes of non-biological internal nutrient loading in shallow lakes. *Hydrobiologia* 610:1-11.

Wilhelm FM, Closs GP and Burns CW (2007) Seasonal diet and amphipod size selection of juvenile common bully, Gobimorphus cotidianus, in a coastal New Zealand lake. *Hydrobiologia* 586: 303-312, doi: DOI 10.1007/s10750-007-0703-9.

Wilhelm FM, Hamann J, Burns CW (2002) Mysid predation on amphipods and *Daphnia* in a shallow coastal lake: prey selection and effects of macrophytes. *Canadian Journal of Fisheries and Aquatic Sciences* 59: 1901-1907.

In: Zooplankton and Phytoplankton
Editor: Giri Kattel, pp. 141-157
ISBN 978-1-61324-508-8

Chapter 7

PHYTOPLANKTON COMPOSITION IN THE FISH FARM AREA: PIGMENT ANALYSES

Vesna Flander-Putrle*
Marine Biology Station, National Institute of Biology, SI-6330 Piran, Fornače 41, Slovenia

ABSTRACT

Fast development of different aquaculture in the world in 90s caused different studies of effects of aquaculture on the environment. Aquaculture activity has a serious impact on environment, e.g. the enrichment of the water column in dissolved organic and inorganic material. This may subsequently affect populations of phytoplankton differentially. In this study we report on short-term changes in the water column in relation to fish feeding and differences along transect from the centre of fish farm towards open waters. We took samples at different sites around the fish cage. We also undertake the enriched enclosures experiment. With HPLC (High Performance Liquid Chromatography) pigment analyses we determine the phytoplankton community structure. Comparing the pigment fingerprints at fish farm area before and after feeding we observed only minor differences. The main phytoplankton group were diatoms. We noticed differences on the profile from the centre of the fish cage outward.

Keywords: Ecosystem approach and Aquaculture, Fish farms, HPLC, pigments, phytoplankton community, northern Adriatic

INTRODUCTION

Aquaculture activity, in particular intense fish farming in open cages, has a serious impact on environment. One of the main environmental concerns associated with fish farming

* e-mail: flander@mbss.org

is the direct discharge of suspended solids and dissolved nutrients into coastal waters and thus enrichment of the water column in dissolved organic and inorganic material. These materials are excreted by the fish, elutriate from uneaten food and derive from other pollution substances (Arzul et al., 1996). This may produce marked changes in the chemical composition of seawater (e.g. dissolved organic components and N/P ratio) that may subsequently affect populations of phytoplankton differentially (Arzul et al., 2001). Mostly they allow or stimulate their growth but they can also inhibit some species. In this way fish farms may change the phytoplankton community in the area. Organic material from fish farms has grater impact on benthic community. Input of dissolved and particulate nutrients from fish farms is highest during summer due to the temperature-dependent increase of metabolism of fish and more intensive feeding.

The rapid development of different types of aquaculture in the world during the 90s brought about different studies of the effects of aquaculture on the environment. The 1992 Earth Summit in Rio de Janeiro addressed the issue that environmental management policies, with their traditional sectoral basis, were not adequately covering the full impacts of human development and exploitation on the environment. There was a concerted move then to incorporate a more holistic approach to policy decision making in regard to sustainable development with an Ecosystem Approach based direction.

This concept has been applied to many Fisheries projects but ECASA (Ecosystem Approach for Sustainable Aquaculture) was one of the first aquaculture based projects to focus on this. In Europe there is approximate 40 % increase of aquaculture activity every 10 years. Not only was this the reason for EU Commission to adopt "A strategy for the Sustainable Development of European aquaculture. COM (2002)511". One of the key aims was to assure continuous environmental growth and to develop specific tools and directions for environmental impact assessment. And that's how the idea for the ECASA project appears (http://www.ecasa.org.uk/index.htm). Our study is part of this project.

In the Bay of Piran (northern Adriatic) we have cage fish farms of sea bass (*Dicentrarchus labrax*). Since the 2000 production has been 50 - 80 tons per year.

As a part of the EU founded project ECASA we want to identify quantitative and qualitative indicators of the effects of aquaculture on the environment and vice versa. In this study we report on short-term changes in phytoplankton pigments in the water column in relation to fish feeding and differences along transect from the centre of fish farm towards open waters. HPLC (High Performance Liquid Chromatography) pigment analyses indicate the phytoplankton community structure. Photosynthetic pigments have proved to be useful biomarkers of the abundance, composition and physiological status of the phytoplankton biomass in the marine environment although they cannot be considered to be fully specific diagnostic markers of individual phylogenetic groups of phytoplankton, and their use should therefore be exercised with caution (Jeffrey et al., 1997). Several studies have demonstrated good agreement between chemotaxonomic assessments of phytoplankton and conventional microscopy (Gieskes and Kraay, 1986; Bidigare et al., 1990; Latasa et al., 1992; Wright et al., 1996; Rodriguez et al., 2002; Carreto et al., 2003). Such a comparison performed in the Gulf of Trieste (Švagelj et al., 1996; Terzić, 1996) confirmed the biomarker approach as a very promising tool to investigate phytoplankton in that area.

The main goal of our research was to study the influence of fish farming on phytoplankton community, and assess the suitability of HPLC – determined pigments as an indicator of water quality in fish farm – influenced area.

Material and Methods

Study Area

The studied area is located in the northernmost part of the Adriatic Sea. In winter the water column is vertically homogeneous, while thermal stratification generally develops from late April through September. The fish farm is situated in the inner part of the semi-enclosed Bay of Piran (Gulf of Trieste, Adriatic Sea). The bay is about 7 km long and 5 km wide and its inner part ends at a salt pan created by the Dragonja River. The depth beneath the fish cages is about 13 m and the sea floor is smooth and muddy with a moderate declination towards the mouth of the bay.

Sampling Strategy

We did three field campaigns at the fish farm area; in October 2005, April 2006 and July 2006. We took samples from 5 m depth. In October 2005 samples were collected at 6 different points of the compass around and in the centre of fish cage (00CC) (Figure 1), approximately 1 hour before feeding of farmed fish (samples marked 0) and three hours after feeding (samples marked A). First ring of sampling sites was 6 m from the centre (00N1, 0EN1, 0ES1, 00S1, 0WS1, 0WN1), and the second was 20 m away (00N2, 0EN2, 0ES2, 00S2, 0WS2, 0WN2) (Figure 1). In addition, we sampled at sites along the profile from the centre of fish cage outwards: 00CC, 6 m (00N1), 20 m (00N2), 1166 m (ZBMA) and 6825 m (ZC0B). In April and July 2006 we sampled in the centre of the mariculture (CESY) and at 4 different outwards directions at 200, 400 and 600 m: north (N200, N400, N600), south (S200, S400, S600), east (E200, E400), west (W200, W400), and at ZBMA and ZC0B (Figure 2). In April 2006 we took samples only 3 hours after feeding and in July 2006 we again took samples before feeding (0) and 3 hours after feeding (A).

Definition of Zones

Three main zones were considered related to point source discharges (CSTT, 1994).

Zone A: local to the farm; water residence times of a few hours; in regions of low or moderate dispersion, particles fall to the sea-bed within this zone, and their footprint effectively defines zone A. In our case in this zone are all the sites at the fish farm area.

Zone B: the scale of small water bodies, such as a fjordic basin that can be treated as laterally homogenous; water residence times of a few days to several weeks, a period sufficiently long for nutrient enrichment to lead to enhanced growth of phytoplankton. In our case Bay of Piran represents such zone: sites 200, 400, 600 m from fish cage, ZBMA, ZBWC, ZBSC and ZBEC.

Zone C: regional scale; water residence times of weeks to months, sufficiently long for complex ecosystem processes to take place; provides the "background" conditions with which waters in potentially effected zone Bs exchange. In our case this is Gulf of Trieste; site ZC0B.

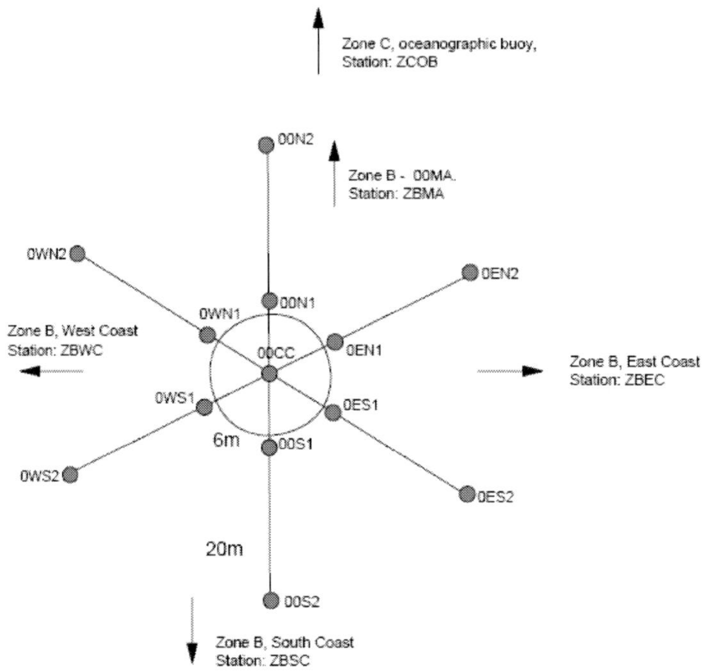

Figure 1. Scheme of sampling sites in October 2005.

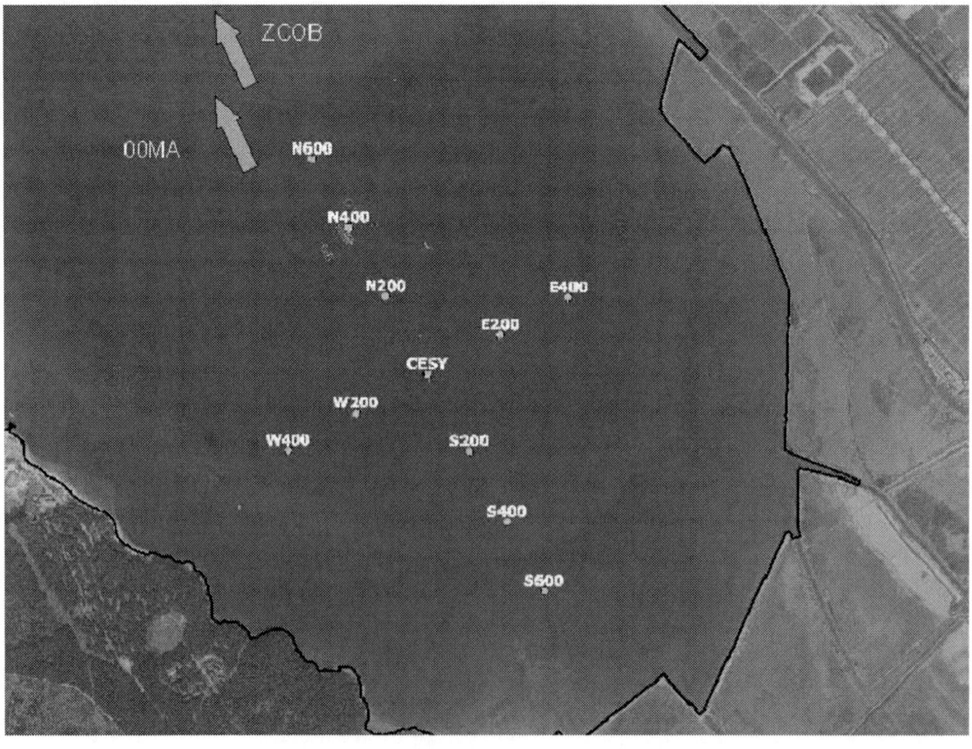

Figure 2. Scheme of sampling sites in April and July 2006.

ENCLOSURES ENRICHMENT EXPERIMENT

Enclosure experiment was undertaken during summer using natural phytoplankton population. We took seawater at the sampling site ZBMA in the central part of Bay of Piran at the depth where maximum chlorophyll *a* concentration above thermocline was detected using CTD fine-scale probe. Natural phytoplankton population was amended by seawater from fish cage area and as a control from area 200 m far from fish cage.

Enclosure A (natural population, control): 8 l seawater from the sampling site ZBMA, pre-filtered through net with mesh size 200 μm.

Enclosure B (amended natural population – 200 m from cage): concentrated seawater (4 l of seawater concentrated to 1 l through Millipore 0.22 μm filter) and enriched seawater from area 200 m far from fish cage (filtrate of 7 l seawater through Millipore 0.22 μm filter).

Enclosure C (amended natural population – cage water): concentrated seawater (4 l of seawater concentrated to 1 l through Millipore 0.22 μm filter) and enriched seawater from fish cage area (filtrate of 7 l seawater through Millipore 0.22 μm filter).

Three 8 l containers were incubated in the sea at 3 m depth. During the experiment we took samples from the containers first, third and fifth day.

HPLC PIGMENT ANALYSIS

Seawater samples (0.7 - 1 l) were filtered through 47 mm Whatman GF/F filters that were immediately frozen until HPLC pigment analysis. Filters were extracted in 90 % acetone using sonication. After centrifugation for 10 min at 4.000 rpm, the supernatant was used for HPLC pigment analysis. An aliquot (300 μl) of clarified acetone extract was mixed with 300 μl 1 M ammonium acetate, and 200 μl of this mixture was injected into a gradient HPLC system. Pigments were analysed using a reverse-phase HPLC method (Mantoura and Llewellyn, 1983; Barlow et al., 1993). The HPLC system was equipped with a reverse phase 3 μm C18 column (Pecosphere, 35x4.5 mm, Perkin Elmer). Solvent A consisted of 80 % methanol and 20 % 1 M ammonium acetate, and solvent B 60 % of methanol and 40 % of acetone. A linear gradient from 0 % B to 100 % B for 10 min was followed by an isocratic hold at 100 % B for 6 min. The flow rate was 1 ml min^{-1}. Chlorophylls and carotenoids were detected by absorbance at 440 nm using an UV/Vis spectrophotometric detector (Spectra Physics, Model UV2000). Degradation products of chlorophyll *a* were detected by measuring fluorescence (420/672 nm) with a spectrofluorimetric detector (Spectra Physics, Model FL2000). To estimate the contribution of various phytoplankton groups, we multiplied the concentrations of individual biomarker pigments with published values of chlorophyll *a* : biomarker pigment ratios (Table 1).

The relative contributions of different phytoplankton groups to total biomass (total chlorophyll *a*) was estimated using the equation: X = K (Cpig/Cchla), where X is the relative contribution of different phytoplankton groups to total biomass, K is the chlorophyll *a* : biomarker pigment ratio for a certain phytoplankton group, Cpig is the concentration of biomarker pigment for that phytoplankton group, and Cchla is the concentration of chlorophyll *a* in the sample.

Table 1. Values of chlorophyll *a* : biomarker pigment ratios (K) in different phytoplankton groups

PHYTOPLANKTON PIGMENT	K	PHYTOPLANKTON GROUP
Fucoxanthin	1.2[a]	diatoms
19'-hexanoyloxyfucoxanthin	1.1[a]	prymnesiophytes
Peridinin	1.5[a]	dinoflagellates
zeaxanthin+lutein	1.7[b]	cyanophytes
19'-butanoyloxyfucoxanthin	1.6[c]	silicoflagellates
Alloxanthin	1.85[d]	cryptophytes
chlorophyll *b*	0.9[a]	green algae (chlorophytes, euglenophytes)

[a] Terzić, 1996), [b] Stransky and Hager, 1970), [c] Everitt *et al.*, 1990), [d] Hager and Stransky, 1970).

Results

October 2005

Changes in nutrients concentration at fish farm area (sampling sites at centre of fish cage, at 6 m and 20 m distance from the centre) we observed before and after feeding. The concentrations of PO_4^{3-}, NH_4^+ and P_{tot} increase (Table 2).

Comparing the pigment fingerprints at fish farm area (sampling sites at centre of fish cage, at 6 m and 20 m distance from the centre) before and after feeding we observed only minor differences in pigment concentrations and phytoplankton groups contribution to the total biomass (Figure 3). The main phytoplankton group were diatoms (67.7 - 77.9 %) followed by Primnesiophytes (13.9 – 19.8 %). The phytoplankton biomass in the centre of fish cage expressed in chlorophyll *a* concentration was 1.06 $\mu g\ l^{-1}$ before feeding and 1.57 $\mu g\ l^{-1}$ 3 hours after the feeding. In the area 6 m from the centre the concentration was 1.03±0.09 $\mu g\ l^{-1}$, and 20 m away 1.24±0.12 $\mu g\ l^{-1}$. The main difference on the profile from the centre of the fish cage outward was a decrease in chlorophyll *a* degradation products concentration.

Table 2. The concentrations of nutrients before and after the feeding in October 2005 sampling at fish farm area (sampling sites at centre of fish cage, at 6 m and 20 m distance from the centre)

	BEFORE THE FEEDING (0)	AFTER THE FEEDING (A)
PO_4^{3-} [$\mu mol\ l^{-1}$]	0.08±0.03	0.14±0.04
NH_4^+ [$\mu mol\ l^{-1}$]	0.58±0.08	0.72±0.15
P_{tot} [$\mu mol\ l^{-1}$]	0.29±0.03	0.35±0.01

The highest concentration of chlorophyll *a* degradation products was measured 6 m (00N1) from the centre of fish cage, followed by location 20 m from centre (00N2) and clear decline in direction further away (ZBMA, ZC0B) (Figure 4).

The main contribution to chlorophyll *a* degradation products was due to concentrations of chlorophyllide *a* and pheophorbide a_1. Similar trend was observed also for pheophorbide a_2 but the concentrations were very low.

April 2006

Since during this sampling the samples were taken only 3 hours after feeding, we cannot compare the concentrations of nutrients before and after feeding.

The main phytoplankton group in area of fish cages (CESY) during this sampling period were Primnesiophytes (32.4 %) followed by diatoms (24.9 %). The phytoplankton biomass expressed in chlorophyll *a* concentration was 0.76 μg l^{-1}.

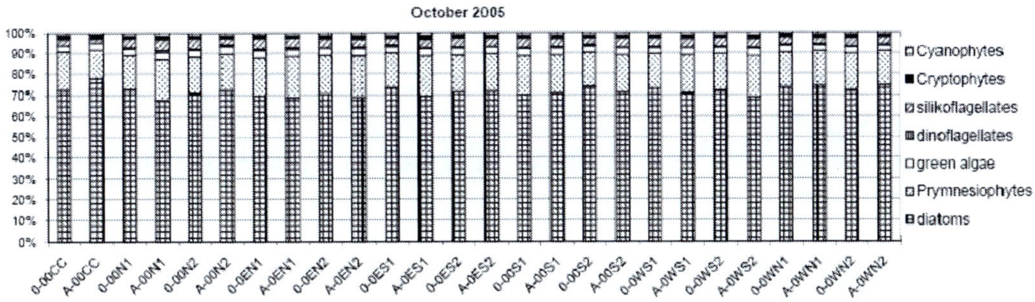

Figure 3. The phytoplankton groups' contribution to the total biomass at the fish farm area during the sampling in October 2005.

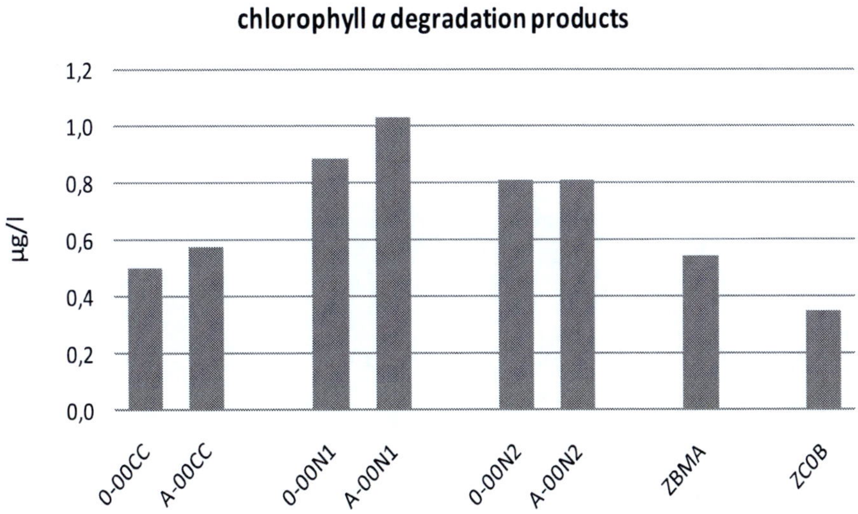

Figure 4. The chlorophyll *a* degradation products concentration distribution along transect from fish farm towards open waters during the sampling in October 2005.

The concentration of chlorophyll *a* degradation products around the fish farm cage was very low (from under detection limit and up to 31 ng l^{-1}), but again with highest values close to fish cage with exception of W200 site where the concentration was higher (70 ng l^{-1}).

July 2006

Likewise in October the main phytoplankton group around the fish cages were diatoms followed by Primnesiophytes (Tab. 3). Greater variability was observed further away from the fish cages (400 m from the fish cage).

The phytoplankton biomass in the centre of fish cage expressed in chlorophyll *a* concentration was 0.32 μg l^{-1} before feeding and 0.23 μg l^{-1} 3 hours after the feeding. In the area 200 m from the centre the concentration was 0.23-0.39 μg l^{-1} (0) and 0.19-0.25 μg l^{-1} (A), and 400 m away 0.22-0.36 μg l^{-1} (0) and 0.18-0.24 μg l^{-1} (A). During this sampling we did not noticed differences on the profile from the centre of the fish cage outward.

Because of the short time between two samplings we can not expect big differences before and after the feeding.

In contrast to situation in October 2005 the concentrations of nutrients in the centre of fish cage decrease (Tab. 4).

Comparisons of chlorophyll *a*, fucoxanthin and 19'-hexanoyloxyfucoxanthin concentrations for all sampling periods in the Zone A (mariculture area), Zone B (Bay of Piran) and Zone C (the control site outside the Bay of Piran, 6825 m from fish cages) were done using ANOVA and shows the difference among sampling periods is significant while among different zones is not (Tab. 5, Figure 5).

Table 3. The phytoplankton groups contribution (%) to the total biomass at the fish farm area during the sampling in July 2006; 0 – before the feeding, A – after the feeding

	Centre of the fish cage		200 m from the fish cage		400 m from the fish cage	
	0	A	0	A	0	A
	%		%		%	
Diatoms	56.0	60.5	61.3-65.7	56.2-59.1	49.5-67.7	47.8-65.0
Prymnesiophytes	22.2	22.3	20.3-22.5	22.2-26.3	17.9-31.1	23.2-27.1
Green algae	4.8	3.3	2.9-4.2	3.2-3.8	3.4-7.6	2.4-4.8
Dinoflagellates	1.0	under detection limit	up to 1.0	1.0-1.6	up to 1.2	up to 1.9
Silicoflagellates	4.9	4.8	3.1-4.4	2.9-4.2	2.9-5.0	2.0-3.9
Cryptophytes	2.2	1.4	0.9-2.7	2.1-4.5	0.8-1.5	0.5-5.3
Cyanophytes	8.8	7.6	4.3-7.2	6.2-7.9	5.0-8.3	5.9-9.2

Table 4. The concentrations of nutrients in the centre of fish cage (CESY) before and after the feeding during July 2006 sampling

	BEFORE THE FEEDING (0)	AFTER THE FEEDING (A)
PO_4^{3-} [μmol l^{-1}]	0.12	0.07
NH_4^+ [μmol l^{-1}]	2.42	0.74
P_{tot} [μmol l^{-1}]	0.40	0.27

Slight decrease of mean values is visible in each zone but it is covered by greater variability (Figure 5).

ENRICHMENT ENCLOSURE EXPERIMENT

The phytoplankton biomass expressed in chlorophyll *a* concentration in different enclosures shows differences on fifth day of the experiment (Figure 6). First day of the experiment the concentration of chlorophyll *a* was 0.32 μg l^{-1}. The highest concentration (0.77 μg l^{-1}) was measured fifth day in enclosure B (enriched with seawater from area 200 m far from fish cage), followed by enclosure C (0.42 μg l^{-1}) and A (0.18 μg l^{-1}) where chlorophyll *a* concentration decrease.

The concentration of chlorophyll *a* degradation products were under detection limit.

Initial composition of phytoplankton biomass was 37 % of prymnesiophytes, 33 % of cyanophytes, 17 % of diatoms, 5 % of silicoflagellates, 4 % of green algae and 3 % of cryptophytes. Among phytoplankton pigments fucoxanthin and 19'-hexanoyloxyfucoxanthin showed more or less the same trend as chlorophyll *a*. 19'-hexanoyloxyfucoxanthin i.e. prymnesiophytes contribute the major part to the phytoplankton biomass (21.1 - 56.9 %), followed by cyanophytes (6.2 – 58.9 %) and diatoms (14.5 – 24.8 %). Third day we observed prevalence of prymnesiophytes (47.8 %) in enclosure A, while in B and C the cyanophytes were the prevailing group (56.0 % and 58.9 %). Fifth day of the experiment prymnesiophytes were the main contributor to phytoplankton biomass in all enclosures (55.7 %, 48.9 % and 56.9 %).

Table 5. Comparisons of chlorophyll *a* (chl *a*), fucoxanthin (fuc) and 19'-hexanoyloxyfucoxanthin (hex) concentrations for all sampling periods in the Zone A (mariculture area), Zone B (Bay of Piran) and Zone C (the control site outside the Bay of Piran, 6825 m from fish cages) done using ANOVA

	Df	Pr(>F)
chl a		
month	2	0,000
zone	2	0,197
Residuals	63	
fuc		
month	2	0,000
zone	2	0,262
Residuals	63	
hex		
month	2	0,000
zone	2	0,085
Residuals	63	

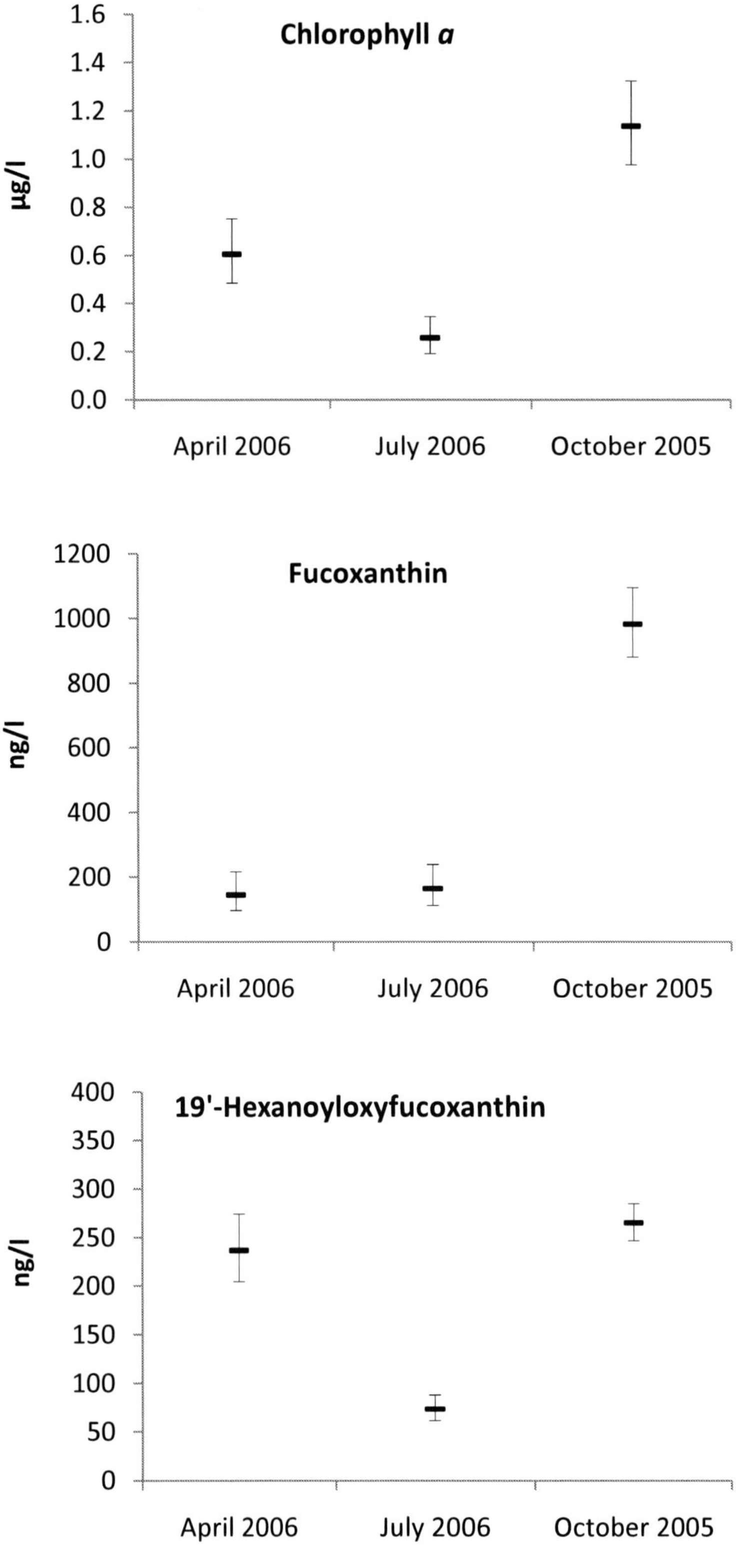
Chlorophyll a
µg/l
1.6
1.4
1.2
1.0
0.8
0.6
0.4
0.2
0.0
April 2006
July 2006
October 2005
Fucoxanthin
ng/l
1200
1000
800
600
400
200
0
April 2006
July 2006
October 2005
19'-Hexanoyloxyfucoxanthin
ng/l
400
350
300
250
200
150
100
50
0
April 2006
July 2006
October 2005

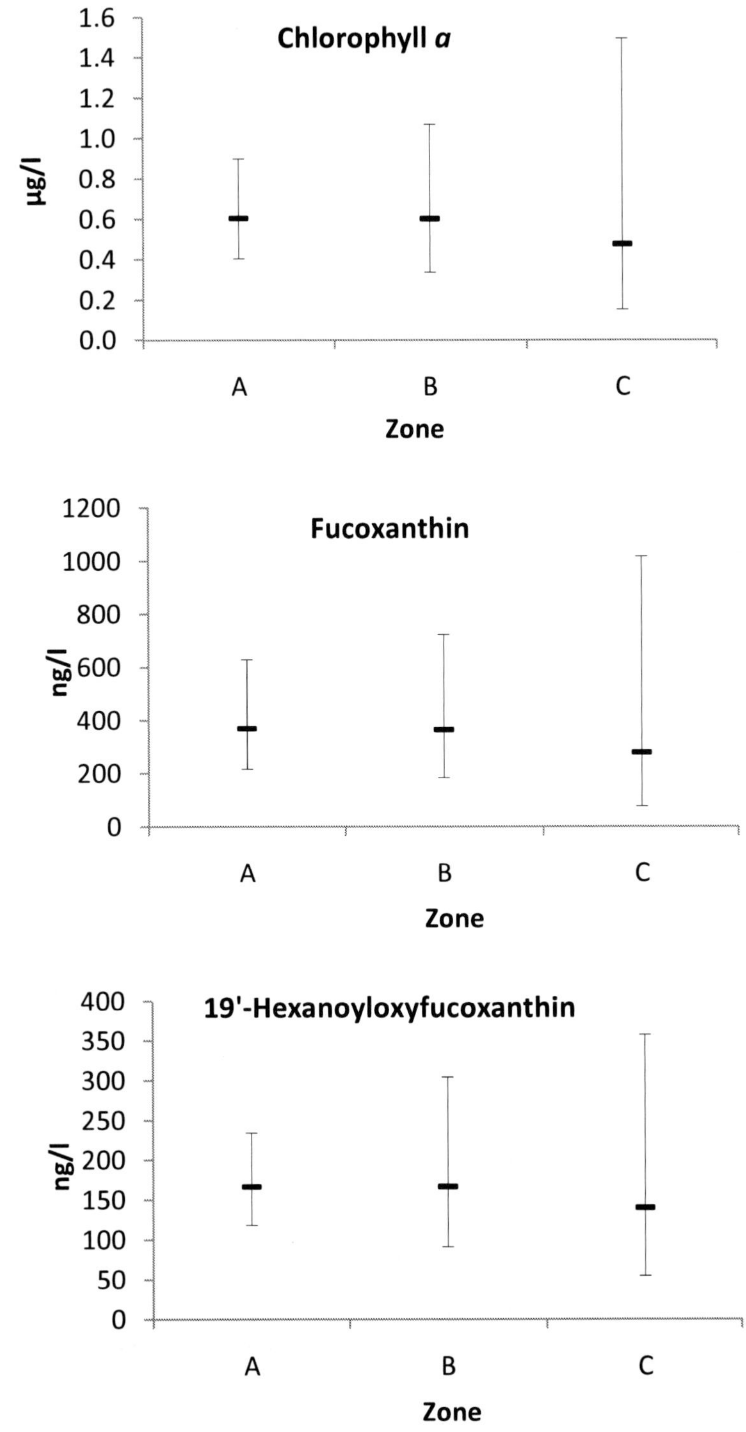

Figure 5. Chlorophyll *a*, fucoxanthin and 19'-hexanoyloxyfucoxanthin concentrations in different sampling periods, and in different zones.

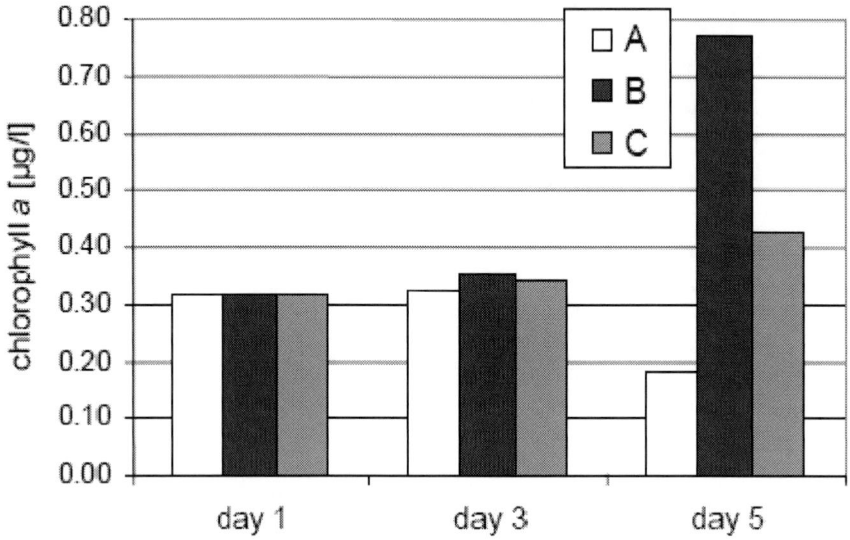

Figure 6. Chlorophyll *a* concentrations in enclosures A (control), B (amended with sea water 200 m from cage) and C (amended with sea water from fish cage) during the enriched enclosures experiment.

DISCUSSION

In temperate regions, such as the Gulf of Trieste, phytoplankton presents seasonal maxima in growth during spring and autumn as a result of nutrient availability due to mixing and other physical processes as well as to adequate light conditions; in contrast, surface nutrient depletion during the summer stratification period results in low phytoplankton growth despite the increased light availability (Pitta et al., 1999). The seasonal plankton dynamics in Gulf of Trieste appear to be strongly related to the flow regime of the main freshwater source, the Soča river (Malej et al., 1995; Mozetič et al., 1998). Peaks of autotrophic biomass were related to extended periods of high freshwater discharge. Annual peaks of phytoplankton biomass in the Gulf of Trieste were mainly due to increases in diatom abundance, particularly evident during the spring and autumn months (Flander, 1999; Mozetič et al., 1998; Flander, 1999). By contrast, the release of nutrients by fish-farms is a continuous process throughout the year, reaching maximal values during summer when high water temperature imposes the need for higher feeding rates. This reflects in phytoplankton population at fish farm area. Measured values of two indicators of the trophic state, Fp ratio (Claustre, 1994) and the trophic index TRIX (Vollenweider et al., 1998), were higher at fish farm area compared to the control site (Flander-Putrle and Malej, 2003) showing influence of the fish farm on the environment. This was more significant during the period of a homogeneous water column (Flander-Putrle and Malej, 2003). During the summer the concentrations of chlorophyll *a* measured at fish farm area were similar to other seasons in contrast to usual decrease of biomass during summer (Flander Putrle, 2003).

The impact of farm-originating nutrients was detected during summer enclosure enrichment experiments. Interestingly amendment with sea water from the centre of cage was stimulating phytoplankton less than sea water from 200 m distance.

Season was a significant factor of variability in the fish farm area. We detect higher differences in phytoplankton abundance (chlorophyll *a*) and other pigment concentrations among different seasons compared to differences among distances from the fish cages. Concentrations of chlorophyll *a*, fucoxanthin and 19'-hexanoyloxyfucoxanthin showed a conspicuous seasonal variation (Figure 5) in all zones. Same pattern with season as the major source of variation for phytoplankton was also observed in Aegean Sea by Pitta et al. (1999). The highest values of fucoxanthin concentration were measured during October 2005 sampling as well as diatom contribution to the total biomass. During October 2005 we also detect a peak of fucoxanthin at the reference site in the Gulf of Trieste (in front of Piran, where oceanographic buoy is situated; 45° 32,90' N, 13° 33,00' E) (Flander-Putrle, 2010). Pitta et al. (2005) in their study found out that most of the differences in geochemical and biological variables of the water column were identified during September when the surface water temperature is higher than during May, thus inducing considerably higher food supply to the fish cages and consequently higher losses of dissolved material to the environment. But the nutrients discharged by fish farming into the water column are not easily detectable at the distances of 2-3 nautical miles from fish farm zone due to rapid advection and diffusion as well as to the rapid uptake of nutrients by plankton organisms and further transfer to higher trophic levels (Pitta et al., 2005). The release of nutrients from fish farming in nutrient-poor systems can have a positive effect on local fisheries with no visible negative change in species composition or biodiversity (Machias et al., 2004). There is an increasing density of wild fish near the fish farm zones. The mechanism involved in this process could be the utilization of nutrients by plankton organisms, coupled with a rapid transfer of these nutrients up the food web without leaving behind significant traces of eutrophication (Machias et al., 2005).

Also in our study performed in the northern Adriatic as well as in the eastern Mediterranean (Pitta et al., 1999; Pitta et al., 2005) no significant response of phytoplankton community to fish farm influence was detected in terms of biomass, community structure or diversity. The differences in phytoplankton groups sampled were so sporadic that they cannot be easily explained (Pitta et al., 2005).

The influence of fish farming on phytoplankton was noticeable higher concentrations of chlorophyll *a* degradation products in the fish farm area. The chlorophyll *a* degradation products are indicators of physiological status of the phytoplankton community. They indicate higher level of decaying, dead and decomposing phytoplankton biomass, and increased grazing. Chlorophyllide *a* and pheophorbide *a* indicate the dying algae (Head et al., 1994). Pheopigments, above all pheophorbide *a*, are also indicators of zooplankton grazing and senescent algae (Jeffrey et al., 1997). Chlorophyllide *a* for example is a biomarker of senescent phytoplankton cells. The presence of chlorophyllide *a*, which was the most abundant among chlorophyll *a* degradation products in the fish farm area, implying negative impacts of fish farming. The significant higher pheopigment concentrations at the fish farm area (< 3 nautical miles from fish farm zone) was also measured in Aegean Sea (Pitta et al., 2005) indicating that there is an intensive grazing on phytoplankton biomass by heterotrophic organisms. They also measured significant higher chlorophyll *a* concentrations in this area.

Three hours are not enough to detect changes in phytoplankton community due to input of organic and inorganic matter after the feeding. And on the other hand this is an open system with normal diurnal migrations of phytoplankton, and currents influence that are possible causes of changes in phytoplankton community composition. The flushing time of a water body can limit the accumulation of biomass within the water body (Gowen et al., 1983). If the exchange time of the water masses is shorter than the generation time of phytoplankton, then no increase of phytoplankton biomass or eutrophication will be observed due to increased input of nutrients from fish farms. The increase of biomass was detected in the mesocosm enrichment experiment with water from fish farm area with no influence of currents and dilution (Flander Putrle, 2003), and of course no grazing effect.

To notice the influence of fish farm to the environment we also perform an experiment in enclosures with enriched water from fish farm area. In contrast to natural conditions the concentrations of chlorophyll *a* degradation products in different enclosures were below detection limit. Due to this we presume the degradation products are result of grazing in natural conditions.

Pitta et al. (2005) also suggested that the effects of fish farming should be studied at wider distances, since the changes in all variables are within the range of oligotrophic conditions and that these were found far away from the farms and may represent a highly diluted pool. In contrast to water column the influence of fish farms is more evident on benthic community (Hall et al., 1990; Holmer, 1991; Angel et al., 1995; Hall et al., 1990; Holmer, 1991; Karakassis et al., 1998; Karakassis et al., 2000).

The fish farms affect the environment. The influence is more evident on benthic than pelagic community since the surrounding water exchange due to currents and hidrometeorological conditions.

Acknowledgments

This study was carried out as a part of the ECASA project (ECosystem Approach for Sustainable Aquaculture) of the European 6th framework program. I would like to thank to the team of field sampling, to Janez Forte for the schemes and to Milijan Šiško for statistical analysis. In particular, I thank to A. Malej for her valuable comments and suggestions.

References

Angel, D.L., P. Krost and H. Gordin (1995): Benthic implications of net cage culture in the oligotrophic Gilf of Aquaba. In: H. Rosenthal, B. Moav and H. Gordin (Eds.). Improving the Knowledge Base in Modern Aquaculture. *Eur. Aquaculture Soc. Spec. Publ.:* 129-173.

Arzul, G., A. Clement and A. Pinier (1996): Effects on phytoplankton growth of dissolved substances produced by fish farming. *Aquat. Living Resour*. 9: 95-102.

Arzul, G., M. Seguel and A. Clement (2001): Effect of marine animal excretions on differential growth of phytoplankton species. *ICES Journal of Marine Science*, 58: 386-390.

Barlow, R.G., R.F.C. Mantoura, M.A. Gough and T.W. Fileman (1993): Pigment signatures of the phytoplankton composition in the northeastern Atlantic during the 1990 spring bloom. *Deep-Sea Res.,* 40(1/2): 459-477.

Bidigare, R.R., J. Marra, T.D. Dickey, R. Iturriaga, K.S. Baker, C.S. Raymond and H. Pak (1990): Evidence for phytoplankton succession and chromatic adaptation in the Sargasso Sea during spring 1985. *Mar. Ecol. Prog. Ser.,* 60: 113-122.

Carreto, J.I., N.G. Montoya, H.R. Benavides, R. Guerrero and M.O. Carignan (2003): Characterization of spring phytoplankton communities in the Rio de La Plata maritime front using pigment signatures and cell microscopy. *Mar. Biol.*, 143: 1013-1027.

Claustre, H. (1994): The trophic status of various oceanic provinces as revealed by phytoplankton pigment signatures. *Limnol. Oceanogr.*, 39(5): 1206-1210.

CSTT (1994): Comprehensive studies for the purposes of Article 6 of DIR 91/271 EEC, the Urban Waste Treatment Directive. Published for the Comprehensive Studies Task Team of Group Coordinating Sea Disposal Monitoring by the Forth River Purification Board, Edinburgh.

Everitt, D.A., S.W. Wright, J.K. Volkman, D.P. Thomas and E.J. Lindstrom (1990): Phytoplankton community compositions in the western equatorial Pacific determined from chlorophyll and carotenoid pigment distribution. *Deep-Sea Res.*, 37: 975-977.

Flander-Putrle, V. and A. Malej (2003): The trophic state of coastal waters under the influence of anthropogenic sources of nutrients (fish farm, sewage outfalls). *Period. Biol.*, 105(4): 359-365.

Flander Putrle, V. (2003): Phytoplankton biomarker pigments as an indicator of the trophic state of the coastal sea. Doctoral dissertation, University of Ljubljana, 158 pp.

Flander-Putrle, V. (2010): Examples of High Performance Liquid Chromatography (HPLC) application in marine ecology studies in the northern Adriatic. *Natura Sloveniae*, 12(1): 5-23.

Flander, V. (1999): Biološke sestavine sluzi v Severnem Jadranu. Magistrsko delo, Univerza v Ljubljani, 96 pp.

Gieskes, W.W. and G.W. Kraay (1986): Analysis of phytoplankton pigments by HPLC before, during and after the mass occurrence of the microflagellate *Corymbellus aureus* during the spring bloom in the open northern North Sea in 1983. Mar. Biol., 92: 45-52.

Gowen, R.J., P. Tett and B.J.B. Wood (1983): Changes in the major dihydroporphyrin plankton pigments during the spring bloom of phytoplankton in two Scottish sea-lochs. *J. mar. biol. Ass.* U.K., 63: 27-36.

Hager, A. and H. Stransky (1970): The carotenoid pattern and the occurrence of the light induced xanthophyll cycle in various classes of algae. V. A few members of Cryptophyceae, Euglenophyceae, Bacillariophyceae, Chrysophyceae and Phaeophyceae. *Archiv für Microbiologie*, 73: 77-89.

Hall, P.O.J., L.G. Anderson, O. Holby, S. Kollberg and M.O. Samuelsson (1990): Chemical fluxes and mass balances in a marine fish cage farm. *I. Carbon. Mar. Ecol. Prog. Ser.,* 61: 61-73.

Head E.J., B.T. Hargrave and D.V. Subba Rao (1994): Accumulation of a phaeophorbide a-like pigment in sediment traps during late stages of a spring bloom: A product of dying algae? *Limnol. Oceanogr.* 39: 176-181.

Holmer, M. (1991): Impacts of aquaculture on surrounding sediments: generation of organic-rich sediments. In: N. De Pauw and J. Joyce (Eds.). Aquaculture and the environment. *Eur. Aquacult. Soc. Spec. Publ.*: 155-175.

Jeffrey, S.W., R.F.C. Mantoura and S.W. Wright, Eds.) (1997): Phytoplankton pigments in oceanography: guidelines to modern methods. Paris, UNESCO Publishing, 661 pp.

Karakassis, I., M. Tsapakis and E. Hatziyanni (1998): Seasonal variability in sediment profiles beneath fish farm cages in the Mediterranean. *Mar. Ecol. Prog. Ser.*, 162: 243-252.

Karakassis, I., M. Tsapakis, E. Hatziyanni, K.-N. Papadopoulou and W. Plaiti (2000): Impact of cage farming of fish on the seabed in three Mediterranean coastal areas. *ICES Journal of Marine Science*, 57: 1462-1471.

Latasa, M., M. Estrada and M. Delgado (1992): Plankton-pigment relationships in the Northwestern Mediterranean during stratification. *Mar. Ecol. Prog. Ser.*, 88: 61-73.

Machias, A., I. Karakassis, M. Labropoulou, S. Somarakis, K.N. Papadopoulou and C. Papaconstantinou (2004): Changes in wild fish assemblages after the establishment of a fish farming zone in an oligotrophic marine system. *Estuarine and Costal Marine Science*, 60(4): 771-779.

Machias, A., I. Karakassis, M. Giannoulaki, K.N. Papadopoulou, C.J. Smith and S. Somarakis (2005): Response of demersal fish communities to the presence of fish farms. *Mar. Ecol. Prog. Ser.*, 288: 241-250.

Malej, A., P. Mozetic, V. Malacic, S. Terzic and M. Ahel (1995): Phytoplankton response to freshwater inputs in a small semi-enclosed gulf (Gulf of Trieste, Adriatic Sea). *Mar. Ecol. Prog. Ser.*, 120: 111-121.

Mantoura, R.F.C. and C.A. Llewellyn (1983): The rapid determination of algal chlorophyll and carotenoid pigments and their breakdown products in natural waters by reverse-phase high-performance liquid chromatography. *Anal. Chim. Acta*, 151: 297-314.

Mozetič, P., S. Fonda Umani, B. Cataletto and A. Malej (1998): Seasonal and inter-annual plankton variability in the Gulf of Trieste (northern Adriatic). *ICES Journal of Marine Science*, 55: 711-722.

Pitta, P., I. Karakassis, M. Tsapakis and S. Zivanovic (1999): Natural vs. mariculture induced variability in nutrients and plankton in the eastern Mediterranean. *Hydrobiologia,* 391: 181-194.

Pitta, P., E.T. Apostolaki, M. Giannoulaki and I. Karakassis (2005): Mesoscale changes in the water column in response to fish farming zones in three coastal areas in the Eastern Mediterranean Sea. Estuarine, *Coastal and Shelf Science*, 65: 501-512.

Rodriguez, F., M. Varela and M. Zapata (2002): Phytoplankton assemblages in the Gerlache and Bransfield Straits (Antarctic Peninsula) determined by light microscopy and CHEMTAX analysis of HPLC pigment data. *Deep-Sea Research II*, 49: 723-747.

Stransky, H. and A. Hager (1970): The carotenoid pattern and the occurrence of the light induced xanthophyll cycle in various classes of algae. IV. Cyanophyceae and Rhodophyceae. *Archiv für Microbiologie,* 72: 84-96.

Švagelj, B., P. Mozetič and S. Terzić (1996): Growth and ecological role of the selected classes of marine phytoplankton. *Annales*, 9(157-166.

Terzić, S. (1996): Biogeokemija autohtone organske tvari u neritičkim područjima Sredozemlja: fotosintetski pigmenti i ugljikohidrati. Doktorska disertacija, Prirodoslovno-matematički fakultet, 177 pp.

Vollenweider, R.A., F. Giovanardi, G. Montanari and A. Rinaldi (1998): Characterization of the trophic conditions of marine coastal waters with special reference to the NW Adriatic Sea: Proposal for a trophic scale, turbidity and generalized water quality index. *Environments*, 9(3): 329-357.

Wright, S.W., D.P. Thomas, H.J. Marchant, H.W. Higgins, M.D. Mackey and D.J. Mackey (1996): Analysis of phytoplankton of the Australian sector of the Southern Ocean: comparisons of microscopy and size frequency data with interpretations of pigment HPLC data using the 'CHEMTAX' matrix factorisation program. *Mar. Ecol. Prog. Ser.*, 144: 285-298.

http://www.ecasa.org.uk/index.htm.

In: Zooplankton and Phytoplankton
Editor: Giri Kattel, pp. 159-179

ISBN 978-1-61324-508-8

Chapter 8

DISPERSAL, CONNECTIVITY, AND LOCAL CONDITIONS DETERMINE ZOOPLANKTON COMMUNITY COMPOSITION AND DYNAMICS IN FOUR MEDITERRANEAN FRESHWATER RESERVOIRS

***Ikbel Sellami*[1,*], *Asma Hamza*[2], *Abderrahmen Bouain*[1], *Lotfi Aleya*[3], *and Habib Ayadi*[1]**

[1]Université de Sfax, Faculté des Sciences de Sfax, Département des Sciences de la Vie. Unité de recherche LR/UR/05ES05 Biodiversité et Ecosystèmes Aquatiques. Route soukra Km 3.5 – BP 1171 – CP 3000 Sfax Tunisie

[2]Institut National des Sciences et Technologie de la Mer, Centre de Sfax BP 1035 Sfax 3018, Tunisie

[3]Université de Franche-Comté, Laboratoire de Chrono-environnement, UMR UFC/CNRS 6249 - Place Leclerc, F-25030 Besançon cedex, France

ABSTRACT

The relative importance of local and regional processes in shaping natural communities within a metacommunity context has been a focus of intense debate in recent years. We examine the variation of the zooplankton communities structure attributable to local (water temperature, dissolved oxygen, pH, suspended matter, nutrients, chlorophyll a, phytoplankton and fish abundances) and regional factors (dispersal and connectivity) in four freshwater reservoirs with (Beni Mtir) and without (Lakhmes, Nabhana, and Sidi Sâad) artificial connections to others reservoirs in Mediterranean area. Fourteen zooplankton species (four copepods: *Copidodiaptomus numidicus*, *Acanthocyclops robustus*, *Acanthocyclops viridis*, *Cyclops strenuous*; four cladocerans: *Bosmina longirostris*, *Ceriodaphnia quadrangula*, *Diaphanosoma brachyurum*, *Daphnia longispina* and six rotifers: *Asplanchna* sp., *Filinia longiseta*,

* Corresponding author: sellamifss@yahoo.fr, Tel: 00 216 (97) 64 21 80, Fax: 00 216 (74) 27 44 37.

Hexarthra mira, Keratella quadrata, Keratella cochlearis, Brachionus urceolaris) were identified in the studied reservoirs. Seasonal changes in the density ($P < 0.01$) and biomass of zooplanktonic organisms ($P < 0.001$) revealed clear differences between the four reservoirs. The results indicated that the zooplankton communities were significantly related to local factors in Lakhmes, Nabhana, and Sidi Saâd reservoirs. However, in Beni Mtir reservoir, connectivity and dispersal with different local conditions determine the zooplankton composition and dynamics. This study demonstrated that artificial connections to other reservoirs alter the metacommunity structure of zooplankton communities.

Keywords: Mediterranean freshwater area, zooplankton communities, local factors, connectivity, dispersal

1. Introduction

The recent development of metacommunity ecology provides a framework for exploring local (environmental conditions, such as physico-chemical variables and biotic interactions) and regional processes (dispersal and connectivity) that affect the assembly and maintenance of communities (Leibold et al., 2004). Zooplankton community structure has been attributed to a number of factors such as the environmental (Leibold and Miller, 2004) and biotic parameters (Tokeshi, 1999; Isari et al., 2007; Yamamoto et al., 2007). Connectivity should also be an important regulator of plankton community. So, connectivity can promote high community diversity by providing new species from the regional pool, but can also limit local diversity by washing out species, setting up local communities for competitive exclusion, and otherwise providing a disturbance that can offset equilibrium conditions (Cottenie and DeMeester, 2005). Empirical surveys of connected and unconnected habitats (Quinn and Harrison, 1988; Harrison, 1999), as well as theoretical modeling (Hastings and Gavrilets, 1999; Amarasekare, 2000; Mouquet and Loreau, 2002) suggested that increased connectance among local communities can lead to lower overall species diversity.

In Tunisia country, some reservoirs are artificially connected through human constructions. Therefore, increases in connectivity can impart community- and even metacommunity-wide effects due to changes in species' dispersal rates; however, to date, no research has addressed these issues in ecosystems that have been artificially connected by human alteration. Thus, we hypothesized that anthropogenic influence on the connectivity of Beni Mtir ecosystems, which provide directly the water to Ghdir El Goulla reservoir, would affect metacommunity dynamics, since the connectivity of ecosystems is expected to influence dispersal rates and induce passive dispersal within metacommunities.

In the current study, we examined the influence local environmental variables and regional factors on zooplankton species richness and community structure in four Mediterranean reservoirs with (Beni Mtir) and without (Lakhmes, Nabhana, and Sidi Sâad) artificial connections to others reservoirs. We used multivariate analyses to evaluate the relative importance of these variables to zooplankton communities.

2. Materials and Methods

2.1. Study Area

Four seasonal sampling in four Mediterranean reservoirs were conducted in spring (May-June), summer (July-August), autumn (September-October) and winter (November-January) (2005-2006) in the deepest area. The Sampling stations of the studied reservoirs were described in Sellami et al. (2009). Morphometrics and hydrobiology of the four reservoirs is shown on Table 1. Beni Mtir is directly connected to Ghdir El Goulla reservoir. A pipeline (135 km) connects the Ghdir El Goulla reservoir to receive water from Beni Mtir reservoir. Lakhmes, Nabhana, and Sidi Sâad are unconnected reservoirs.

2.2. Sampling Methods

Water samples were taken with a limnological bottle (a 5-liter Van Dorn bottle) at every metre of depth (from surface, 5 and 10 m deep). Water temperature, dissolved oxygen concentration (mg l^{-1}) and pH were measured using a multiparameter probe (Multi 340i/SET). The concentration of the suspended matter (mg l^{-1}) was determined by measuring the dry weight of the residue after filtration through a Whatman GF/C membrane. Nutrient concentrations were analyzed by standard colorimetric techniques with an automatic BRAN and LUEBBE type 3 analyzer and were expressed as mg l^{-1}. Air temperature, evaporation and precipitation were obtained from the meteorology service of Tunisia.

Table 1. Morphometric and hydrologic characteristics of the studied reservoirs.

	Beni Mtir	*Lakhmes*	*Nabhana*	*Sidi Saâd*
Longitude	36°43'10''N,	36°N,	36°03'N,	35° 31'40''N,
Latitude	08°44'10''E	09°22'E	09°54'E	09° 59' 30''E
Location	Ain Draham	Siliana	Kairouan	Kairouan
Construction (year)	1953	1966	1966	1981
Area (ha)	310.6	102	532	1800
Catchment area (Km^2)	103	131	855	8650
Volume (10^6 m^3)	57.6	14	86	209
Annual mean air temperature (°C)	16.7	18.1	20.1	22.4
Annual mean evaporation (mm)	90.2	129.8	147.2	145.1
Annual mean precipitation (mm)	123.3	46.9	28.8	27.2
Uses	hydroelectricity drinking water	irrigation	irrigation drinking water protection against the violent floods	irrigation protection against the violent floods

For phytoplankton determination, water samples were sedimented in Utermöhl chambers (1958) after fixation with a Lugol solution, and identification and counting of the different populations were later performed with an Olympus inverted microscope.

Sub samples (0.5 l) for quantification of chlorophyll *a*, were filtered using Whatman GF/C filters (1.2 µm pore size filter and 25 mm- diameter) and extracted in 90% acetone (Lorenzen, 1967). The spectrophotometric absorption was measured at 750 and 663 nm. Fish density was obtained from the ministry of agriculture of Tunisia.

Samples for qualitative analysis of zooplankton were taken by pulling the plankton net, mesh size 55 µm, through water from bottom to surface. The samples for quantitative zooplankton analysis were taken by straining 20 liters integrated water through a 55 µm mesh Juday plankton net and preserved with 4% formalin and coloured with Bengal Pink. The zooplankton was enumerated and counted under a binocular microscope type leica in Dolffus chambers (Paterson, 1993). Identification of species was made according to Dussart (1969), Koste (1978), Stella (1982), Margaritora (1985), Pourriot and Francez (1986), Korovchinsky (1992). Zooplankton was presented in terms of abundance (ind l^{-1}) and biomass (µg l^{-1}).

Zooplankton biological diversity (H'), evenness (J) and Simpson index (D) were calculated according to Eq. (1) (2) (3) (Shannon and Weaver, 1949; Pielou, 1966; Steel et al., 1997):

$$H' = -\sum_{i=1}^{S} \frac{n_i}{N} \times \log_2 \frac{n_i}{N} \tag{1}$$

$$J = \frac{H'}{\log_2 S} \quad (2) \tag{2}$$

where ni is the average density of i species and N is the average density of the entire community, S is the total species.

$$1\text{-}D = 1 - \Sigma\,(Ni \times N^{1})^2 \tag{3}$$

Ni is the density of i species, N is the number of species.

A canonical correspondence analysis or CCA (Ter Braak, 1995) was used to ascertain the relationships between zooplankton, phytoplankton and environmental variables along the study period. An independent one-way ANOVA was used to compare physicochemical variables and zooplankton density and biomass between reservoirs and between months.

3. Results

3.1. Physico-Chemical Parameters

In Beni Mtir reservoir, the water temperature varied between 8.1 and 26.6 °C (mean ± s.d. = 19.2 ± 5.3°C), showing a seasonal pattern with the highest values in September at surface after a dry season, the lowest values in January during a cooler period along the water column (Figure 1a). The decrease is abrupt in the zone between 5 and 10 m particularly during summer and the early autumn, because the thermocline, which occurs at these depths, is developed at the end of the spring and prevents effectively the vertical mixing. In Lakhmes reservoir the water temperature ranged from 7.9 ºC in the winter season (in January) to 27.3 ºC in the summer season (in July), with an average of 21.5 ± 6.3°C (Figure 1b). The temperature was decreased in the layer of 5-10 m depth; the highest differences between surface and the depth of 10 m were 3 °C in November. The temperature profiles between Nabhana and Sidi Saâd reservoirs are very similar, the water temperature fluctuated from 9.7 to 29 °C (mean ± s.d. = 22.1 ± 6.0°C) and from 8.4 to 28.4 °C (mean ± s.d. = 22.3 ± 6.0°C) respectively (Figure 1c and d). In the two reservoirs a seasonal variation was observed with a highest values in summer season (in August at a depth of 5m) and lowest values in the winter season (in January at a surface and at the depth of 10m). There are not remarkable differences of temperature with depth in these two reservoirs.

The dissolved oxygen shows a similar trend in Beni Mtir (mean ± s.d. = 4.2 ± 2.1 mg l^{-1}) and Lakhmes (mean ± s.d. = 3.6 ± 1.7 mg l^{-1}) reservoirs (Figure 1e and f). The highest dissolved oxygen levels were recorded in May (7.3 mg l^{-1} - 5.3 mg l^{-1} respectively) and June (6.7 mg l^{-1} - 6.6 mg l^{-1} respectively). In Nabhana reservoir, dissolved oxygen varied from 3.0 mg l^{-1} (In January at a depth of 5m) to 8.2 mg l^{-1} (In June at surface) (mean ± s.d. = 4.6 ± 1.4 mg l^{-1}) and its vertical profiles showed that the water column was always well oxygenated in June, averaging 8 mg l^{-1} (Figure 1g). In Sidi Saâd reservoir, dissolved oxygen profile showed a decrease with depth, the highest dissolved oxygen observed was 7.2 mg l^{-1} (In May at the surface) while the lowest was 2.3 mg l^{-1} (In July at a depth of 10m) (mean ± s.d. = 4.6 ± 0.8 mg l^{-1}) (Figure 1h). The oxycline becomes established in spring at 5m, with a maximum vertical difference of 2.1 mg l^{-1} in early summer in July. In late summer, autumn and winter season ddissolved oxygen concentrations were distributed with small differences between the surface and bottom.

In Beni Mtir, Lakhmes and Sidi Saâd reservoirs, pH values were always on the alkaline side and varied from 8.1 to 10 (mean ± s.d. = 8.7 ± 0.6), 7.6 to 8.4 (mean ± s.d. = 8.1 ± 0.2) and 8.0 to 8.6 (mean ± s.d. = 8.3 ± 0.1) respectively (Figure 2a,b and c). In Nabhana reservoir, pH was around neutral and varied between 6.4 and 8.5 (mean ± s.d. = 8.1 ± 0.7) (Figure 2d). pH values were characterized by the lowest variability.

Concentrations in suspended matter ranged from 23.5 mg l^{-1} in January at the surface to 72.8 mg l^{-1} in November at a depth of 5m (mean ± s.d. = 47.0 ± 15.0 mg l^{-1}) in Beni Mtir (Figure 2e). The vertical distribution in suspended matter concentrations is the same in Lakhmes (mean ± s.d. = 32.5 ± 23.6 mg l^{-1}) (Figure 2f) and Nabhana (mean ± s.d. = 29.3 ± 23.8 mg l^{-1}) (Figure 2g). In January suspended matter concentrations were higher throughout the water column, a maximum was observed at a depth of 5m (97.0 and 104.3 mg l^{-1} respectively). In Sidi Saâd reservoir suspended matter concentrations varied between 14.0 mg l^{-1} in July at a depth of 10m to 45.4 mg l^{-1} in July at a depth of 5m (mean ± s.d. = 27.0 ± 5.2 mg l^{-1}) (Figure 2h).

N/P: DIN (DIN = NO_2^- + NO_3^- + NH_4^+) to DIP (DIP = PO_4^{3-}) ratio varied from 1.62 to 10.54 (mean ± s.d. = 4.36 ± 2.27) in Beni Mtir (Figure 3a), from 1.63 to 138.09 (mean ± s.d. = 17.31 ± 26.44) in Lakhmes (Figure 3b), from 2.45 to 13.78 (mean ± s.d. = 7.50 ± 3.66) in Nabhana (Figure 3c), from 0.29 to 148.05 (mean ± s.d. = 18.64 ± 33.09) in Sidi Saâd (Figure 3d).

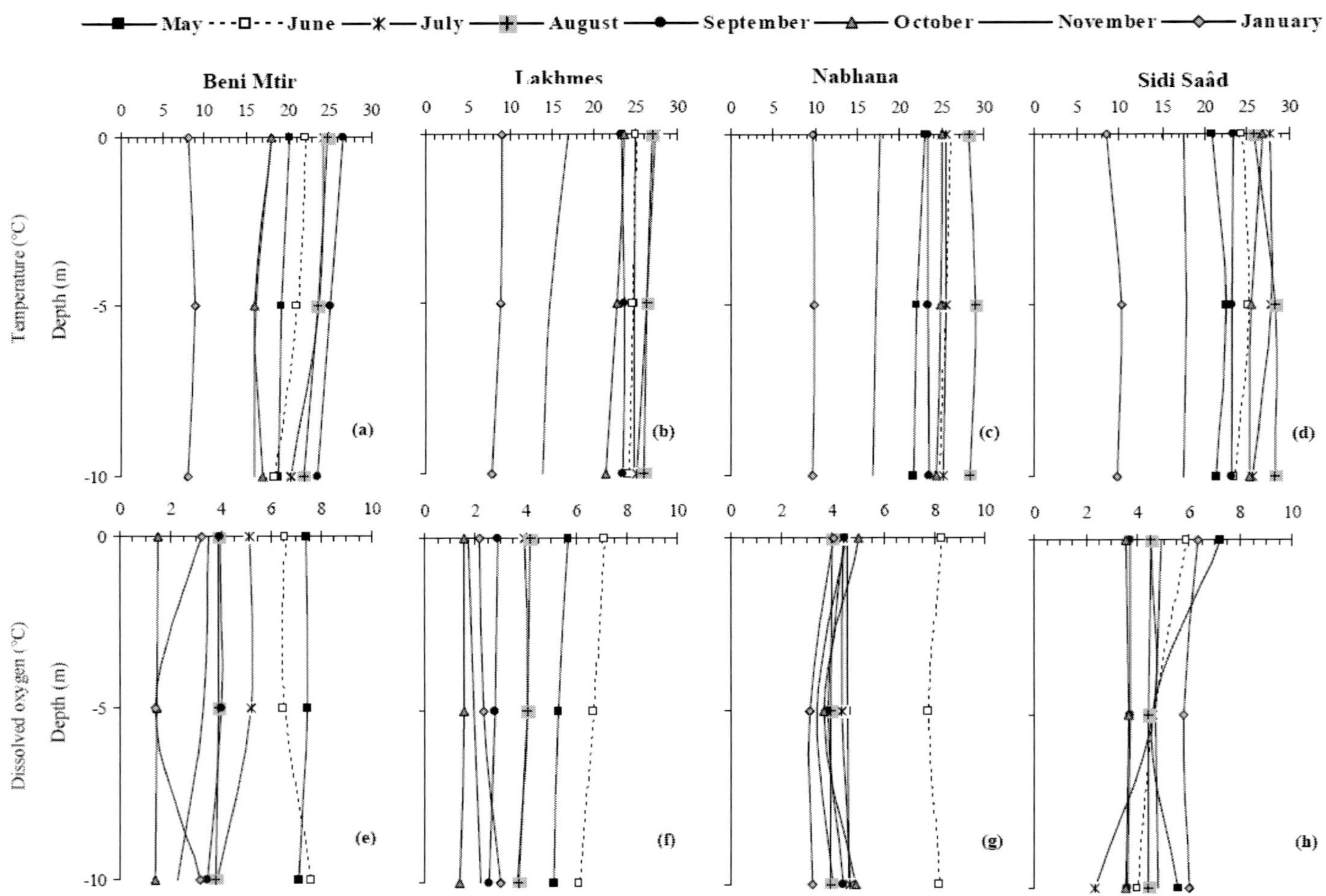

Figure 1. Spatial and temporal variation of the water temperature (a, b, c and d) and dissolved oxygen (e, f, g and h) along the water column in Beni Mtir, Lakhmes, Nabhana and Sidi Saâd reservoirs.

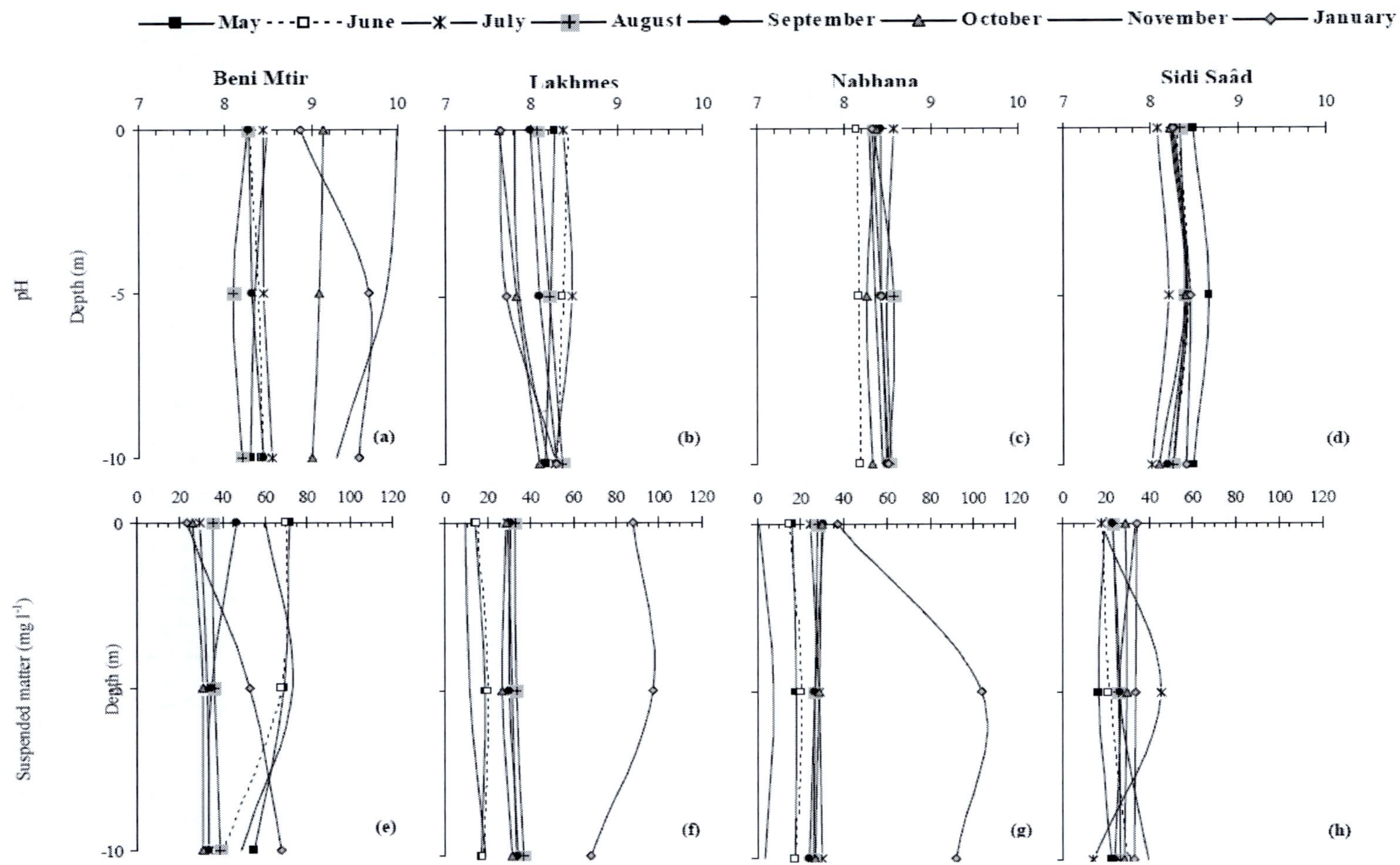

Figure 2. Spatial and temporal variation of the pH (a, b, c and d) and suspended matter (e, f, g and h) along the water column in Beni Mtir, Lakhmes, Nabhana and Sidi Saâd reservoirs.

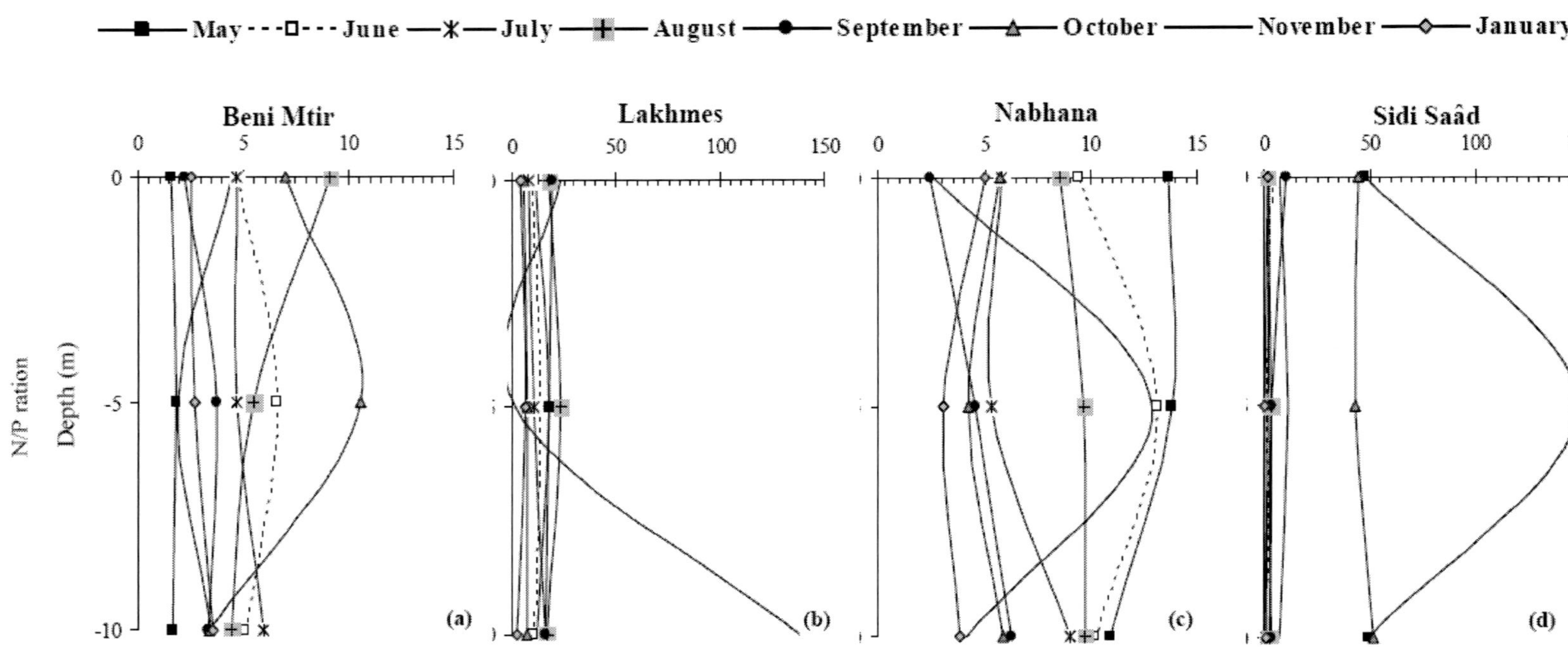

Figure 3. Spatial and temporal variation of the N/P ration (a, b, c and d) along the water column in Beni Mtir, Lakhmes, Nabhana and Sidi Saâd reservoirs.

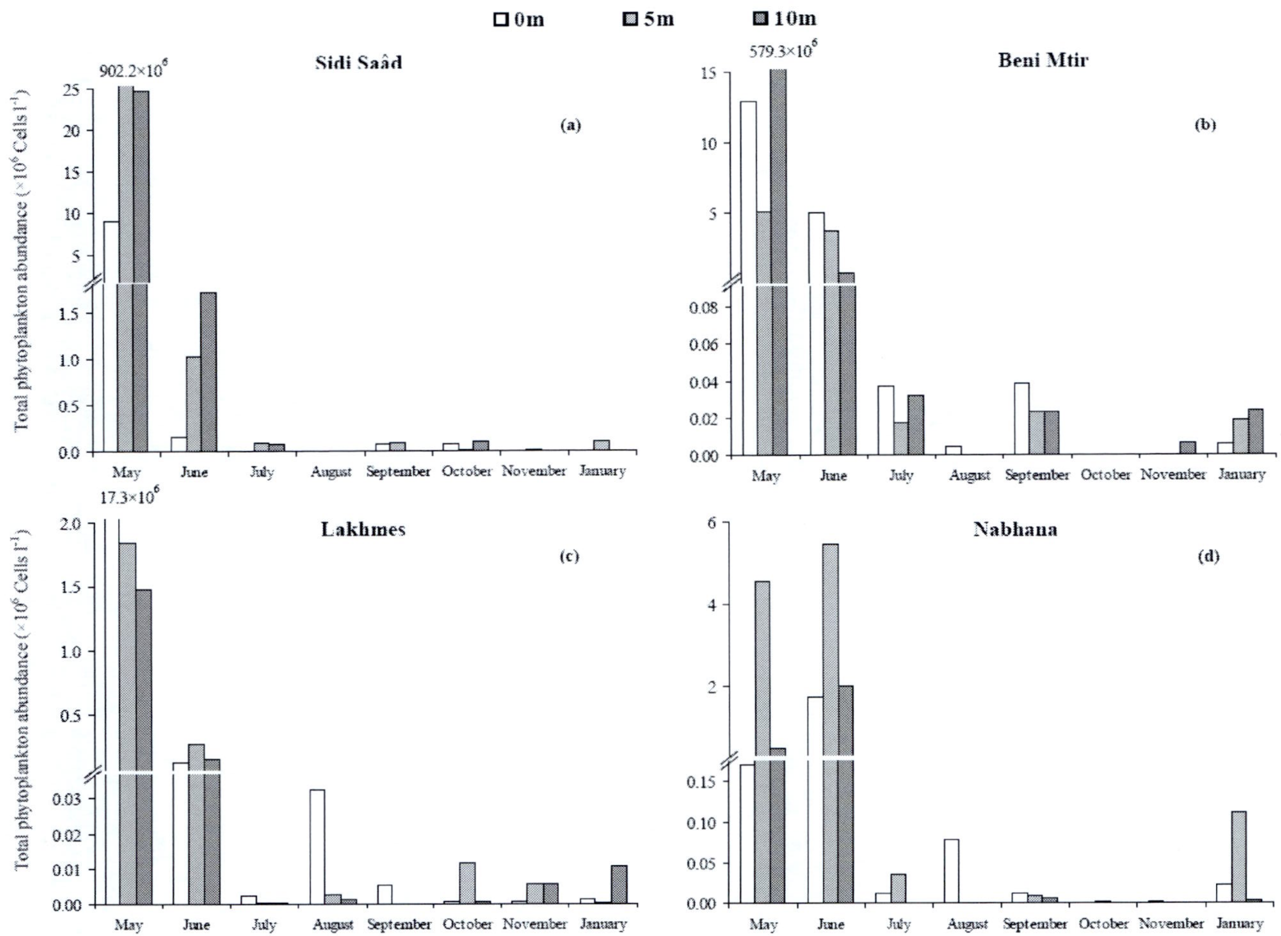

Figure 4. Spatial and temporal variation of total phytoplankton abundance along the water column in Sidi Saâd (a), Beni Mtir (b), Lakhmes (c) and Nabhana (d) reservoirs.

The averages were higher than the Redfield ratio (16) in Lakhmes and Sidi Saâd reservoirs, which suggests potential P limitation. Contrary, in Beni Mtir and Nabhana reservoirs the averages were less than (16), which suggests potential N limitation.

3.2. Phytoplankton and Zooplankton Communities Structure

In Sidi Saâd and Beni Mtir reservoirs, phytoplankton density was higher ($3.9 \times 10^7 \pm 1.8 \times 10^8$ cells l^{-1}) and $2.5 \times 10^7 \pm 1.1 \times 10^8$ cells l^{-1}) respectively compare to Lakhmes and Nabhana ($0.8 \times 10^6 \pm 3.5 \times 10^6$ cells l^{-1}) and ($6.5 \times 10^5 \pm 1.5 \times 10^6$ cells l^{-1}) respectively (Figure 4a, b, c and d) . Large notable phytoplankton populations were found in May and June with a highest peak in Sidi Saâd (9.0×10^8 cells l^{-1} in May at a depth of 5m) and Beni Mtir reservoirs (5.7×10^8 cells l^{-1} in May at a depth of 10m) (Figure 4a, b). Cyanobacteriae were the most numerically abundant at all study reservoirs (99% Nabhana, 99% Beni Mtir, 98% Lakhmes and 78% of the total phytoplankton abundance in Sidi Saâd) (Figure 5a, b c and d).

During the sampling periods, the zooplankton community consisted of 4 copepods species (*Copidodiaptomus numidicus*, *Acanthocyclops robustus*, *Acanthocyclops viridis*, *Cyclops strenuus*), 4 cladocerans (*Bosmina longirostris*, *Ceriodaphnia quadrangula*, *Diaphanosoma brachyurum*, *Daphnia longispina*) and 6 rotifers (*Asplanchna* sp., *Filinia longiseta*, *Hexarthra mira*, *Keratella quadrata*, *Keratella cochlearis*, *Brachionus urceolaris*) in the studied reservoirs (Table 2).

Zooplankton was significantly higher in abundance and in biomass in Lakhmes reservoir (144.3 ± 106.9 ind l^{-1}; 848.5 ± 601.4 µg l^{-1}) than in Nabhana (75.8 ± 36.1 ind l^{-1}; 376.5 ± 194.5 µg l^{-1}), Beni Mtir (54.7 ± 35.5 ind l^{-1}; 439.5 ± 329.0 µg l^{-1}) and Sidi Saâd (34.4 ± 29.2 ind l^{-1}; 97.4 ± 83.3 µg l^{-1}) reservoirs. The maximum density of zooplankton was recorded in September (4.8×10^2 ind l^- at a depth of 5m), June (2.1×10^2 ind l^- at a depth of 5m), July (1.5×10^2 ind l^- at a depth of 5m) and August (1.3×10^2 ind l^- at a depth of 10m) in Lakhmes, Nabhana, Beni Mtir and Sidi Saâd reservoirs respectively (Figure 6a, b, c and d). Significant differences were found in the density ($P < 0.01$) and in the biomass ($P < 0.001$) of zooplankton between the four reservoirs. Copepods were the most abundant group averaging 82% in Nabhana, 67% in Beni Mtir, and 52% in Lakhmes 60% of the total zooplankton abundance in Sidi Saâd reservoirs (Figure 7a, b, c and d).

The species *Copidodiaptomus numidicus* was the most abundant in Lakhmes, Nabhana and Beni Mtir reservoirs ($p < 0.001$) (Table 2). The species *Acanthocyclops viridis* ($p < 0.001$) and *Cyclops strenuus* ($p < 0.01$) were found only in Sidi Saâd and Beni Mtir respectively. The species *Acanthocyclops robustus* was found in Lakhmes, Nabhana and Sidi Saâd reservoirs ($p < 0.01$, $p < 0.05$). *Bosmina longirostris* was present in Lakhmes, Nabhana reservoirs ($p < 0.001$). *Ceriodaphnia quadrangula* ($p < 0.05$, $p < 0.01$) and *Filinia longiseta* ($p < 0.01$) were absent in Sidi Saâd reservoirs. Species richness at the four reservoirs was low with a range of between 11 in Lakhmes and 6 in Sidi Saâd reservoir. Significant differences between the species were found ($p < 0.05$). In general, species richness, evenness and diversity increased as average abundance increased (Table 2).

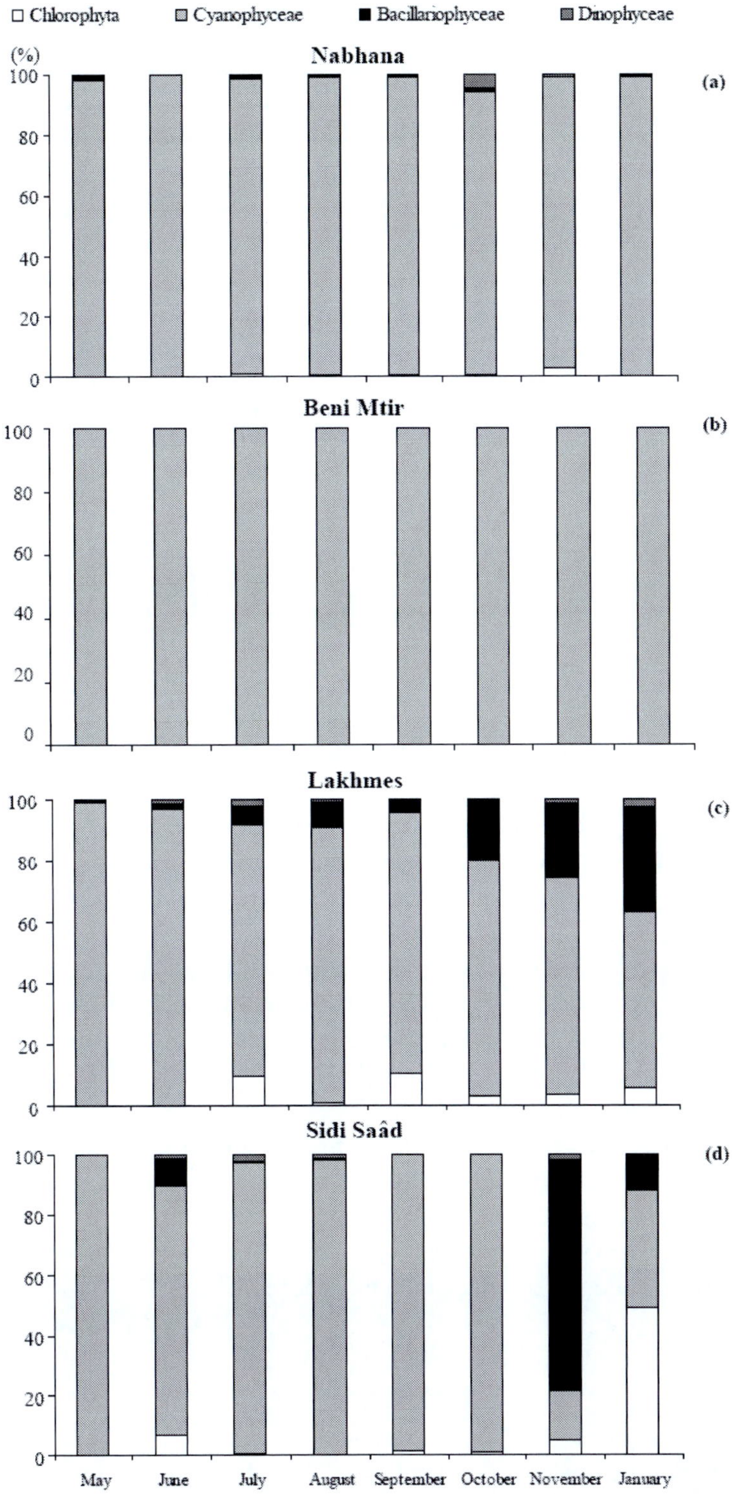

Figure 5. Spatial and temporal variation of the phytoplankton groups' abundance in Nabhana (a), Beni Mtir (b), Lakhmes (c) and Sidi Saâd (d) reservoirs.

Table 2. Species of zooplankton (%) found in the four reservoirs

	Lakhmes		Nabhana		Beni Mtir		Sidi Saâd		*P*	
	Density	Biomass	Density	Biomass	Density	Biomass	Density	Biomass	Density	Biomass
Copepoda										
Copidodiaptomus numidicus	62.3	83.1	66.0	84.0	66.5	58.5			***	***
Acanthocyclops robustus	0.8	0.6	0.4	0.4			5.1	12.7	**	*
Acanthocyclops viridis							55.0	64.4	***	***
Cyclops strenuus					0.6	0.5			**	**
Cladocera										
Bosmina longirostris	6.0	1.1	0.1	0.03					***	***
Ceriodaphnia quadrangula	9.4	3.7	5.7	2.2	16.3	16.0			*	**
Diaphanosoma brachyurum	8.8	1.1	22.3	5.5	5.3	1.7	37.9	22.3	*ns*	*ns*
Daphnia longispina	0.9	3.2			8.9	22.7			*ns*	*ns*
Rotifera										
Asplanchna sp.	4.4	4.8	4.7	7.9	0.03	0.03			*ns*	*ns*
Filinia longiseta	6.7	2.0	0.1	0.01	2.3	0.5			**	**
Hexarthra mira	0.01	<0.01	0.6	0.1	0.05	<0.01	0.4	0.1	*ns*	*ns*
Keratella quadrata	0.6	0.3	0.1	0.04	0.01	<0.01	0.1	0.1	*ns*	*ns*
Keratella cochlearis	0.1	0.0	0.1	0.01					*ns*	*ns*
Brachionus urceolaris							1.5	0.5	*ns*	*ns*
Species richness (S)	11		10		9		6			
Shannon and Weaver (H')	0.79-2.03	1.6 ± 0.4	0.4-2.16	1 ± 0.6	0.42-1.51	0.9 ± 0.5	0.9-1.83	1.3 ± 0.4	*	*
Evenness (J)	0.28-0.64	0.5 ± 0.1	0.25-0.77	0.4 ± 0.2	0.0-0.84	0.5 ± 0.3	0.47-0.92	0.6 ± 0.2	*ns*	*ns*
Simpson index (D)	0.26-0.67	0.5 ± 0.1	0.12-0.72	0.4 ± 0.2	0.0-0.56	0.3 ± 0.2	0.34-0.69	0.3 ± 0.2	*ns*	*ns*

* $p < 0.05$; ** $p < 0.01$; *** $p < 0.001$; *ns*: not significant.

The distribution of the chlorophyll *a*, the phytoplankton and the zooplankton abundances is illustrated in Figure 8 a, b, c and d. The abundance of zooplankton were negatively correlated with phytoplankton abundance (r=-0.718, $p<0.05$) and the chlorophyll *a* concentration (r=-0.436, $p<0.05$) (Figure 9).

The zooplankton were related to physical and chemical factors in the four reservoirs (Table 3). Correlations between the zooplankton community and the physical and chemical factors differed between the 2 types of reservoirs (Table 3). All of the factors were more strongly correlated for unconnected reservoirs (Lakhmes, Nabhana and Sidi Saâd) than for the connected reservoirs (Beni Mtir).

The first two CCA axes explain as a whole 85.82% of the total variance (Figure9). The first axis explaining 52.78% of the total variance separates Lakhmes and Nabhana Sidi saâd reservoirs. The second axis explaining 33.04% of the total variance is associate Beni Mtir reservoir presenting heavy precipitation. In this later reservoir, precipitation could be important for species immigration and community succession.

4. Discussion

Our results indicated that the communities found among unconnected ponds were strongly related to environmental factors. Thus, patterns of zooplankton species occurrence and abundance were correlated with the ecosystems and local environmental conditions.

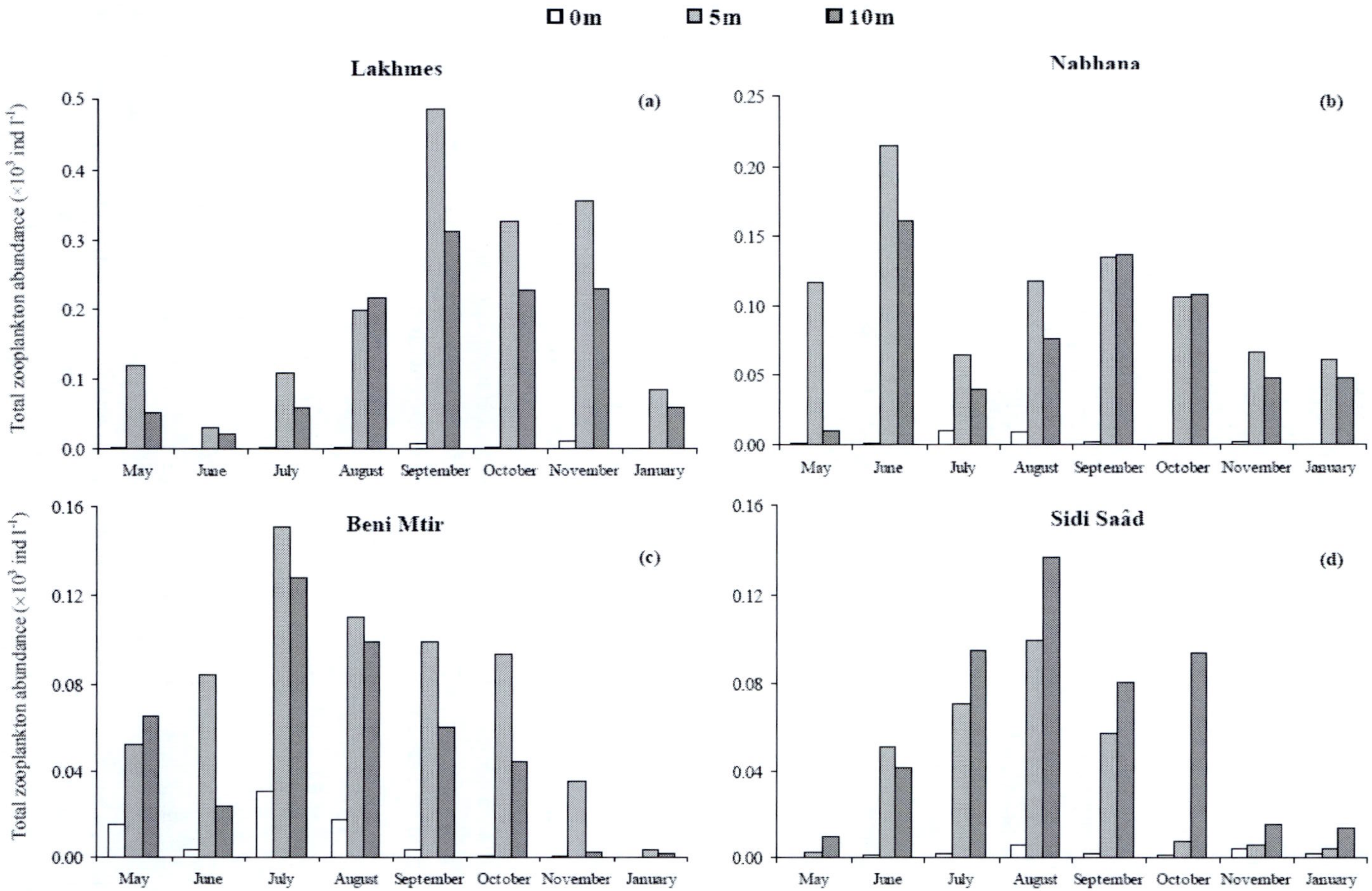

Figure 6. Spatial and temporal variation of the total zooplankton abundance along the water column in Lakhmes (a), Nabhana (b), Beni Mtir (c) and Sidi Saâd (d) reservoirs.

In contrast, the communities among connected reservoir (Beni Mtir) were weakly related to all of the factors. In this reservoir, the connectivity of ecosystems also strongly affected dispersal within the zooplankton metacommunity as well as local ecological conditions. Thus, from a species-sorting perspective, ecosystem connectivity strongly affected the metacommunity status as well as the influence of important environmental factors within the local community (Leibold et al., 2004). These findings are also supported by previous studies (Cottenie and De Meester, 2004; Thompson and Townsend, 2006; Doi et al., 2010).

Zooplankton in the four reservoirs follows a seasonal pattern, being more abundant in summer and autumn than in winter. The maximums densities of zooplankton were recorded in September, June, July and August in Lakhmes, Nabhana, Beni Mtir and Sidi Saâd reservoirs respectively. Positives correlations were found between zooplankton and temperature (Table 3). Contrary, the lowest densities were found in January in Nabhana and Beni Mtir reservoirs, in May in Sidi Saâd and in June in Lakhmes reservoir. In Beni Mtir reservoir the intense precipitation (in January) can wash out the entire populations.

Previous studies have revealed numerous important local conditions that affect zooplankton diversity: lake area (Allen et al., 1999; Dodson et al., 2000), lake morphology (Keller and Conlon, 1994), productivity (Dodson et al., 2000), nutrients (Jeppesen et al., 2000), water quality (Jeppesen et al., 2000; Cottenie et al., 2001), and interspecific competition (Shurin, 2000). Connectivity also can have an important effect on zooplankton community structure (Cottenie and De Meester, 2004; Vanschoenwinkel et al., 2007; Doi et al., 2010). Species richness was much higher at Lakhmes (11), Nabhana (10) and Beni Mtir (9). Sidi Saâd reservoir exhibited the lowest richness (6 species). The reservoirs Lakhmes, Nabhana and Sidi Saâd trophic status as mesotrophic. Contrary, Beni Mtir was an oligotrophic reservoir. The diversity of zooplanktons varies according to the trophic state of freshwater bodies (Jeppesen et al., 2002). An ecosystem generally becomes poor at high trophic levels (Kira, 1993) with few species and a low structural complexity. In Beni Mtir reservoir, connectivity may affect species diversity. The same results were found by previous study showed that species richness decreased as connection increased (Baranyi et al., 2002; Forbes and Chase, 2002). Species richness was not positively correlated with the surface area of the water bodies. These results were not corroborating with previous studies (Shurin et al., 2000; Tavernini et al., 2009). Species–area relationships are well documented in zooplankton (Dodson et al., 2000), and large aquatic systems are usually considered to have higher immigration and lower extinction rates than small biotopes (Angermeier and Schlosser, 1989).

In natural ecosystems, the biomass of zooplankton is determined by both phytoplankton (bottom-up effect) (Maher et al., 2000, 2009) and zooplanktivores (top-down controls) (Liu et al., 2007). The abundance of zooplankton were negatively correlated with phytoplankton abundance ($r=-0.718$, $p<0.05$) and the chlorophyll *a* concentration ($r=-0.436$, $p<0.05$). In all reservoirs the zooplankton were dominated by copepods and the phytoplankton were dominated by cyanobacteriae. Previous results showing that copepods were less affected by cyanobacteriae, maintaining high densities throughout the bloom (Ferrão-Filho et al., 2002; Zhao et al., 2008). A low zooplankton density and biomass in the mesotrophic Sidi Saâd reservoir resulted from high fishing pressure. Fish are well known to impact zooplankton community structure (Brooks and Dodson, 1965). However, the control of zooplankton biomass by fish feeding in oligotophic Beni Mtir reservoirs is negligible.

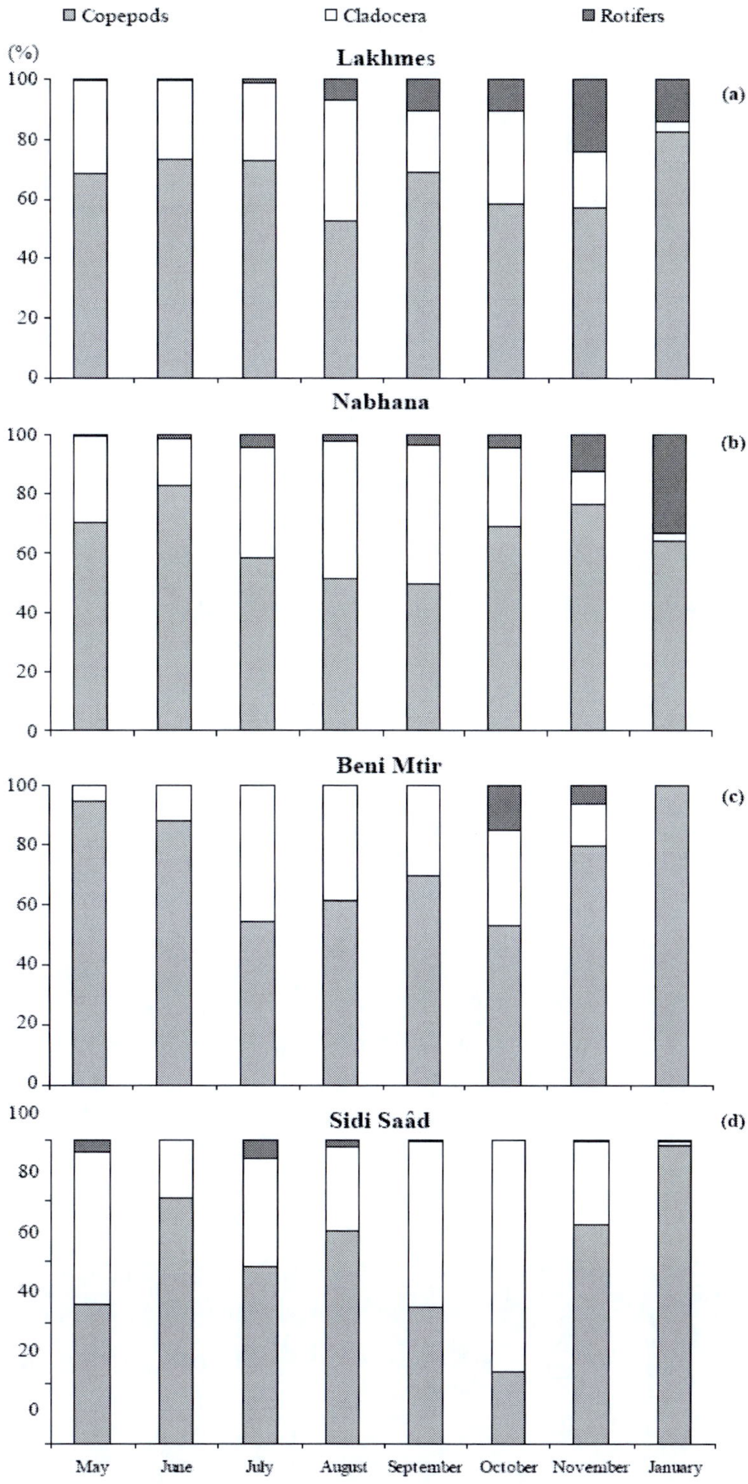

Figure 7. Spatial and temporal variation of the zooplankton groups' abundance in Lakhmes (a), Nabhana (b), Beni Mtir (c) and Sidi Saâd (d) reservoirs.

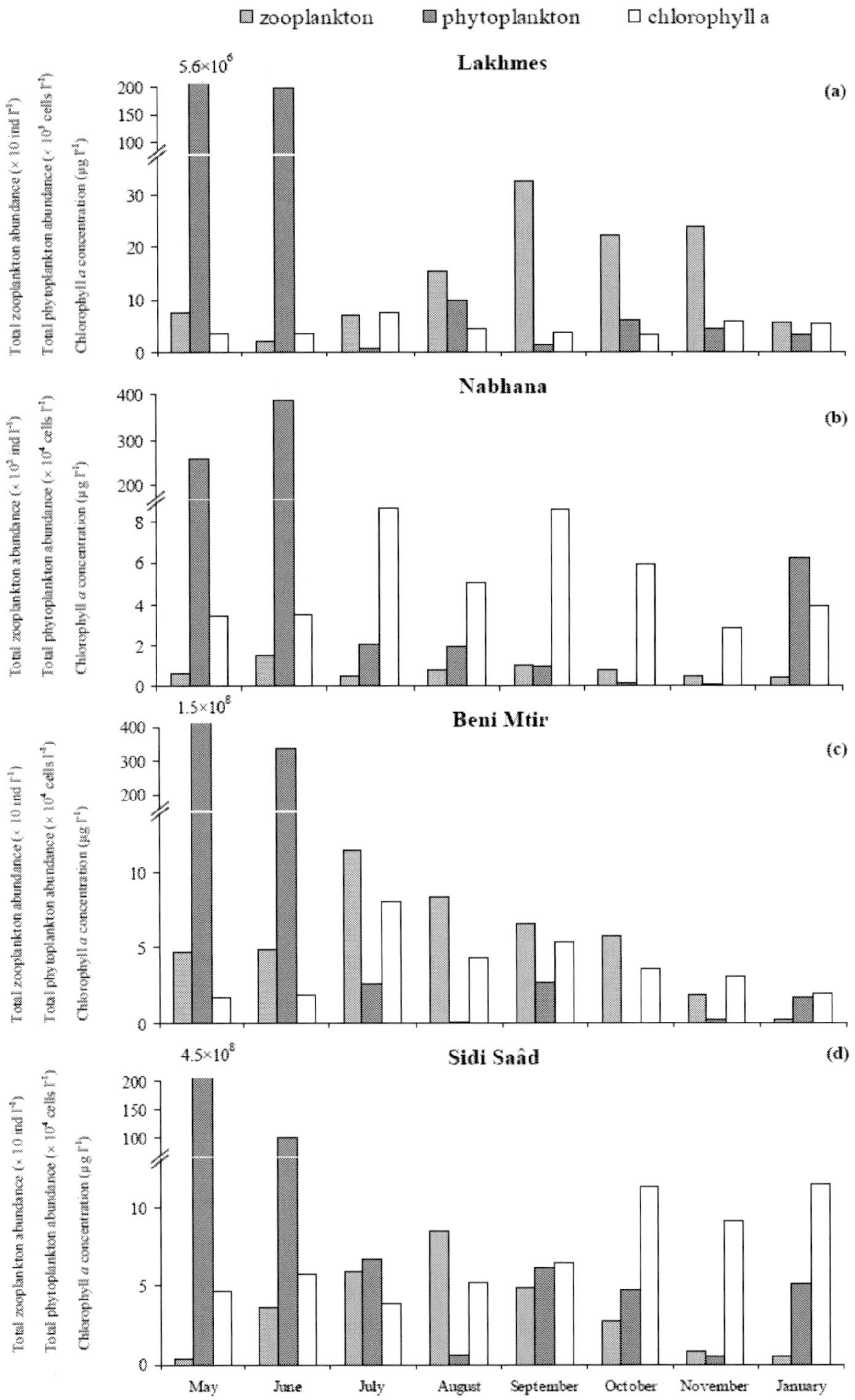

Figure 8. Spatial and temporal variation of the total phytoplankton, total zooplankton abundances and chlorophyll a concentrations in Lakhmes (a), Nabhana (b), Beni Mtir (c) and Sidi Saâd (d) reservoirs.

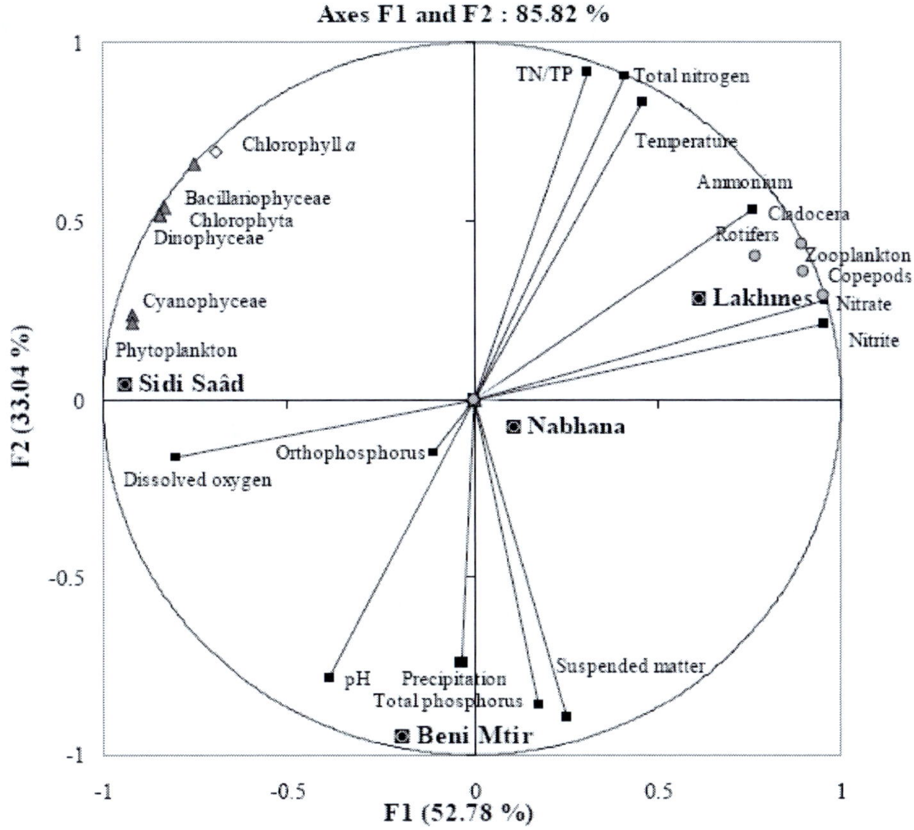

Figure 9. Canonical Correspondence Analysis (CCA) (Axis I and II) on mean values of environmental and biological variables in the four reservoirs (Lakhmes, Nabhana, Beni Mtir and Sidi Saâd).

Table 3. Pearson's correlation coefficients between zooplankton and physical-chemical factors

		Total Zooplankton	Copepods	Cladocera	Rotifers
Lakhmes	Temperature	**0.816***	0.568	0.309	-0.345
	Ammonium	**0.894****	**0.894****	0.687	**0.807***
	Orthophosphorus	0.372	0.291	0.129	**0.820***
Nabhana	Temperature	0.505	0.364	**0.707*****	**- 0.892*****
	Dissolved oxygen	**0.939****	**0.939*****	0.145	-0.416
	Suspended matter	-0.291	-0.302	-0.309	**0.827****
Beni Mtir	Temperature	**0.821***	**0.902***	0.526	-0.075
	pH	**-0.709***	**-0.877***	-0.474	0.311
	Suspended matter	-0.282	-0.362	**-0.759***	-0.427
Sidi Saâd	Temperature	**0.727***	0.617	**0.726***	0.499
	Dissolved oxygen	-0.341	-0.341	**-0.869****	-0.312
	pH	-0.548	-0.428	-0.581	**-0.721***

Seasonal dynamics of zooplankton could also explained by invertebrates' predation and/or resource competition (Isari, 2007). An inverse relationship between *Daphnia longispina* and rotifer abundance could be found in Beni Mtir and Lakhmes reservoirs. Yoshida et al. (2003) attributed bottom-up forces being more important than crustaceans in regulating the rotifer density to the lack of large- sized *Daphnia*. Previous study (Fussmann, 1996; Conde-Porcuna, 2000; Wang et al., 2009) suggested that exploitative competition was the main mechanism through which cladocerans constrain rotifer populations.

Conclusion

In our study, the diversity and distribution zooplankton were influenced by an interaction between environmental and biotic factors in unconnected reservoirs. In Beni Mtir reservoir, our results also indicated that the connectivity and dispersal with different local conditions determine the zooplankton composition and dynamics. To conserve local communities, attention must be paid to the connectivity of systems to accurately estimate metacommunity dynamics within natural ecosystems.

Acknowledgments

This study was supported by the University of Sfax (Tunisia) and the National Institute of Marine Sciences and Technologies (INSTM) (Tunisia). We acknowledge two anonymous reviewers for their valuable comments on the manuscript.

References

Allen, A. P., Whittier, T. R., Kaufmann, P. R., Larsen, D. P., O'Connor, R. J., Hughes, R. M., Stemberger, R. S., Dixit, S. S., Brinkhurst, R. O., Herlihy, A. T. and Paulsen, S. G. (1999). Concordance of taxonomic richness patterns across multiple assemblages in lakes of the northeastern United States. *Canadian Journal of Fisheries and Aquatic Sciences*, *56*, 739-747.

Amarasekare, P. (2000). The geometry of coexistence. *Biological Journal of Linnean Society 71*, 1-31.

Angermeier, P. L. and Schlosser, I. J. (1989). Species-area relationship for stream fishes. *Ecology*, *70*, 1450-1462.

Baranyi, C., Hein, T., Holarek, C., Keckeis, S. and Schiemer, F. (2002). Zooplankton biomass and community structure in a Danube River floodplain system: effects of hydrology. *Freshwater Biology*, *47*, 473-82.

Brooks, J. L. and Dodson, S. I. (1965). Predation, body size, and composition of plankton. *Science*, *150*, 28-35.

Conde-Porcuna, J. M. (2000). Relative importance of competition with *Daphnia* (Cladocera) and nutrient limitation on Anuraeopsis (Rotifera) population dynamics in a laboratory study. *Freshwater Biology*, *44*, 423-430.

Cottenie, K. and De Meester, L. (2004). Metacommunity structure: synergy of biotic interactions as selective agents and dispersal as fuel. *Ecology*, *85*, 114-119.

Cottenie, K., Michels, E., Nuytten, N., De Meester, L. (2001). Zooplankton community structure and environmental conditions in a set of interconnected ponds. *Hydrobiologia*, *442*, 339-350.

Cottenie, K. and De Meester, L. (2005). Local interactions and local dispersal in a zooplankton metacommunity. pp. 189-211. In Holyoak, M., Leibold, M. A. and Holt R. D. (eds.) Metacommunities: Spatial Dynamics and Ecological Communities. The University of Chicago Press, Chicago, IL, USA.

Dodson, S. I., Arnott, S. E., and Cottingham, K. L. (2000). The relationship in lake communities between primary productivity and species richness. *Ecology*, *81*, 2662-2679.

Doi, H., Chang, K. H., Nakano, S. (2010). Dispersal, connectivity, and local conditions determine zooplankton community composition in artificially connected ponds. *Aquatic Biology*, *10*, 47-55.

Dussart, B. (1969). Les copépodes des eaux continentales d'Europe occidentale. Tome II : Cyclopoides et biologie quantitative, N. Boubée et Cie, Paris, 292 pp.

Ferrão-Filho, A. S., Domingos, P. and Azevedo, S. M. F. O. (2002). Influences of a *Microcystis aeruginosa* KOTZ NG bloom on zooplankton populations in Jacarepaguli Lagoon (Rio de Janeiro, Brazil). *Limnologica*, *32*, 295-308.

Forbes, A. E. and Chase, J. M. (2002). The role of habitat connectivity and landscape geometry in experimental zooplankton metacommunities. *Oikos*, *96*, 433-440,

Fussmann, G. (1996). The importance of crustacean zooplank- ton in structuring rotifer and phytoplankton communities: an enclosure study. *Journal Plankton Research*, *18*, 1897-1915.

Harrison, S. (1999). Native and alien species diversity at the local and regional scales in a grazed California grassland. *Oecologia*, *121*, 99-106.

Hastings, A. and Gavrilets, S. (1999). Global dispersal reduces local diversity. *Proc. R. Soc. Lond. B.*, *266*, 2067-2070.

Isari, S., Psarra, S., Pitta, P., Mara, P., Tomprou, M. O., Ramfos, A., Somarakis, S., Tselepides, A., Koutsikopoulos, C. and Fragopoulu, N. (2007). Differential patterns of mesozooplankters' distribution in relation to physical and biological variables of the northeastern Aegean Sea (eastern Mediterranean). *Marine Biology*, *151*, 1035-1050.

Jeppesen, E., Jensen, J. P., Søndergaard, M., Lauridsen, T. and Landkildehus, F. (2000). Trophic structure, species richness and biodiversity in Danish lakes: changes along a phosphorus gradient. *Freshwater Biology*, *45*, 201-218.

Jeppesen, E., Jensen, J. P. and Sondergaard, M. (2002). Response of phytoplankton, zooplankton and fish to re-oligotrophication: an 11-year study of 23 Danish lakes. *Aquatic Ecosystems Health and Management*, *5*, 31-43.

Keller, W. and Conlon, M. (1994). Crustacean zooplankton communities and lake morphometry in precambrian shield lakes. *Canadian Journal Fish Aquatic Science*, *51*, 2424-2434.

Kira, T. (1993). Major environmental problems in world lakes. In: De Bernardi, R., Pagnotta, R., Pugnetti, A., (Eds), Strategies for Lake Ecosystems Beyond 2000. *Memorie dell'Istituto Italiano di Idrobiologia*, *52*, 1-7.

Korovchinsky, N. M. (1992). Sididae et holopediidae (Crustacea: Daphniiformes).Guides to the identification of the Microinvertebrates of the continental of the word 3. SPB Academic Publishing, 85pp.

Koste, W. (1978). Rotatoria. Die Rädertiere Mitteleuropas. Ein Bestimmungswerk begründet von Max Voigt. Borntrager, Stuttgart, (1), Textband, 673pp., (2), Tafelband, 234pp.

Leibold, M. A. and Miller, T. E. (2004). From metapopulations to metacommunities. In: Hanski, I., Gaggiotti, O. E. (eds) Ecology, genetics, and evolution of metapopulations. Elsevier Academic Press, San Diego, CA, pp. 133-150.

Leibold, M. A., Holyoak, M., Mouquet, N. , Amarasekare, P., Chase, J. M., Hoopes, M. F., Holt, R. D., Shurin, J. B., Law, R., Tilman, D., Loreau, M. and Gonzalez A. (2004). The metacommunity concept: a framework for multi-scale community ecology. *Ecology Letters*, *7*, 601-13.

Liu, Q. G., Chen, F. Y., Li, J. L. and Chen, L. Q. (2007). The food web structure and ecosystem properties of a filter-feeding Chinese carps dominated Chinese deep lake ecosystem. *Ecology Modelling, 203*, 279-289.

Mahar, M. A., Jafri S. I. H., Leghari, S. M. and Khuhawar, M. Y. (2009). Seasonal periodicity of phytoplankton and chlorophyll content in lake Manchar. *Pakistan*, *41*(*2*), 871-884.

Mahar, M. A., Jafri, S. I. H. and Leghari S. M. (2000). Studies on water chemistry and fish production of Manchar lake, Dadu, Sindh, Pakistan. *Pakistan Journal Biology Science*, *3*(*12*), 2151-2153.

Margaritora, F. G. (1985). Cladocera. Fauna d'Italia, Bologna, 399pp.

Mouquet, N. and Loreau, M. (2002). Coexistence in metacommunities: the regional similarity hypothesis. *American Naturalist*, *159*, 420-426.

Paterson, M. (1993). The distribution of micro-crustacea in the littoral zone of a freshwater lake. *Hydrobiologia*, *263*, 173-183.

Pielou, E. C. (1966). The measurement of diversity in different types of biological collections. *Journal of Theoretical Biology*, *13*, 131-144.

Pourriot, R. and Francez, A. J. (1986). Rotifères. Introduction pratique à la systématique des organismes des eaux continentales françaises. *Bulletin Mensuel Société Linnéenne Lyon*, 5-37.

Quinn, J. F. and Harrison, S. P. (1988). Effects of habitat fragmentation and isolation on species richness: evidence from biogeographic patterns. *Oecologia*, *75*, 132-140.

Sellami, I., Guermazi, W., Hamza, A., Alyea, L., Ayadi, H. (2010). Seasonal dynamics of zooplankton community in four Mediterranean reservoirs in humid area (Beni Mtir: North of Tunisia) and semi arid area (Lakhmes, Nabhana and Sidi Saâd: Center of Tunisia). *Journal of Thermal Biology*, *35*, 392-400.

Shannon, C. E., Weaver, G. (1949). The Mathematical Theory of Communication. university of illinois Press, urbana, Chicago, Il.

Shurin, J. B., Havel, J. E., Leibold, M. A., Pinel-Alloul, B. (2000). Local and regional zooplankton species richness: a scale-independent test for saturation. *Ecology*, *81*, 3062-3073.

Steel, R. G. D., Torrk, J. H., Dickey, D. A. (1997). Principks and procedures of statistics a biometrical appmach, Third edition, McGraw-Hill Compagnies, Inc., New-York, 666 pp.

Taverninia, S., Primiceriob, R., Rossettia, G. (2009). Zooplankton assembly in mountain lentic waters is primarily driven by local processes. *Acta oecologica*, *35*, 22-31.

Ter Braak, C. J. F. (1995). Ordination. In: Jongman, R. H. G., Ter Braak, C. J. F., Tongeran, O. F. R. (Eds.), Data Analysis in Community and Landscape Ecology. Cambridge University Press, Cambridge, pp. 91-631.

Thompson, R., Townsend, C. (2006). A truce with neutral theory: local deterministic factors, species trait and dispersal limitation together determine patterns of diversity in stream invertebrates. *Journal Animal Ecology*, *75*, 476-484.

Tokeshi, M. (1999) Species coexistence: ecological and evolutionary perspectives. Blackwell Publishing, Oxford.

Utermöhl, H. (1958). Zur Vervollkommung der quantitativen phytoplankton methodik. mitteilungen internationale Vereinigung für theoretische und angewandte. *Limnology*, *9*, 1-38.

Vanschoenwinkel, B., De Vries, C., Seaman, M. and Brendonck, L. (2007). The role of metacommunity processes in shaping invertebrate rock pool communities along a dispersal gradient. *Oikos*, *116*, 1255-1266.

Wang, S., Xie, P. and Geng, H. (2009). The relative importance of physicochemical factors and crustacean zooplankton as determinants of rotifer density and species distribution in lakes adjacent to the Yangtze River, China. Limnologica doi:10.1016/j. limno.2009.03.001.

Yamamoto, N., Yokoyama, J. and Kawata, M. (2007). Relative resource abundance explains butterfly biodiversity in island communities. *Proc Natl Acad Sci USA*, *104*, 10524-10529.

Yoshida, T., Urabe, J. and Elser, J. (2003). Assessment of ''top- down'' and ''bottom-up'' forces as determinants of rotifer distribution among lakes in Ontario. *Canadian Ecology Research*, *18*, 639-665.

Zhao, J., Ramin, M., Cheng, V. AND Arhonditsis, G. B. (2008). Plankton community patterns across a trophic gradient: The role of zooplankton functional groups. *Ecology Modelling*, *213*, 417-436.

In: Zooplankton and Phytoplankton
Editor: GiriKattel, pp. 181-192

ISBN 978-1-61324-508-8

Chapter 9

DAPHNIA PULICARIA HIJACKED BY *VIBRIO CHOLERAE*: ALTERED SWIMMING BEHAVIOUR AND PREDATION RISK IMPLICATIONS

***Ai Nihongi*[1], *Joshua J. Ziarek*[2], *Takeyoshi Nagai*[3], *Marco Uttieri*[4,*], *Enrico Zambianchi*[4], *and J. Rudi Strickler*[1]**

[1]Great Lakes WATER Institute, University of Wisconsin-Milwaukee, 600 East Greenfield Avenue, Milwaukee, Wisconsin, 53204 USA

[2]Department of Biochemistry, Medical College of Wisconsin, 8701 Watertown Plank Road, Milwaukee, Wisconsin 53226, USA

[3]Department of Ocean Sciences, Tokyo University of Marine Science and Technology, Minato-ku Konan 4-5-7, Tokyo 108-8477, Japan

[4]Department of Environmental Sciences, University of Naples "Parthenope", Centro Direzionale di Napoli – Isola C4, 80143 Napoli, Italy

ABSTRACT

Parasites hijack hosts and alter the latters' behaviour so as to enhance their dispersion and transmission to new hosts and new environments. *Daphnia*, as one of the major zooplankton members in freshwater environments, are targets of parasitic fungi, bacteria, and even waterborne human pathogens. Although the effects of infestation on *Daphnia* have been studied for over 120 years, scarce information is available regarding behavioural consequences. Conspicuous swimming behaviour of *Daphnia* can increase predation rates by visual predators such as fish. Therefore, if the swimming behaviour of *Daphnia*are modified by infestation, their predation risk could increase. Here we observed the effect of *Vibrio cholerae* infestation on the swimming performances of *Daphnia pulicaria* in dark and light conditions to simulate the risk of detection by visual predators. Our results show that *D. pulicaria*both with and without infestation by *V. cholerae* display similar swimming patterns in the dark. However, in light conditions, *D. pulicaria* infected by *V. cholerae* swim faster and travel more convoluted trajectories than those without infestation. Our study is the first to indicate that a microbial

[*]Corresponding Author: Marco Uttieri.

infestation can modify the swimming behaviour of *D. pulicaria* and that, consequently, infected *D. pulicaria* could experience more extensive predation by fish.

INTRODUCTION

The crustacean cladoceran*Daphnia* is a dominant zooplankton species in freshwater communities (Hebert, 1977), its biology and ecology being an important element of the functioning of the pelagic web. *Daphnia*'s behaviour reflects dynamical adaptations to numerous environmental factors (Sommer et al., 1986; Ziarek et al., 2011) and it also provides important details to understand the mechanisms regulating the interactions with other organisms.

Daphnia are not exempt from parasites; fungal infections of *Daphnia* were first described by Metschnikoff (1884). Since then, an increasing number of reports regarding the effects of parasites on the life history and population dynamics of *Daphnia*has accumulated. However, no information addressing the effects of parasites on the behaviour of *Daphnia* is available. Literature evidence on other organisms indicate that infected hosts behave differently from uninfected individuals and are subject to higher mortalityrates due to conspicuous motions (e.g., clams: Hulscher, 1973; amphipods: Bethel and Holmes, 1977; sand shrimp: Helluy and Thomas, 2003) or throughmanipulated suicide (e.g., pea aphids: McAllister and Roitberg, 1987; grasshopper: Thomas et al., 2002). Following this line of reasoning, we hypothesize that the behaviour of *Daphnia* might be modified by parasites similarly to other aquatic and terrestrial animals. To test our assumption, we investigated the swimming behaviour of *D. pulicaria*, one of the main zooplankton species living in the Allequash Lake (Wisconsin, USA), infected by the parasite *Vibrio cholerae*.

*V. cholerae*serogroup O1 is the causative agent of the endemic disease cholera. It is considered an autochthonous inhabitant of estuarine and brackish waters (Tamplin et al., 1990). *V. cholerae* strains can attach to numerous phyto- and zooplankton species and can be a threat to their biology and ecology (Tamplin et al., 1990: Epstein, 1993). In particular, *V. cholerae* adheres to the zooplankton chitin exoskeleton and eggs by producing a surface pilus called the mannose-sensitive hemagglutinin (MSHA) (Chiavelli et al., 2001). The bacillus can thus benefit from a source of carbon and nitrogen, while at the same time being more efficiently dispersed in the environment (Pruzzo et al., 2008).

To evaluate any pathogen-induced behavioural modification, we observed the swimming behaviour of *D. pulicaria*infected by *V. cholerae* in dark and light conditions mimicking the alternation of day and night luminous fields. We filmed a total of twenty*Daphnia* for 15 min continuously: tencontrols, not infected by *V. cholerae*, and ten infected animals were exposed to each condition. We measured the velocity fluctuations and the spatial complexity of each *D. pulicaria* swimming trajectory to quantify its swimming behaviour. The swimming of *Daphnia* species involves a cycle of hop and stop movements while they propel themselves through water. The kinetic energy provided by such swimming behaviour might have certain distributions in frequency space that could reflectthe characteristic behaviour or swimming response of *Daphnia* to the imposed stimulations. Spectral analysis was performed on the swimming velocity of *D. pulicaria*to investigate the variations of the velocity spectra caused by the infestation. Fractal dimension provided a synthetic description of the level of space-occupancy of the trajectories in the different tested conditions.

Our results are the first to indicate that *Daphnia* can be behaviourally manipulated through *V. cholerae* infestation. Infected specimens in the presence of light describe more convoluted tracks with a peculiar power spectrum signature, indicative of a behaviour differing from that displayed by uninfected individuals. Preliminary assessments of associated predation risk indicate that visual planktivorous fish would preferentially select *D. pulicaria* infected with *V. cholerae*, suggesting a possible mechanism for introduction into higher trophic organisms.

METHODS

Daphnia pulicaria–Sampling and Rearing

Daphnia pulicaria Forbes 1893 were sampled from Allequash Lake (46.041°N, 89.628°W) in Wisconsin (USA) in April 2005. A hole was cut in the frozen ice cover of the lake, and a plankton net (100 µm mesh size) was gently let down into it. Water temperature was 3°C. Adult *D. pulicaria* were then brought to the laboratory (University of Wisconsin-Milwaukee, Great Lakes WATER Institute, Milwaukee, Wisconsin, USA), were they went through an acclimation period of some days until the water temperature was raised to 22°C. They were then maintained in culture in 1 l beakers filled with artesian-well water, and were fed daily with a 5 ml mixture of *Ankistrodesmusfalcatus*, *Chlamydomonasreinhardtii* and *Selenastrumcapricornutum*.

Vibriocholerae

Vibrio cholerae was provided by Dr. Andrew Camilli (Tufts University). The serotype used for our study was El Tor O1 Inaba; it was provided in a 2 ml tube containing 0.85 ml of *V. cholerae* in Luria-Bertani (LB) medium and 0.15 ml of 100% glycerol at -80°C. Two days before observations, the frozen *V. cholerae* were inoculated into 2 ml LB medium with 100 µg/ml streptomycin and incubated for 18 h at 30°C. The resulting growth culture contained approximately 10^9 cells/ml, and it was then diluted to a concentration of 10^6*V. cholerae* cells/ml, a dose considered infectious for humans (Huq et al., 1996; Chowdhury et al., 1997).

Prior to observations, 30 adult *D. pulicaria* females without eggs were sorted and kept for 24 h in the dark in 3 l of moderately-hard reconstituted water (MHRW) (U.S. Environmental Protection Agency, 2003) containing 10^6*V. cholerae* cells/ml and 10 ml of *Chlamydomonas*. Preliminary SEM investigations (our personal observations) showed that exposure of *D. pulicaria* to *V. cholerae* resulted in the attachment of the pathogen to the body, the feeding appendages and the gut of the cladoceran. A time-dependent effect was also noticed, with exposures lasting 24-48 h producing greater infestations than shorter time (1 and 12 h). The *D. pulicaria* treated in this way were considered as infected by *V. cholerae*, and their swimming behaviour analysed as discussed in the following sections. As a control, the motion of *D. pulicaria* adults not exposed to the pathogen but sorted directly from the rearing culture was analysed for comparative purposes. In the following, cladocerans exposed to *V. cholerae* will be indentified as "infected *D. pulicaria*", in contrast to the control organisms, which will be termed "uninfected *D. pulicaria*".

Experimental Set-up and Observation Protocol

We individually observed a total of twenty randomly picked *D. pulicaria*: ten uninfected and ten infected by *V.cholerae*. Each single animal was preliminarily kept in a paper cup containing ~100 ml of artesian-well water enriched with 2 ml of *Chlamydomonas*. The animals were exposed to the water for at least 2h before observation. The animals and their water were then transferred to a 80.3 ml experimental aquarium (72×93×12 mm) where they acclimated for 20 minutes. The vessel was then placed in the observation apparatus (Figure 1) where it underwent an additional 5 minute acclimation.

Each *D. pulicaria* was observed for 15 minutes, using an IR-CCD camera (XC-E150, SONY, Japan) connected to a camcorder (DCR-HC40, SONY, Japan) at a frame rate of 30 frames/s. The observation period allowed studying the behaviour of the cladocerans as induced by the experimental conditions for a prolonged time, rather than focusing on the instantaneous behavioural responses (Ziarek et al., 2011). Filming was conducted at a water temperature of 22°C (laboratory temperature of 22-23°C), a value typical of summer conditions in the Allequash Lake (Ziarek et al., 2011). All observations were performed with two near-infrared diode lights (~980 nm) positioned at a distance of 20 cm and at 45° to the vessel. A visible-light filter was mounted to the lens of the IR-CCD camera. Each trial was saved to a MiniDV tape for analysis, and transferred as an AVI file (720×480 pixels) onto a computer drive. The movie was then modified and digitised frame-by-frame using the automatic tracking software VidAna, provided by Dr. Michael Hofmann (University of Bonn).

Daphnia are well known for their photoadaptive behaviour (e.g., Smith and Baylor, 1953) and for performing diel vertical migrations (e.g., Ringelberg, 1999), and consequently the experimental protocol was scheduled with several provisos. All experiments were carried out over a 12 h period, from 4 p.m. of the starting day to 4 a.m. of the following one, to minimize the effects of diel vertical migrations. Sorted *D. pulicaria* (both uninfected and infected) were randomly exposed to light and dark conditions to account for the day/night changes in the luminous fields experienced by the animals. Consequently, four conditions were tested: uninfected *D. pulicaria* in light (UL) and in the dark (UD), and infected *D. pulicaria* in light (IL) and in the dark (ID).

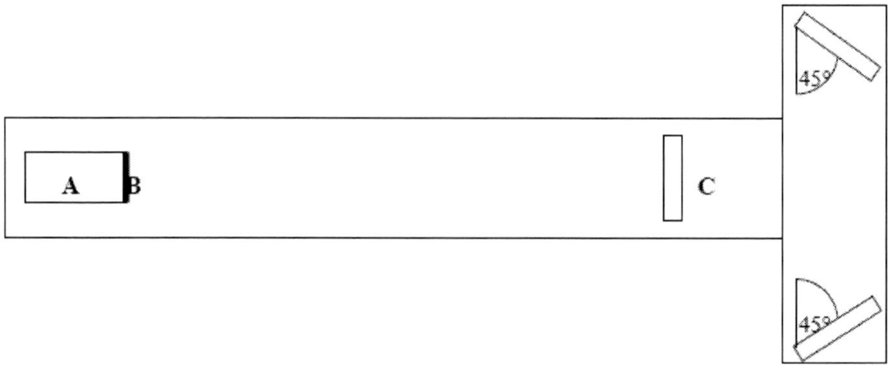

Figure 1. Schematic view of the filming apparatus used for *D. pulicaria* filming (A: IR-CCD camera; B: visible-light filter; C: experimental aquarium; D: near-infrared lights).

For light conditions, standard fluorescent bulbs were placed approximately 2 m above the aquarium. In dark conditions, only the two near-infrared diode lights were on. Each individual was tested in both light and dark conditions. This protocol allowed investigating how uninfected and infected *D. pulicaria* reacted to different light stimuli. To prevent any turbulence, the entire set-up was installed on a solid-steel ballast, while a plastic plug covered the top of the aquarium.

Data Analysis

For each animal, we obtained a track of ~27,000 xy coordinates. As in Ziarek et al. (2011), each trajectory was characterised in terms of three different metrics: mean swimming speed, velocity power spectral density and fractal dimension.

The swimming speed (*V)* was calculated from two positions in two dimensions using the formula:

$$V = \frac{1}{T}\sqrt{\left(x_{i+1} - x_i\right)^2 + \left(y_{i+1} - y_i\right)^2}$$

where$T$ was the sampling rate, i.e. 1/30 seconds between two successive positions. We measured the velocity fluctuations and the spatial complexity of each swimming trajectory to quantify the swimming behaviour. In order to reduce the noise contained in the raw data, a preprocessing was carried out using 25, 50 and 75 data point running means.Tukey's test (Sokal and Rohlf, 1995) was applied to the results to compare the differences between the tested conditions.

The swimming of *Daphnia* involves a cycle of hops and stops while they propel themselves through water. The kinetic energy provided by such swimming behaviour might have certain distributions in frequency space that could depend on the zooplankton behaviour or on the response to the imposed stimulation. Spectral analysis was performed on the swimming velocities of zooplankton to investigate whether there were significant differences in particular frequency ranges caused by the infection or by the illumination stimulus. Power spectra were estimated using the Welch method (Welch, 1970) by dividing the data into several segments and applying the Hanning window before the estimation of power spectra. The spectra were then averaged to construct a smoothed spectrum.

The convolution and space-fillingness of the swimming trajectories were evaluated using the fractal dimension D (Mandelbrot, 1967) using a modified box-counting method (Uttieri et al., 2005); a recent review of the applications of fractal concepts to zooplankton swimming behaviour can be found in Uttieri and Zambianchi (in press). Fractal dimension was calculated using a linear regression of the number of boxes occupied vs. the size of the cell with a dyadic sequence of box values (Buczkowski et al., 1998). Prior to the analysis, all tracks were checked for the presence of an intermittent pattern (data not shown) indicative of a fractal structure (Cole, 1995). In addition, a calibration of the method in relation to the dimension of the available domain was carried out to calculate a correction factor (Uttieri et al., 2005). A one-way ANOVA (Zar, 1984) was used to compare the fractal estimates obtained for the different experiments, returning a P value for the null hypothesis that the values were statistically different. The boxplot method (Tukey, 1977) was used to visualise the statistical results. In addition, the statistical robustness of fractal estimates obtained with linear regressions was verified by computing the coefficient of regression r^2 (Sokal and Rohlf, 1995).

RESULTS

During their motion, both uninfected and infected *Daphniapulicaria* moved according to the typical hop and sink behaviour, with an alternation of active jumps and a motionless sinking in the fluid. Uninfected and infected *D. pulicaria* behaved differently under light and dark conditions (Figure 2): in the presence of the light stimulus, uninfected cladocerans (UL) moved mostly vertically displaying positive phototaxis, while in the dark the motion was almost exclusively horizontal for both uninfected (UD) and infected *D. pulicaria*(ID). When the light was switched on, *D. pulicaria* infected by *Vibriocholerae* (IL) aggregated at the horizontal boundary (either the top or the bottom of the vessel), with a movement pattern drastically different than that reported for UL organisms.

The swimming speeds recorded for uninfected and infected *D. pulicaria* (Table I) fell in the interval 2.70-3.66 mm/s, with some differences depending on the experimental tested condition. In all comparisons, the Tukey's test showed that there was no statistical difference among treatments (P > 0.4).

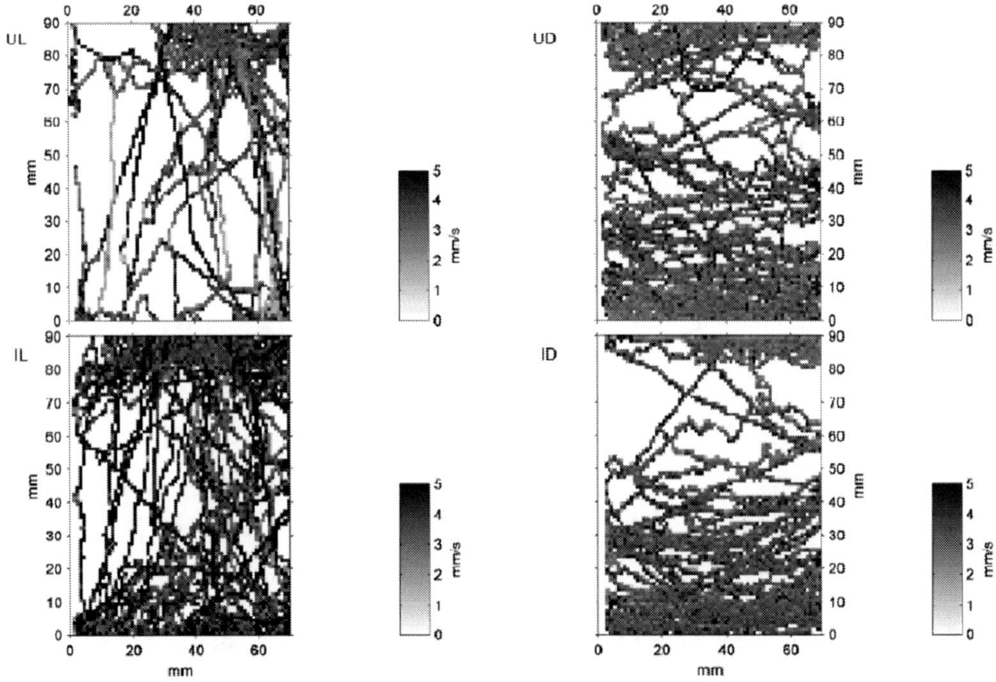

Figure 2. Swimming trajectories of *D. pulicaria* in the four conditions tested (UL: uninfected and light; UD: uninfected and dark; IL: infected and light; ID: infected and dark).

Table I. Mean swimming speed (± standard deviation) of *Daphnia pulicaria* in all the conditions tested in the experiments

Experimental Condition	Swimming Speed (mm/s) (mean ± standard deviation)
UL	2.70 ± 0.51
UD	3.58 ± 0.35
IL	3.66 ± 1.10
ID	3.33 ± 1.10

UL: uninfected *D. pulicaria*, light conditions; UD: uninfected *D. pulicaria*, dark conditions; IL: infected *D. pulicaria*, light conditions; ID: infected *D. pulicaria*, dark conditions.

Figure 3 shows velocity power spectra for the four tested conditions: in particular, Figure 3a displays power spectra relative to infected and uninfected organism velocities for light conditions and Figure 3b the same in dark conditions. As for the comparison between dark and light conditions, remarkable differences can be seen in the range of low frequencies of motion: in light conditions that portion of the spectrum was well populated, while in the dark slower motion was evidently less active.

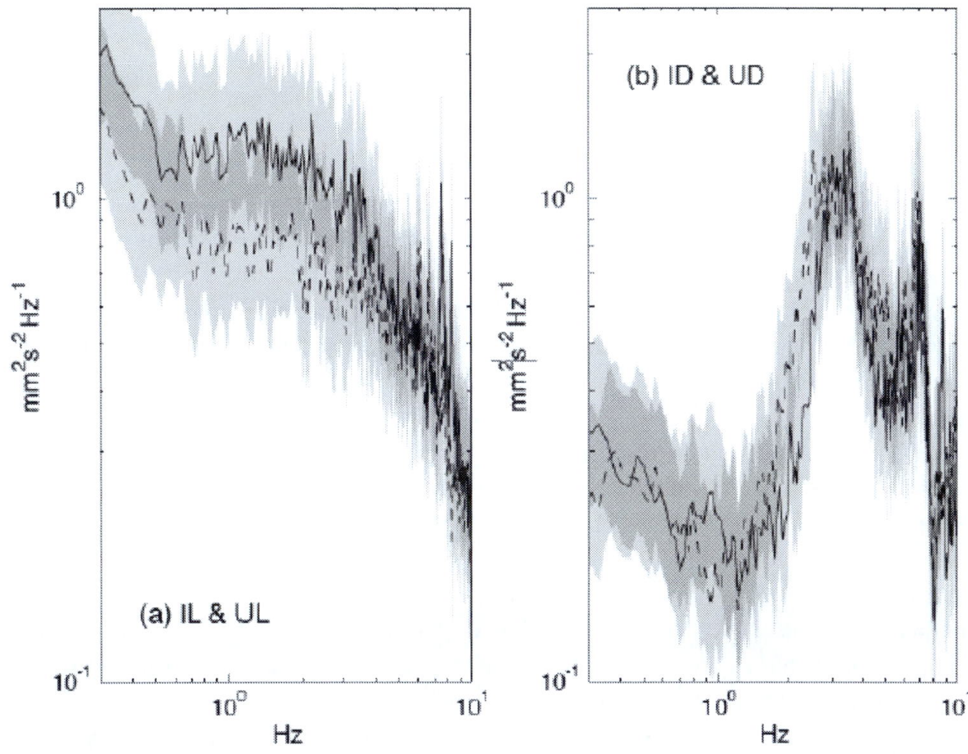

Figure 3. Power spectrum density of swimming velocities in light (a) and dark (b) conditions. The dashed lines indicate the average power spectrum density of uninfected *D. pulicaria* and the solid lines indicate the average power spectrum density of infected *D. pulicaria*. Shading is 95% confidence interval. Dark shading part is the overlapping part of the confidence interval between infected and uninfected D. pulicaria.

Moreover, in dark conditions the differences between the motion of infected and unifected organisms were practically negligible, as shown qualitatively by the spectra. Both spectra relative to dark conditions showed major peaks in the region between 3 and 3.5 Hz, at 7 and 10 Hz. The spectrum of ID velocities also showed two peaks at 6 and 8.8 Hz that were not visible in the UD spectrum.

On the other hand, in light conditions there were strong differences between the power spectra relative to infected and uninfected organism velocities, as furthermore shown by the basic statistics reported in Table I: infected organisms showed a more energetic motion over the whole range of frequencies. The range of low frequencies was clearly different in the two spectra. Major peaks for both components may be identified at 3.5, 4 and 7.5 Hz; the IL spectrum also showed strong peaks at 5 and 10 Hz, while the UL spectrum displayed an individual 6 Hz peak, but generally speaking all peaks in the IL spectrum were definitely more pronounced than their UL counterparts.

The trajectories of uninfected *D. pulicaria* attained statistically different values of fractal dimension D with respect to the other three conditions (Table II) (Figure 4). In particular, in UL trials D was lower than in UD (P=0.01), indicating a different strategy of exploration of the environment. Infected cladocerans instead recorded similar fractal dimensions in both IL and ID conditions, with values significantly similar to those of UD (Figure 3). In all conditions, the estimates of fractal dimension were associated with high r^2 values (r^2>0.99), indicating a good reliability of the estimates.

DISCUSSION

The movement of individuals is an important source of information about the autoecology (Strickler, 1977) and the niche separation (Strickler, 1985) of zooplanktonic species. The movement behaviour of an individual affects the modalities by which it interacts with the environment and, more importantly, with the other organisms present in the fluid, be they prey, mates or predators. The motion behaviour must balance between different needs and aims, and it must be flexible enough to allow for the adaptation of the dynamics to external changes.

The issue of infection is also of ecological importance. Zooplankton organisms are exposed to a wide range of infective agents (e.g., fungi, bacteria, viruses), which can impact both their biology and physiology. In freshwater systems, the cladoceran*Daphniapulicaria* is one of the main actors of the functioning of pelagic web. Among its different potential pathogens, in this work we have investigated for the first time the effects of *Vibriocholerae* infections on *D. pulicaria* from a behavioural perspective. Our goal was to evaluate if infected animals behaved differently from uninfected specimens, with potential consequences on the interactions between *D. pulicaria* and possible predators.

In the dark, the velocity fluctuations of the control and the infected animals both contain one major peak at ~3 Hz in frequency space. A peak conserved in both spectra reveals that control and infected *D. pulicaria* swimming motions periodically punctuate particular variances, or in other words all animals maintain certain rhythms in their swimming behaviour. Furthermore, the complexity of the overall motion path was similar between uninfected and infected animals in the absence of light.

Table II. Mean fractal dimension (± standard deviation) of *Daphnia pulicaria* in all the conditions tested in the experiments

Experimental Condition	Fractal Dimension D (mean ± standard deviation)
UL	1.44 ± 0.12
UD	1.62 ± 0.07
IL	1.65 ± 0.04
ID	1.65 ± 0.10

UL: uninfected *D. pulicaria*, light conditions; UD: uninfected *D. pulicaria*, dark conditions; IL: infected *D. pulicaria*, light conditions; ID: infected *D. pulicaria*, dark conditions.

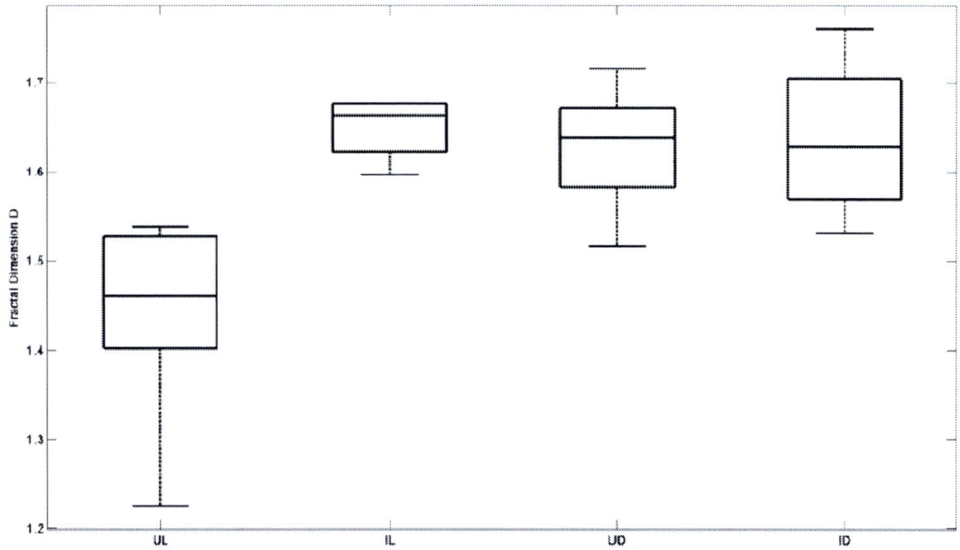

Figure 4. Fractal dimension values D of uninfected (UL) and infected (IL) *D. pulicaria* in light treatment and of uninfected (UD) and infected (ID) *D. pulicaria* in dark treatment.

In light conditions, uninfected *D. pulicaria* adjusted their swimming behaviour by decreasing their swimming velocity and the convolution of their motion pattern, while infected cladoceransdisplayed statistically more convoluted trajectories.

Daphnia exhibiting higher velocities increase the probability of encountering food and mates(Gerritsen and Strickler, 1977); moreover, increased motion complexity in response to greater prey concentrations effectively maximizes nutrient uptake (Seuront et al., 2004). Unfortunately, more active behaviour also leads to greater visibility and encounter rates with predatory animals (Uttieri et al., 2007); therefore, zooplankton survival depends on effectively transitioning between motions that maximize encounter with prey and avoidance of predators. Our data demonstrate that when the environment is dark, *D. pulicaria* may express more active swimming behaviour to maximize food allocation without additionalpredator risk. On the other hand, in light conditions individuals have to suppress any “appealing” behaviour so as to minimize predation.

It is generally accepted that the main visual predators of *Daphnia* species,planktivorous fish, select their prey depending on their motion patterns (Brooks and Dodson, 1965; Ware,

1973; Zaret, 1980; Brewer and Coughlin, 1996; Strickler et al., 2005).Strickler et al.(2005) used computer-animated zooplankton to demonstrate that fish preferentially select spinning *Daphnia* over hop-sink ones, their movement seemingly being more attractive. A velocity spectral analysis of the *Daphnia* used by Strickler et al. (2005) indicates that the magnitude of variance, in all frequency ranges, is significantly greater during spinning behaviour than hop-sink motion (F-test; P> 0.05, df = 4 at each frequency point), similarly to what reported here for *D. pulicaria* infected by *V. cholerae*. In the presence of light, uninfected *D. pulicaria* suppress the visually appealing behaviour by reducing motion complexity, whereas infected specimens maintain a potentially more attractive movement.

Preliminary assessments of associated predation risk indicate that visual planktivorous fish preferentially selected *D. pulicaria* infected with *V. cholerae*. Adolescent rainbow trout (*Oncorhynchusmykiss*) responded to infected *D. pulicaria* at a distance three times greater than their reactive distance to control (uninfected) animals (our personal observation). This evidence suggests that infected individuals are more susceptible to predation, while healthier specimens may have higher chances of surviving and reproducing. Following this line of thought, infection by *V. cholerae* could play a role in the population dynamics of *Daphnia* population. In addition, since zooplankters can act as a reservoir for *V. cholerae* (a single individual carrying up to 10^4 pathogens; Colwell and Huq, 1994) and can be preyed upon by fish and shellfish, which are in turn eaten by humans (Huq et al., 1986), the infection of *Daphnia* can be a key node in the spread of the infectious pathogen.

Our study shows that the behaviour of *D. pulicaria* can be hijacked by *V. cholerae* and raises the questions: what would happen to fish as a consequence of feeding on infected *D. pulicaria*; are they the next targets of the hijackers; what are the mechanisms underlying the behavioural alterations and how do *V. cholerae* physically associate with *Daphnia*?

Acknowledgments

The authors would like to thank Dr.AndewCamilli (Tufts University School of Medicine) for the *Vibrio cholerae* strains and the sharing of his microbe culture and animal infestation technical knowledge. We would also like to recognize the generous donation of the VidAna tracking software by Dr. Michael Hofmann (University of Bonn).

References

Bethel, W. M.; Holmes, J. C. *Can. J. Zool.* 1977, 55, 110-115.

Brewer, M. C.; Coughlin, J. N. In Zooplankton: Sensory Ecology and Physiology; Lenz, P. H.; Hartline, D. K.; Purcell, J. E.; Macmillan, D. L. (Eds.); Gordon and Breach: Amsterdam, 1996; pp 425-434.

Brooks, J. L.; Dodson, S. I. *Science*. 1965, 150, 28-35.

Buczkowski, S.; Kyriacos, S.; Nekka, F.; Cartilier, L. *Pattern Recogn.* 1998, 31, 411-418.

Chiavelli, D. A.; Marsh, J. W.; Taylor, R. K. *Appl. Environ. Microbiol.* 2001, 67, 3220-3225.

Chowdhury, M. A. R.; Huq, A.; Xu, B.; Madeira, F. J. B.; Colwell, R. R. *Appl. Environ. Microbiol.* 1997, 63, 3323-3326.

Cole, B. J. *Anim. Behav.* 1995, 50, 1317-1324.

Colwell, R. R.; Huq, A. In *Vibrio cholerae* and Cholera: Molecular to Global Perspectives; Wachsmuth, I. K.; Olsvik, O.; Blake, P. A. (Eds.); American Society for Microbiology: Washington, 1994; pp 117-133.

Epstein, P. R. *Biosystems*. 1993, 31, 209-211.

Gerritsen, J.; Strickler, J. R. *J. Fish.Res. Board Can.* 1977, 34, 73-82.

Hebert, P. D. N. *J. Anim. Ecol.* 1977, 46, 399-409.

Helluy, S.; Thomas, F. *Proc. R. Soc. Lond.* B. 2003, 270, 563-568.

Hulscher, J. B. *J. Sea Res.* 1973, 6, 141-156.

Huq, A.; Huq, S. A.; Grimes, D. J.; O'Brien, M. O.; Chu, K. H.; Capuzzo, J. M.; Colwell, R. R. *Appl. Environ. Microbiol.* 1986, 52, 586-588.

Huq, A.; Xu, B.; Chowdhury, M. A. R.; S., I. M.; Montilla, R.; Colwell, R. R. *Appl. Environ. Microbiol.* 1996, 62, 2508-2512.

Mandelbrot, B. B. *Science.* 1967, 156, 636-638.

McAllister, M. K.; Roitberg, B. D. *Nature.* 1987, 328, 797-799.

Metschnikoff, I. I. *Arch. Pathol. Anat. Ph.* 1884, 96, 177-195.

Pruzzo, C.; Vezzulli, L.; Colwell, R. R. *Environ. Microbiol.*2008, 10, 1400-1410.

Ringelberg, J. *Biol. Rev.* 1999, 74, 397-423.

Seuront, L.; Brewer, M. C.; Strickler, J. R. In Handbook of Scaling Methods in Aquatic Ecology - Measurements, Analysis, Simulation; Seuront, L.; Strutton, P. G. (Eds.); CRC Press: Boca Raton, 2004; pp 333-359.

Smith, F. E.; Baylor, E. R. *Amer. Nat.* 1953, 87, 49-55.

Sommer, U.; Gliwicz, Z. M.; Lampert, W.; Duncan, A. *Arch. Hydrobiol.* 1986, 106, 433-471.

Sokal, R. R.; Rohlf, F. J. Biometry; W. H. Freeman and Company: New York, 1995; pp 880.

Strickler, J. R. Limnol.*Oceanogr.* 1977, 22, 165-170.

Strickler, J. R.In Physiological Adaptations of Marine Animals.Symposia of the Society for Experimental Biology;Laverack, M. S. (Ed.); Company of Biologists, Ltd.: Cambridge, 1985; pp 459-485.

Strickler, J. R.; Udvadia, A. J.; Marino, J.; Radabaugh, N.; Ziarek, J.; Nihongi, A. *Sci. Mar.* 2005, 69 (Suppl. 1), 111-124.

Tamplin, M. L.; Gauzens, A. L.; Huq, A.; Sack, D. A.; Colwell, R. R. *Appl. Environ. Microbiol.* 1990, 56, 1977-1980.

Thomas, F.; Schmidt-Rhaesa, A.; Martin, G.; Manu, C.; Durand, P.; Renaud, F. *J. Evolution. Biol.* 2002, 356-361.

Tukey, J. W. Exploratory Data Analysis; Addison-Wesley: Reading, MA, 1977; pp 688.

U. S. Environmental Protection Agency. Standard Operating Procedure for Moderately Hard Reconstituted Water, #204; SoBran Inc.: Cincinnati, 2003; pp 5.

Uttieri, M.; Zambianchi, E. In Classification and Application of Fractals: New Research;Mitchell, E. W.; Murray, S. R. (Eds.); Nova Science Publishers Inc.: New York, in press

Uttieri, M.; Zambianchi, E.; Strickler, J. R.; Mazzocchi, M. G. *Ecol. Model.* 2005, 185, 51-63.

Uttieri, M.; Cianelli, D.; Strickler, J. R.; Zambianchi, E. *J. Theor.Biol.* 2007, 247, 480-491.

Ware, D. M. *J. Fish.Res. Board Can.* 1973, 30, 787-797.

Welch, P. D. IEEE *T. Acoust.Speech.* 1970, 15, 70-73.

Zar, J. H. Biostatistical Analysis; Prentice Hall: Englewood Cliffs, NJ, 1984; pp 718.

Zaret, T. M.In Evolution and Ecology of Zooplankton Communities; Kerfoot, W. C. (Ed.); Special Symposium III American Society of Limnology and Oceanography. University Press of New England: Hanover, 1980; pp 594-603.

Ziarek, J. J.; Nihongi, A.; Nagai, T.; Uttieri, M.; Strickler, J. R. *Hydrobiologia.* 2011, 661, 317-327.

In: Zooplankton and Phytoplankton
Editor: Giri Kattel, pp. 193-217
ISBN: 978-1-61324-508-8

Chapter 10

BODY SIZE VERSUS RATE PARAMETERS OF ZOOPLANKTON AND PHYTOPLANKTON: EFFECTS ON AQUATIC ECOSYSTEM

Santanu Ray[1], Sandip Mandal[2], Joyita Mukherjee[1], Madhumita Roy[1], Sudipto Mandal[1] and Surjya Kumar Saikia[1]

[1]Ecological Modelling Laboratory, Department of Zoology,
Visva-Bharati University, Santiniketan, India
[2]Centre for Cellular and Molecular Biology, Hyderabad, India

ABSTRACT

Body size is an important parameter in ecological processes, at both individual and community levels suggest that density of the biomass in any ecosystem is dependent on the community structure that is reflected in body sizes of the constituent organisms. Model behaviour and dynamics are mainly regulated by physiological rate parameters of the living state variables. The physiological processes of living organisms show an allometric relationship to body sizes and the calculation follow algebraic rules that are homologus to those used with logarithms. The role of body size in planktonic system has significant contribution in the system dynamics. In this article an alternative approach has been developed in two generic models of planktonic systems and in this approach the rate processes vary according to the body sizes of planktons following allometric principle and these are incorporated in the dynamic model. In the first model different body sizes of phytoplankton and zooplankton are used and their rate parameters particularly the phtotosynthetic rate of phytoplankton and grazing rate of zooplankton vary accordingly. Optimization tool has been used to determine the perfect combination of phytoplankton and zooplankton to obtain the maximum fish yield in different trophic conditions such as oligotrophic, mesotrophic and eutrophic conditions. Model result shows in different trophic conditions the combinations of phytoplankton and zooplankton are different.

In any ecosystem the equilibrium condition may gradually turn into a chaotic situation for different reasons. In the later part of the article a three species (phytoplankton, zooplankton and fish) model is proposed. Rate parameters are changed according to the change of the size of the organisms. The model is run in different conditions with different sizes of zooplankton by increasing the grazing rate and consequently decreasing the half-saturation constant of this organisms following allometric principles. The system exhibits different states (equilibrium point – stable limit cycle – doubling and ultimately chaos) by gradual increment of zooplankton grazing rate and decrease of half-saturation constant.

INTRODUCTION

The size of an organism affects virtually all aspects of its physiology and ecology (La Barbera, 1986). The works of Hemmingsen (1950, 1960) and Schmidt-Nielsen (1975) on the physiology of body size, Gould (1966) on its phylological ramifications, and a general theoretical model by McMahon (1973) collectively have served to bring the importance of body size as a biological variable forcibly to the attention of biologists. The recent explosive growth of both theoretical and empirical studies on the influence of size on all aspects of biology and evolution, and the extension of such studies from a focus on physiology and functional morphology to broader ecological characteristics have resulted in a spate of books on the subject (Calder, 1984, Peters, 1983, Schmidt-Nielsen, 1984). As a result, the field of allometry, the study of relationships between organism size and function, has rather suddenly become a prominent focus in ecology and evolutionary biology.

Body size is an important parameter in ecological processes, at both individual and community levels (Peters, 1983). Dickie et al. (1987) suggest that density of the biomass in any ecosystem is dependent on the community structure that is reflected in body sizes of the constituent organisms. Size distributions of different organisms of the system provide a holistic ecosystem description that facilitates ecosystem comparisons in time and space (Gaedke, 1993). Of different morphological, behavioral and life-history features, body size seems to be the most important factor responsible for vulnerability of any system dynamics (Gliwicz and Umana, 1994). Biomass size spectra provide an important contribution to the search for generalization in pelagic ecosystem (Gaedke, 1992). Model behaviour and dynamics are mainly regulated by physiological rate parameters of the living state variables (Tang, 1995). The physiological processes of living organisms show an allometric relationship to body sizes and the calculation follow algebraic rules that are homologus to those used with logarithms (Peters, 1983; Platt, 1985). The role of body size in planktonic system has significant contribution in the system dynamics (Harbacek, 1962). Commonly panktonic ecosystems are described by compartmental models, and each compartment represents a trophic level or taxonomic group and such models are primarily descriptive, because the most important components of the ecosystem are represented by compartments and their interactions are described by linking compartments. (Moloney and Field, 1991). However, when on the basis for dynamic simulation modeling these models are used and numerous problems are encountered. Unrealistic lumping of all plankton sizes with widely different rates of physiological parameters is very common in the use of present dynamic model. (Moloney et al., 1986). To overcome this problem Anderson et al., (1987) suggested the increase of compartments, but this approach adds the complexity of the model and most of the cases the output is far from reality. Therefore an alternative approach has been developed in a generic model of planktonic systems and in this approach the rate processes vary according to the body sizes of planktons and these are incorporated in the dynamic model. This approach was first used by Lebedeva, 1971 and latter Radtke and Straskraba (1980), Plat et al., (1981), Moloney and Field (1991), Ray et al. (2001a, 2001b) and Mandal et al., (2006, 2007) successfully used this concept in phytoplankton and zooplankton and applied in their models. In this present article two models have been focused, in the first model the same approach has been used in both phytoplankton and zooplalnkton and incorporated into the dynamic model. Different rate processes of phytoplankton and zooplankton have been calculated on the basis of allometric relationship with their body sizes as proposed by Peters, 1983 and applied in the model. This model describes the structure and functioning of a size based simulation model of phytoplankton and zooplankton in presence of fish during different varying nutrient conditions by optimizing the exergy, a thermodynamic based ecological goal function. Exergy is defined as the work a system can perform when it is brought into equilibrium with the environment or another well defined reference state (Jørgensen et al., 2000). (see detailed about exergy, Ray et al. 2001a). Under oligotrophic, mesotrophic and eutrophic conditions the overall system dynamics by selecting continuous size classes of phytoplankton and zooplankton have been tested in this first model of planktonic system.

In the second model the movement of planktonic system dynamics from ordered to chaotic state has been studied by changing two important parameters viz., half saturation constant and grazing rate of zooplankton following allometric principle by incorporating the changes of body sizes of zooplankton. Hastings and Powell (1991), Chattopadhyay and Sarkar (2003) used half saturation constant of zooplankton as key parameter in their model to study the system dynamics from order to chaos. To handle single parameter particularly half saturation constant is not very realistic. May (1987) showed that the system produce chaotic dynamics for plausible combination of parameters values. In the second model, different body sizes of zooplankton have been considered and accordingly the growth rate and half saturation constant have changed. Due to gradual reduction of the size of zooplankton the specific growth rate and half saturation constant of zooplankton are increased and decreased respectively and the system dynamics moves from ordered state to chaotic condition.

Physiological Rate Parameters of Phytoplankton and Zooplankton in Respect to Body Size

For parameterization of the physiological processes of phytoplankton and zooplankton the deductions have been made from general ecological allometric principle of body size by using logarithmic scale (log 10) of this size (Peter, 1983). Different authors (Sheldon et al., 1972, Kerr, 1974, Radtke and Straskraba, 1980 and Tang, 1995) used cell volume as a measure of size for the scaling of allometric relatinships with physiological processes. The same procedure has been followed in this present work that relies on empirically established allometric relationships between individual cell volume and metabolic processes. For selecting the cell volume of phytoplankton and zooplankton the literature have been surveyed and seen that normally the phytoplankton cell volume ranges between log 1 (μm^3) and log 7 (μm^3) and zooplankton size comprises between the smaller log 1 (μm^3) and the larger log 4 (μm^3). The cell volumes of phytoplankton and zooplankton in this present work have been expressed as V_P and V_z respectively.

Physiological Rate Parameters of Phytoplankton

Size dependent phytoplankton growth rate was observed and studied experimentally by different authors in different environments (Williams, 1965, Eppley and Sloan, 1966, Tilzer, 1972, Findlay, 1972, and Smith et al., 1999). On the basis of the observations of these authors for maximum gross photosynthetic activity (P_{max}), is estimated:

$$P_{max}\,(V_p) = 3.0 - 0.3 \log V_p \tag{1}$$

Photosynthesis is temperature dependent and the values in Eq. (1) hold true at a temperature of 20°C, and the dependence of temperature (T) is exponential: thus

$$P_{max}\,(V_{p,}\,T) = (P_{max}\,(V_p) / 6.05)\exp(0.09\,T) \tag{1a}$$

Yurista (1999) studied that different phytoplankton species according to their sizes have varied respiration rate (Resp), the observation results showed that in the smaller species of phytoplankton the respiratory rate is four times higher than the larger species. This results approximately in a logarithmic dependence of the form

$$Resp\,(V_p) = 0.02 - 0.002 \log V_p \tag{2}$$

On the basis of physical laws of particle sizes the self-shading of phytoplankton is size dependent. According to the law of Kiefer and Austin (1974), Radtke and Straskraba (1980) proposed the following equation:

$$Ext\,(V_p) = 0.12\,V_p^{-0.33} \tag{3}$$

Natural mortality rate of phytoplankton (Mort) is also size dependent, the rate being high in smaller but low in larger sized phytoplankton (Findlay, 1992). The mortality rate is estimated by the following equation:

$$Mort\,(V_p) = 0.03 - 0.001 \log V_p \tag{4}$$

Eppley et al., (1969) suggested a logarithmic increase of half saturation constant (K_{sp}) of phytoplankton for nitrogen according to its size. It is assumed that a similar relationship for phosphorus and in agreement with previous simulation using the values between 5 and 75 and propose the following equation.

$$K_{sp}\,(V_p) = 10 \log V_p - 5 \tag{5}$$

Table 1. Parameter symbols, their description, values and units

Symbol	Meaning	Value	Unit
Iopt	Optimum surface radiation for photosynthesis	300	cal cm $^{-2}$ day^{-1}
Is	Surface solar radiation	280 + 210 SIN (Time 58.1)	cal cm $^{-2}$ day^{-1}
Time			day
Ec	Extinction coefficient	0.2	m^{-1}
T	Temperature	12 + 10 SIN (Time 58.1)	^{0}C
Time			day
Ze	Depth of epilimnion	4	m
Q	Discharge rate of water		m^3 day^{-1}
V	Volume of epilimnion		m^3
Q/V		10^{-2}	
Fin	Concentration of the inflow of N in mesotrophic / oligotrophic / eutrophic conditions	20 / 5 / 100	mg N m^{-3}
P_{max} (V_p T)	Maximum photosynthetic activity of P		day^{-1}
Resp (V_p)	Respiration rate of P		day^{-1}
Mort (V_p)	Mortality rate of P		day^{-1}
K_s (V_p)	Saturation constant of P for N		mg N m^{-3}
Growth (V_z)	Maximum growth rate of Z		day^{-1}
r_{rz}	Assimilation rate of Z	0.6	day^{-1}
Resp (V_z)	Respiration rate of Z		day^{-1}
Mort (V_z)	Mortality rate of Z		day^{-1}
K_s (V_z)	Saturation constant of Z for P		mg N m^{-3}
μzf	Maximum predation rate of F on Z	0.14	day^{-1}
μff	Maximum selfpredation rate of F	0.08	day^{-1}
r_{rf}	Respiration rate of F	0.0015	day^{-1}
m_{rf}	Mortality rate of F	0.001	day^{-1}
a_{rf}	Assimilation rate of F	0.5	day^{-1}
s_{zf}	Predation preference of F on Z	2/3	dimensionless
s_{ff}	Self predation preference of F	1/3	dimensionless
K_{sf}	Saturation constant of F for Z and F	70	mg N m^{-3}
Din	Concentration of D in inflow	20	mg m^{-3}
d_S	Settling rate of D	0.01	day^{-1}
d_D	Remineralization rate of D to N	0.3	day^{-1}

Physiological Rate Parameters of Zooplankton

Blueweiss et al. (1978), Ahrens and Peters (1991) and Gillooly (2000), noticed that zooplankton growth rate varies according to their size, maximum recorded in smaller species and minimum in larger species. On the basis of this observation and on the basis of calculation by Peters (1983) the following equation is proposed:

$$\text{Growth}\ (V_z) = 0.715 - 0.13 \log V_z \quad (6)$$

Wetzel (1975) noticed that the respiratory rate of zooplankton is size dependent, maximum in smaller species and minimum in larger species. According to the estimate of Peters (1983) the following equation is assumed:

$$\text{Resph}\,(V_Z) = 0.0728 - 0.0104 \log V_z \quad (7)$$

Nyholm (1975) reported the size dependent relationship of zooplankton mortality rate, here also he noticed the smallest value in larger species and largest value in smaller species. The rate has been calculated by the following equation:

$$\text{Mort}\,(V_Z) = 0.235 - 0.058 \log V_z \quad (8)$$

Half saturation constant of zooplankton grazing on phytoplankton (K_{sz}) was studied by Smith (1977) and it is logarithmically increasing according with body size. Straskraba and Gnauck (1985) proposed that it varies for phosphorus from 10 to 80 and therefore the following equation is proposed:

$$K_{SZ} = 15 \log V_Z - 5 \quad (9)$$

Equations of First Model

All of the above size dependent parameters are incorporated in to a dynamic model of phosphorus (N), phytoplankton (P), zooplankton (Z), fish (F) and detritus (D). The model equations are presented below and the parameter description their values and units are given in table 1.

$$\frac{dN}{dt} = \left(\frac{Q}{V}\right)(Fin - N) + d_D D - \frac{P_{\max}(V_p T)LePN}{\left[N + K_{sp}(V_p)\right]} \quad (10)$$

Le = light effect on photosynthesis

$$= \frac{I}{\left[(Iopt)\exp\left(1 - \frac{1}{Iopt}\right)\right]}$$

$$I = Is\exp\left[\left(-Ec + ExtV_p\right)Ze\right]$$

$$\frac{dP}{dt} = \frac{P_{\max}(V_p T)LePN}{\left[N + K_{sp}(V_p)\right]} - \operatorname{Re}sp(V_p)P - \frac{Growth(V_z)ZP}{\left[P + K_s(V_z)\right]} \quad (11)$$

$$\frac{Growth(V_z)ZP}{[P+K_s(V_z)]} = \text{grazing of Z on P}$$

$$\frac{dZ}{dt} = (ar_z)\frac{[Growth(V_z)ZP]}{[P+K_s(V_z)]} - \frac{(\mu_{zf}\mathrm{Fs}_{zf}Z)}{(s_{ff}F + s_{zf}Z + K_{sf})} - Mort(V_z)Z - \mathrm{Re}\,sp(V_z)Z \quad (12)$$

$$\frac{(\mu_{zf}\mathrm{Fs}_{zf}Z)}{(s_{ff}F + s_{zf}Z + K_{sf})} = \text{predation of F on Z}$$

$$\frac{dF}{dt} = a_{rf}\frac{(\mu_{zf}\mathrm{Fs}_{zf}Z)}{(s_{ff}F + s_{zf}Z + K_{sf})} - m_{rf}F - r_{rf}F - (1-a_{rf})\frac{(\mu_{ff}Fs_{ff}F)}{(s_{ff}F + s_{zf}Z + K_{sf})} \quad (13)$$

$$\frac{(\mu_{ff}Fs_{ff}F)}{(s_{ff}F + s_{zf}Z + K_{sf})} = \text{self predation of F}$$

$$\frac{dD}{dt} = \frac{Q}{V}Din + (1-ar_z)\frac{[Growth(V_z)ZP]}{[P+K_s(V_z)]} + MortV_pP + \mathrm{Re}\,sp(V_p)P + (1-a_{rf})\frac{(\mu_{zf}\mathrm{Fs}_{zf}Z)}{(s_{ff}F + s_{zf}Z + K_{sf})} \quad (14)$$
$$+ Mort(V_z)Z + \mathrm{Re}\,sp(V_z)Z + (1-a_{rf})\frac{(\mu_{ff}Fs_{ff}F)}{(s_{ff}F + s_{zf}Z + K_{sf})} + m_{rf}F + r_{ff}F - d_sD - d_DD$$

EQUATIONS OF SECOND MODEL

The basic mathematical model is represented as a set of three ordinary differential equations describing the change of phytoplankton (P), zooplankton (Z) and fish (F) over time.

$$\frac{dP}{dT} = R_0P\left(1 - \frac{P}{K_0}\right) - \frac{C_1A_1PZ}{B_1+P}$$

$$\frac{dZ}{dT} = \frac{A_1PZ}{B_1+P} - \frac{A_2ZF}{B_2+Z} - D_1Y \quad (15)$$

$$\frac{dF}{dT} = \frac{C_2A_2ZF}{B_2+Z} - D_2F$$

To characterize interference between phytoplankton and zooplankton population, Holling types I, II and III functional responses are considered to study the behavior of the system. For determination of exact combination of parameters which are controlling the behavior of the system the number of parameters are reduced by choosing $p = P/K_0, z = C_1Z/K_0$,

$f = C_1F / C_2K_0, t = R_0T$. Making these substitutions in model (1) and after simplification it yields the following system of equations:

$$\frac{dp}{dt} = p(1-p) - \frac{a_1 pz}{1+b_1 p}$$

$$\frac{dz}{dt} = \frac{a_1 pz}{1+b_1 p} - \frac{a_2 zf}{1+b_2 z} - d_1 z \qquad (16)$$

$$\frac{df}{dt} = \frac{a_2 zf}{1+b_2 z} - d_2 f$$

The non-dimensional parameters are defined as

$$a_1 = K_0 / R_0 B_1, b_1 = K_0 / B_1, a_2 = A_2 C_2 K_0 / C_1 R_0 B_1, b_2 = K_0 / B_2 C_1, d_1 = D_1 / R_0, d_2 = D_2 / R_0$$

The observation made in this study are limited to the problem and the system under consideration, which is general aquatic ecosystem model, it leads towards chaos from ordered state.

During different run of the model the parameter values (growth rate and half saturation constant) are changed by following the equations (6) and (9) in the way that zooplankton size in the system are gradually shifted from larger size to smaller size. According to the change of the size of the zooplankton both parameters are changed following allometric principles. In this model a_1 = Growth (V_z) (growth rate of zooplankton) and b_1 = K_{sz} (half saturation constant of zooplankton grazing on phytoplankton) of equations (6) and (9) respectively.

RESULTS

The first model is based on empirically established allometric relationships between body size of phytoplankton and zooplankton and metabolic processes, suggesting that body size reveals interplay between physiological properties of individual organisms and community structure. Equations (1) to (9) represent the physiological rate parameters of phytoplankton and zooplankton as functions of their body sizes. The linearization used allows a more detailed numerical representation, but it is to be stressed that in most instances the curves are highly nonlinear. At low V_p and V_z there is in most cases a steep drop in the parameter values, and at high V_p and V_z, the changes in the rate parameters with V_p and V_z are leveled out. A change of V_p and V_z by one order of magnitude will have a rather pronounced effect on process rates at low V_p and V_z, but much smaller effect at high V_p and V_z.

All the size dependent parameters of phytoplankton and zooplankton are substituted to the five compartment dynamic model and because the parameters vary in different size classes of phytoplankton and zooplankton, the model results vary accordingly. In this model it is considered for a homogeneous epilimnion of temperate lake. This type of model with different size classes of phytoplankton was used earlier by Dvořáktová (1974), Straškraba (1976, 1979) and Radtke and Straškraba (1980), Ray (2001a, 2001b). The parameters are separated into two categories: those specifying the state variables; and those characterizing the water body, its driving variables and site constants.

As mentioned earlier that the implication of body sizes of phytoplankton (V_p1 to V_p7 and zooplankton (V_z1 to V_z4) is tested. The model is run for all the size combinations of phytoplankton and zooplankton by using the computer package STELLA Version 6.0 Research of High Performance Company. All runs are started with identical initial conditions. For standard run the input rate of phosphorus 20 mg m^{-3} (mesotrophic condition) is used. The annual average biomass of all three living state variables (N, P and Z) for size classes of phytoplankton ranges from V_p1 to V_p7 and zooplankton from V_z1 to V_z4 (with unit step of both P and Z) are given in Table 2. The water body to which the simulations correspond is a through flowing temperate mesotrophic lake, with an average retention time of 100 days, phosphorus concentration in the inflow of 20 mg m^{-3} , and an extinction depth (Steel, 1973) corresponding to 4 m. This is equally valid for the shallow epilimnion of a very cloudy (e.g. bog like) pond as well as for a deeply mixed but transparent lake.

Table 2. Average (365 days) biomass of phytoplankton, zooplankton and fish and corresponding exergy value in normal nutrient (mesotrophic) condition

	V_z1	V_z2	V_z3	V_z4
V_p1 P	8.34	13.46	14.21	12.26
Z	3.89	2.75	2.74	5.31
F	1.64	0.91	0.82	2.52
E	7254	5439	5633	9606
V_p2 P	8.42	14.78	15.22	13.75
Z	2.42	3.12	2.98	5.72
F	1.35	0.87	0.97	3.61
E	6172	5572	5843	9738
V_p3 P	9.32	16.87	17.61	14.97
Z	1.78	4.64	2.94	5.98
F	0.82	0.78	1.01	3.83
E	5241	4971	6239	9878
V_p4 P	10.14	17.03	15.23	14.68
Z	2.66	3.69	4.76	5.77
F	0.57	0.58	0.98	3.71
E	4233	4695	6198	9782
V_p5 P	14.92	17.98	16.77	13.32
Z	3.56	4.11	4.86	5.17
F	0.19	0.35	0.78	3.20
E	2912	3743	6104	9684
V_p6 P	23.74	18.13	17.86	18.63
Z	0.81	4.87	4.72	4.25
F	0	0.03	0.27	2.01
E	1093	2623	5234	8789
V_p7 P	28.89	19.88	18.93	17.97
Z	0.92	1.04	4.34	3.07
F	0	0	0.03	1.76
E	613	2312	2584	6305

In this model light and temperature dependent photosynthesis and the effect of extinction coefficient as well as self-shading of phytoplankton on photosynthesis is incorporated. The temperature of water body varies from low 2° C to high 22° C and surface solar radiation varies from 70 cal m^{-2} d^{-1} to 490 cal m^{-2} d^{-1} annually. For the variation of temperature and surface solar radiation sine function in the respective equations is used. These two parameters have immense effects on photosynthetic rate of phytoplankton and this rate is more or less directly proportional to these parameters and therefore maximum photosynthesis is recorded in summer and minimum recorded in winter. For these two parameters an annual cycle of photosynthetic rate of all the size classes of phytoplankton is observed. As the zooplankton growth rate is mainly dependent on phytoplankton and in turn fish growth rate on zooplankton, annual cycles (stable limit cycles) of all the state variables in almost all size combinations of phytoplankton and zooplankton are noticed.

Two basic features are reflected in the results. (1) Direct size effects, including both changes in the gains in population as well as in the losses. Smaller phytoplankton grows rapidly than larger phytoplankton under comparable conditions but larger phytoplankton is able to utilize low phosphorus concentrations more favourably. Larger phytoplankton is less grazed by zooplankton and also less predation pressure is noticed for larger zooplankton. Larger sizes of both phytoplankton and zooplankton have relatively smaller respiration and mortality rates and large phytoplankton produces less self-shading, (2) Indirect effects, due to feedback in the phytoplankton, zooplankton, fish, detritus and phosphorus system.

For low V_p the growth rate is high, which results in the rapid utilization of phosphorus and also causes the increase of high zooplankton and fish biomass. In turn these consequences both control phytoplankton, zooplankton and fish levels in such a way that fluctuation of all the living state variables are smoothed, and the peak does not reach a large value. With increasing V_p in spite of decreasing growth rate of phytoplankton the population increases and in most of the cases in this condition the biomass of zooplankton and fish are also increased. In nature many authors showed that this type of low growth rate which is particularly evident in springs, when with increasing V_p it takes the phytoplankton more time to take advantage to the improved light and temperature conditions, in spite of increasing phosphorus concentration and low control from zooplankton grazing. The retarded phytoplankton pulse is reached by a steep exponential increase, which is higher for great V_p due to less control from nutrient limitation and from zooplankton. At V_p 7, there is only a small, late biomass peak, and zooplankton are unable to reproduce at a reasonable level. The phosphorus concentration in the water body nearly reaches the inflow concentration.

Although the smaller zooplankton (V_z1) having maximum growth rate but in association with all size combinations of phytoplankton the results indicate that these small sized zooplankton only can be well adapted with the smaller phytoplankton (V_p1). But with the increasing of phytoplankton size, the biomass of phytoplankton and zooplankton begin to oscillate and the intensity of oscillation directly related with the increased size of phytoplankton. And in these situations the fish biomass is gradually decreased and remains in the system when phytoplankton size is selected up to V_p5 and beyond that the fish is eliminated from the system. Whenever slightly larger zooplankton V_z2 is selected maximum fish biomass is recorded along with proportional biomass of phytoplankton and zooplankton in combination with V_p2. This size of zooplankton shows maximum phytoplankton and zooplankton biomass in combination with larger phytoplankton size but here fish can not survive. The simulation results also indicate that zooplankton size V_z3 gives maximum fish biomass with relative proportion of phytoplankton and zooplankton in association with phytoplankton size V_p3, but further increment of phytoplankton size causes the reduction of fish biomass. In our simulation run when largest zooplankton V_z4 is selected the maximum fish biomass with relative amount of zooplankton and phytoplankton is obtained in association with phytoplankton size V_p3 and here also fish biomass is decreased with further increase of phytoplakton size.

From different runs and corresponding observations the results show that the comparatively more fish biomass with good proportion of zooplankton and phytoplankton is recorded for the size classes of V_p1 and V_z1, V_p2 and V_z2, V_p3 and V_z3, but maximum fish biomass is obtained for V_p3 and V_z4. In all possible size combinations the corresponding exergy values are tested. Greater exergy values are coincided with the above results. The highest average value 9787 is obtained for size combination of V_p3 and V_z4. The other exergy values are 7254 for V_p1 and V_z1, 5572 for V_p2 and V_z2 and for the size combination of V_p3 and V_z3 the average exergy is 6239.

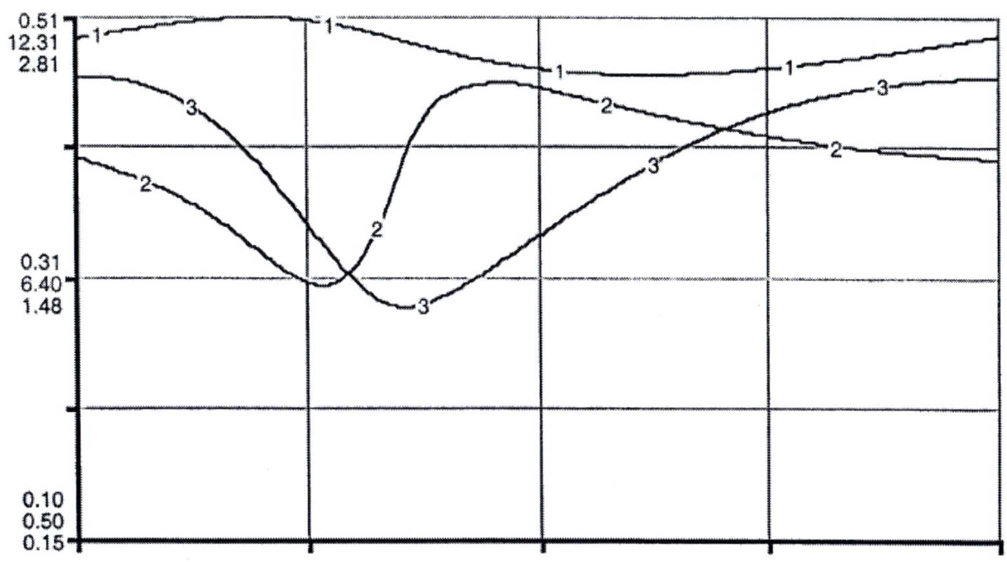

Figure 1a. Biomass of fish (1), phytoplankton (2) and zooplankton (3) in oligotrophic condition, x – axis showing the time (0 – 365 days) and y – axis showing the concentration of biomass (mg phosphorus m^{-3}), from upper to lower the values representing fish, phytoplankton and zooplankton biomass.

During optimization of exergy, the continuous selection of size classes of phytoplankton and zooplankton instead of particularly selected size classes has been used. First the model is run in mesotrophic conditon (nutrient 20 mg m^{-3}) but the model is tested in oligotrophic (nutrient 5 mg m^{-3}) and eutrophic condition (nutrient 100 mg m^{-3}) too. In mesotrophic condition two peaks of exergy are obtained, the smaller peaks for smaller phytoplankton V_p1 and smaller zooplankton V_z1 sizes, while the large peak with largest zooplankton V_z4 and medium sized phytoplankton $V_p3.12$. And here the largest exergy value is obtained (Figs. 1b and 3). In oligotrophic condition the largest fish biomass with relative proportional biomass of zooplankton and phytoplankton is recorded for smaller phytoplankton size $V_p1.48$ and V_z4 which also corresponds by highest exergy value (Figs. 1a and 2). During eutrophic condition the optimization results show that the large zooplankton V_z4 is best adapted with comparatively with larger phytoplankton $V_p5.25$ and this combination gives maximum fish biomass and the highest exergy value (Figs. 1c and 4). Different physiological rate parameters of phytoplankton and zooplankton in different nutrient conditions where maximum exergy values obtained are shown in table 3 and 4 respectively.

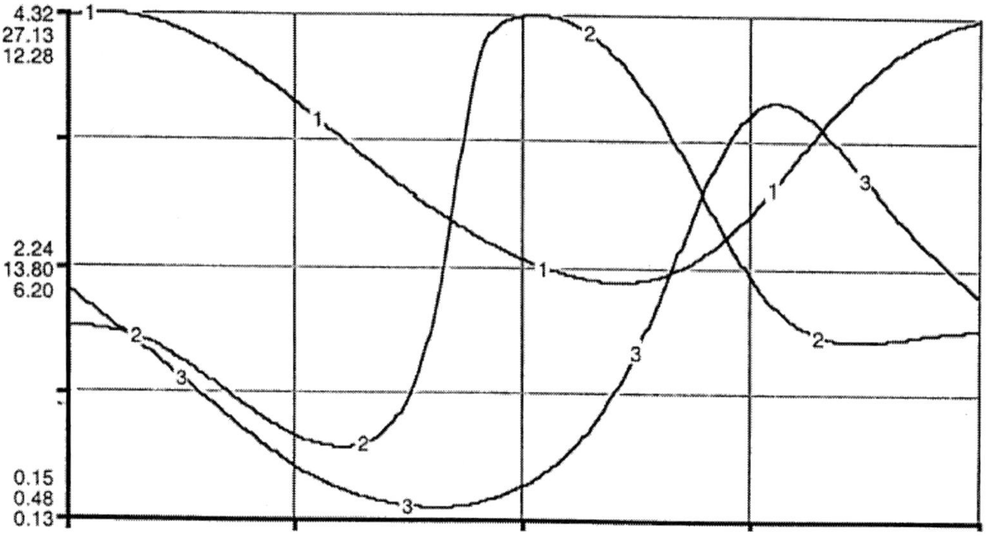

Figure 1b. Biomass of fish (1), phytoplankton (2) and zooplankton (3) in mesotrophic condition, x – axis showing the time (0 – 365 days) and y – axis showing the concentration of biomass (mg phosphorus m^{-3}), from upper to lower the values representing fish, phytoplankton and zooplankton biomass.

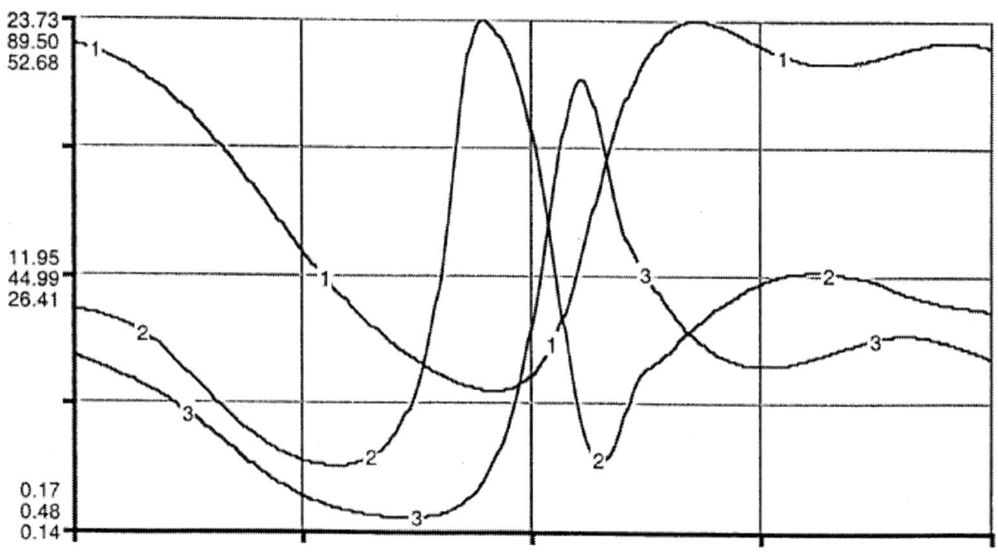

Figure 1c. Biomass of fish (1), phytoplankton (2) and zooplankton (3) in eutrophic condition, x – axis showing the time (0 – 365 days) and y – axis showing the concentration of biomass (mg phosphorus m^{-3}), from upper to lower the values representing fish, phytoplankton and zooplankton biomass.

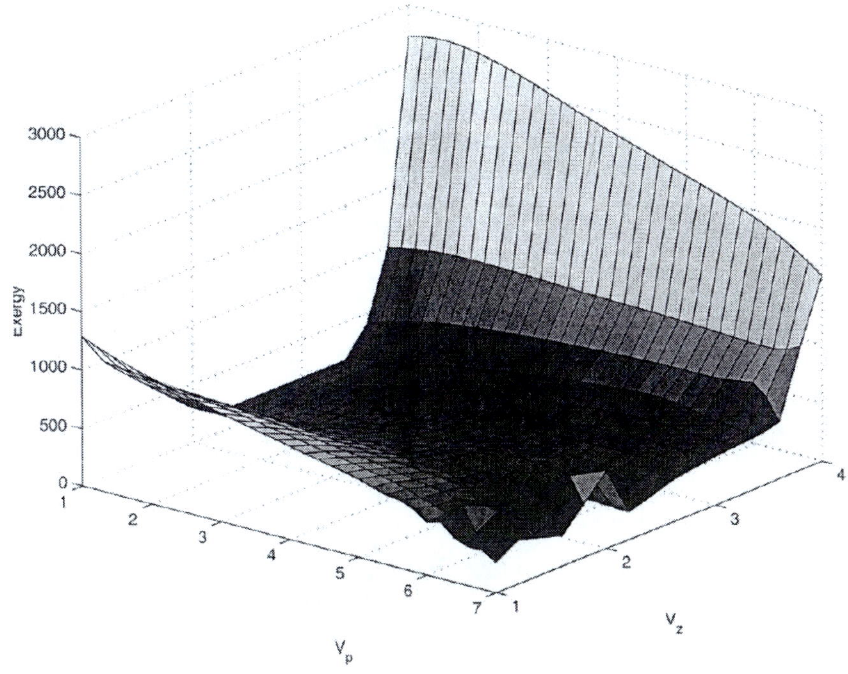

Figure 2. Optimization of exergy in between different size classes of phytoplankton and zooplankton in a time period of 365 days during oligotrophic condition.

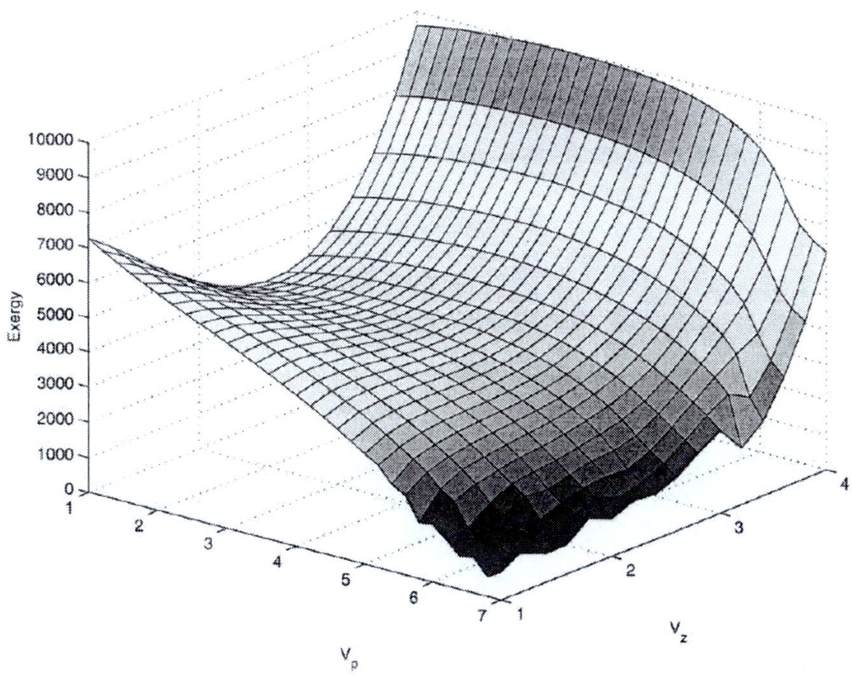

Figure 3. Optimization of exergy in between different size classes of phytoplankton and zooplankton in a time period of 365 days during mesotrophic condition

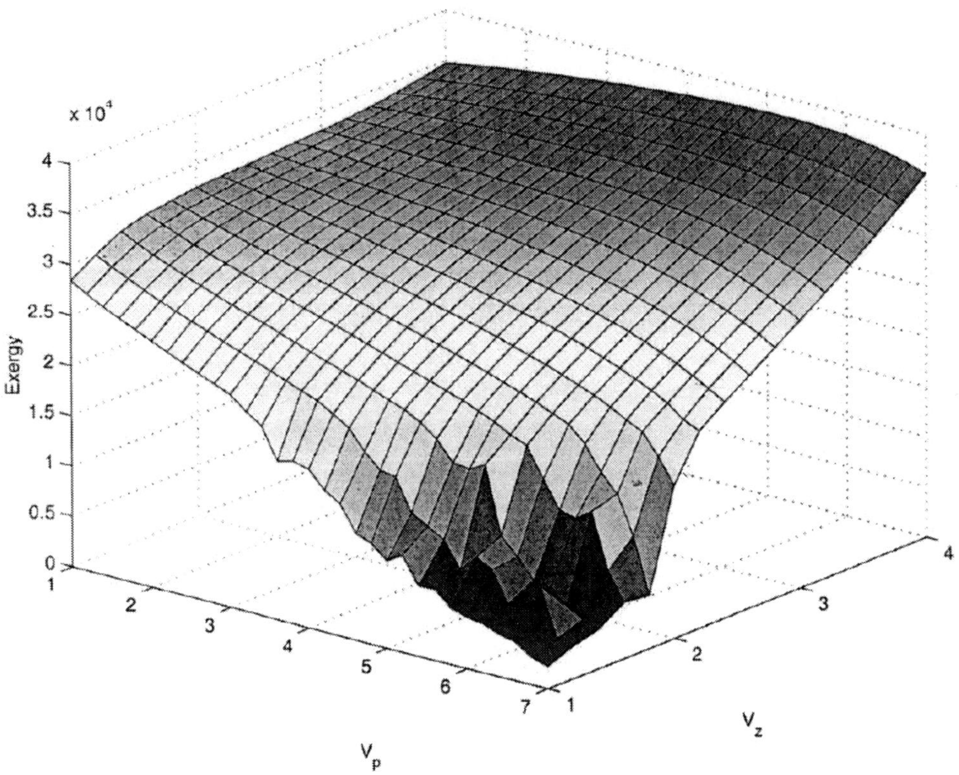

Figure 4. Optimization of exergy in between different size classes of phytoplankton and zooplankton in a time period of 365 days during eutrophic condition.

Table 3. Physiological rate parameters of phytoplankton at different nutrient conditions where maximum exergy values are obtained through optimization

	Mesotrophic condition - 20 mg phos m^{-3} (V_p = log 3.12)	Oligotrophic condition - 5 mg phos m^{-3} (V_p = 1.48)	Eutrophic condition - log100 mg phos m^{-3} (V_p = log 5.25)
Phytoplankton growth rate	2.064 (at 20 ^{0}C) d^{-1} 1.01 (at 12 ^{0}C) d^{-1} 0.41 (at 2 ^{0}C) d^{-1} 2.47 (at 22 ^{0}C) d^{-1} 1.182 d^{-1} (average)	2.556 (at 20 ^{0}C) d^{-1} 1.25 (at 12 ^{0}C) d^{-1} 0.51 (at 2 ^{0}C) d^{-1} 3.06 (at 22 ^{0}C) d^{-1} 1.464 d^{-1} (average)	1.425 (at 20 ^{0}C) d^{-1} 0.7 (at 12 ^{0}C) d^{-1} 0.28 (at 2 ^{0}C) d^{-1} 1.71 (at 22 ^{0}C) d^{-1} 0.818 d^{-1} (average)
Phytoplankton respiration rate	0.0138 d^{-1}	0.017 d^{-1}	0.009 d^{-1}
Phytoplankton mortality rate	0.027 d^{-1}	0.03 d^{-1}	0.011 d^{-1}
Phytoplankton self shading	0.011	0.039	0.002
Phytoplankton half saturation constant	26.2	9.8	47.5

Table 4. Physiological rate parameters of zooplankton where maximum exergy values are obtained through optimization

	In all nutrient conditions Zooplankton size (V_p 4)
Zooplankton growth rate	0.195 d^{-1}
Zooplankton respiration rate	0.031 d^{-1}
Zooplankton mortality rate	0.003 d^{-1}
Zooplankton half saturation constant	55

From the overall results it is clear that in all nutrient conditions the system gives maximum exergy value when large sized zooplankton is selected but with changing nutrient conditions the associated phytoplankton size changes. From the above results it is clear that size as a function of phosphorus in accordance with the observation and it is supported by early works (Ahrens and Peters, 1991b; Dickie et al.,1987; Jørgensen, 1986, 1992c; Radtke and Straškraba, 1980).

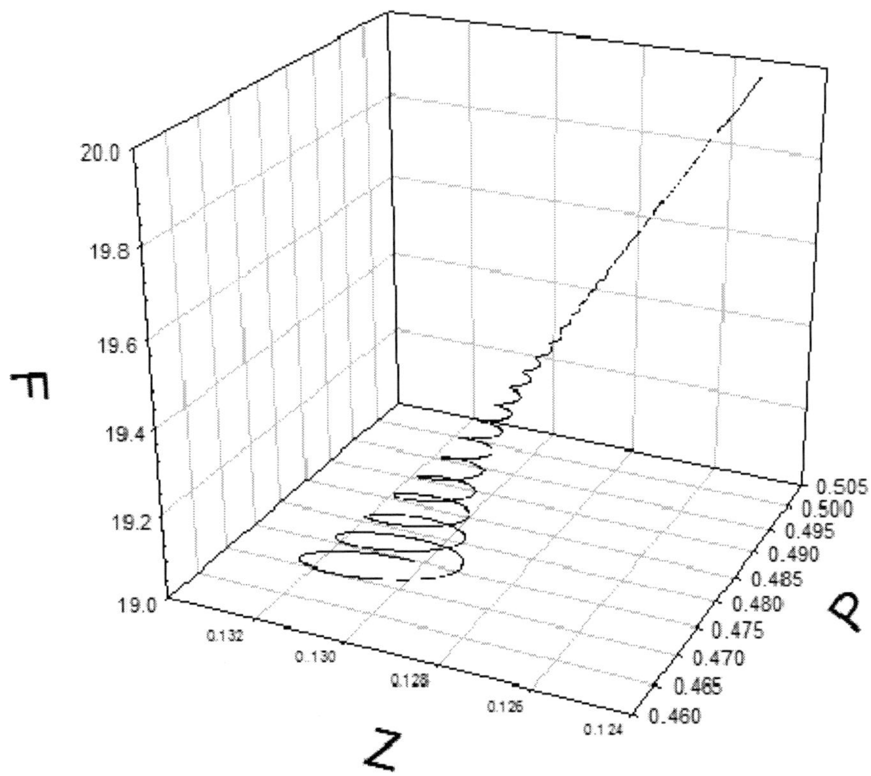

Figure 5. Phase portrait of system, with $a_1 = 5.0$, $a_2 = 0.1$, $b_2 = 2.0$, $d_1 = 0.4$, $d_2 = 0.01$, $\Theta = 0$ and $b_1 = 0.5$, showing convergent oscillation and ultimately settles to equilibrium point.

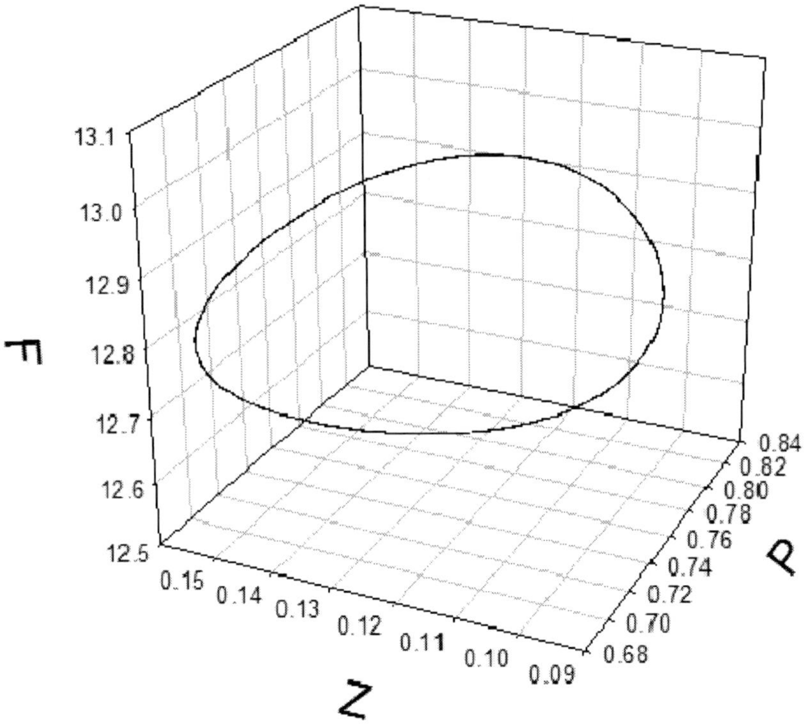

Figure 6. Phase portrait of system with b_1= 2.2 (all other parameter values are same as fig. 1.), showing limit cycle.

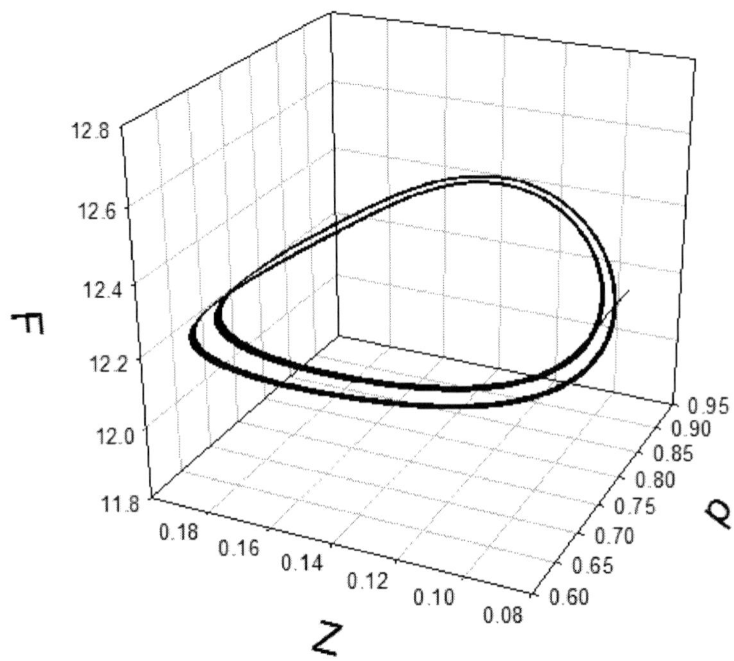

Figure 7. Phase portrait of system with b_1= 2.3 (all other parameter values are same as fig. 1.), showing doubling period.

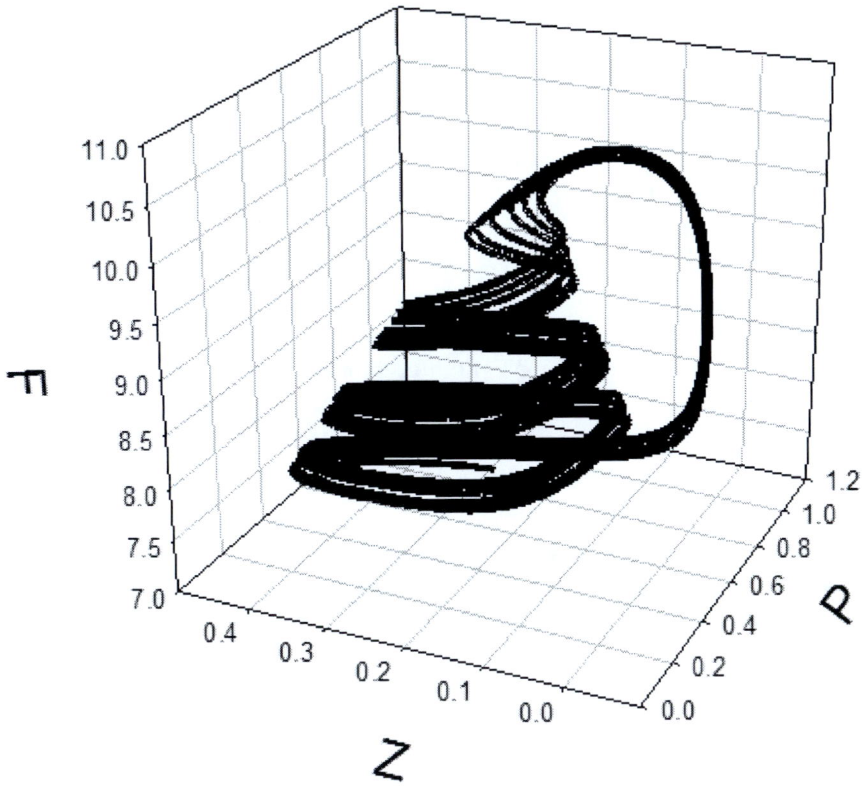

Figure 8. Phase portrait of system with b_1= 3.0 (all other parameter values are same as fig. 1.), showing chaos.

Equation of second model is numerically solved by using modified fourth order Runge-Kutta method considering the parameter values of Hastings and Powell (1991) except growth rate and half saturation constant of zooplankton. In the model run gradually the body size of zooplankton has been decrease by incorporating the gradual increment and decrement of specific growth rate and half saturation constant of zooplankton following the equations of (6) and (9) respectively. By considering the different body sizes of zooplankton in the model equilibrium point, stable limit cycle, doubling and finally chaos (Figs. 5, 6, 7, and 8 respectively) are observed in different run. When model shows equilibrium or stable point the phase portrait of the system is considered with $a_1 = 0.29, a_2 = 0.1, b_2 = 2.0, d_1 = 0.4, d_2 = 0.01$ and $b_1 = 3.41$ (corresponding zooplankton body volume $V_z^{3.26}$), phase portrait of the system during limit cycle with $a_1 = 0.34, b_1 = 2.94$ (corresponding zooplankton body volume $V_z^{2.87}$), and $a_1 = 0.38, b_1 = 2.56$ (corresponding zooplankton body volume $V_z^{2.55}$) in case of doubling period and finally at the time of chaos the value of $a_1 = 0.42, b_1 = 2.20$ (corresponding zooplankton body volume $V_z^{2.24}$). The aim is to observe the movement of system dynamics from equilibrium to chaos by changing two rate parameters corresponding with body size.

DISCUSSION

Size dependence of metabolic processes is a well-known phenomenon in the animal and plant kingdom that is usually been expressed in allometric terms. In this work these relationships are expressed by means of linear functions of log cell or body volume of phytoplankton and zooplankton. Many authors previously observed the size dependent effects on metabolic rates of phytoplankton and zooplankton in the different aquatic system (Eppley and Sloan, 1966; Taguchi, 1976; Harris, 1986). The size dependence of different parameters for phytoplankton and zooplankton has been considered in a number of recent papers, suggesting that a considerable improvement might be achieved by incorporating this recent knowledge into the model. New data concerning growth and photosynthesis rates suggest no change to the basic idea (Taguchi, 1976; Smith et al., 1999; Mostajir et al., 1999). Observations in nature using 14C autoradiography agree with ideas derived from the analysis of cultures (Nikulina and Gutelmacher, 1974; Desortová, 1976). One problem concerns the more exact definition of the interactions between gross growth, respiration and the effects of nutrient limitation of growth. The organization of most experiments is not adequate to cover these parameter interactions, and the independent treatment of variables related to size can lead to serious over under estimation. For instance, the marked size dependence of growth rates might be a consequence of dependence on the other two factors (respiration and nutrient limitation), and their additive inclusion would then be superfluous.

Steele and Frost (1977) treated self-shading using an approach similar to present work. However, a detailed theoretical analysis indicates a much less pronounced effect on size on Ext (V_p) than that modelled here. This is due to changes in the optical properties of phytoplankton of different volumes, which are ignored (this consideration was not taken into account either by Steele and Frost). In addition, the optical properties of phytoplankton groups differ depending on the proportion of photosynthetic pigments, and different cell shapes show different attenuation laws. Unfortunately, the absolute values obtained by Kirk (1975, 1976) seem to be underestimated by about one half in comparison with direct observations for particulate wavelength.

Pasciak and Gavis (1974), Falco et al. (1975), Liu (1976) and Gavis (1976) observed that nutrient uptake is strongly transport limited and hence dependent on cell volume or body volume. Direct proportionality to cell or body diameter for ks values (phosphorus) was obtained by Nyholm (1975). However, Radtke and Straškraba (1980) showed that the half saturation constant is not an ideal parameter for comparing different species and conditions, as it is determined by all the variables affecting photosynthesis or growth of organisms of higher trophic levels. The quantity termed "affinity" (Button, 1978), related to the initial slope of the photosynthesis – nutrient curve, should provide a more representative basis for comparison, but few evaluations have been made up to date. Implicit in the transport limitation theory is a dependence on the gradient access external and internal nutrient concentration. The two stage internal pool models (e.g. Nyholm, 1975, 1978; Jørgensen, 1976) therefore provide a more adequate description of nutrient uptake and growth limitation, particularly for non-steady-state solutions.

Regarding respiration the equation results in almost the same values as the formulation used by Steele and Frost (1977) when recalculated to 10^0C. On the other hand, Laws's (1975) equation point to an extremely steep drop in respiration rate with size; also the absolute values are not comparable between this study and that of Steele and Frost (1977). Humphrey (1975) experimentally studied the photosynthesis: respiration ratio in 11 species of marine phytoplankton, but an evaluation of the size dependence from his data is not simple because pH and the population sizes were also varied. Smith (1977) studied the effect of motility on respiration losses.

Analysis of the light dependence of phytoplankton indicates that light sensitivity parameters indicate that these parameters should also be a function of V_p, although this could be a consequence of other effects (Radtke and Straškraba, 1980).

It is extremely necessary to point out that all the physically driven, size dependent phytoplankton and zooplankton properties, however much information they allow to drive, should be supplemented by differences of biological origin, of which present understanding is very limited. It is very probable that the inclusion of such differences in the present work would increase our degree of understanding of the modelled process, or would at least reduce the scatter between theory and observations which is found in most direct comparisons.

Considering the above critical evaluations it can be distinguished between those that were considered more or less adequately in the framework of contemporary ideas (P_{max} ($V_{p,}$ T), Growth (V_z), Resp (V_p), Resp (V_z), Mort (V_p), Mort (V_z)) and those for which considerable details might have been added (Ext (V_p), K_s (V_p), K_s (V_z)). Also, there are areas in which a few additional parameters such as filtration rate of zooplankton, sinking rate of phytoplankton may be included.

It is mentioned earlier that different goal functions, which were used by different authors in different ecological perspectives, Here to investigate the best system organization, the exergy as a goal function is used in the first model and thereby exergy is optimized. It is defined as the work a system can perform when it is brought into equilibrium with the environment or another well defined reference state (Jørgensen et al., 2000). The question may arise which parameters such as growth rate, mortality rate, respiration rate, half-saturation constant and so on are the most favourable for the species of the system. The concept of “most favourable” can be measured by the survival of the species, that is, the biomass of the species. In complex ecosystems all trophic levels struggle for survival and they can not all get the highest possible biomass, as they are dependent on each other. The sum of all organisms in a system can not be used as measure of the survival of the ecosystem, because different species have different functions and so have different importance for the ecosystem (Jørgensen and DeBernardi, 1997). To overcome this problem exergy is used as a goal function in the first model. The exergy calculation does not account for the information embodied in the network, but as the number of possible links is much lower than the number of possible amino acid sequences, the information of the network is much lower than the information embodied in the organisms. Calculation of the absolute values of exergy is not needed but only the relative changes of exergy due to changes of the properties are required. According to Jørgensen and DeBernardi (1997) the idea behind the use of exergy as goal function by development of models with dynamic structure is expressed as follows: the combination of properties of all the organisms in the ecosystem that gives the highest exergy also expresses the combination of properties that is best fitted to the prevailing condition to offer the highest extent of the survival of the ecosystem or to give the best resistance of the ecosystem against decomposition to an inorganic lifeless soup. It can therefore be described the shifts in properties as a result of changed conditions by a current change of the parameters in accordance with the combination that gives the highest exergy.

In the first model, it is observed that during different nutrient conditions the highest value of exergy coincides with the highest fish biomass with adequate proportion of zooplankton and phytoplankton. From this observation it is clear that in all types of nutrient conditions the large sized zooplankton always yields the best fish production but in association with different sizes of phytoplankton with changing the nutrient conditions. In oligotrophic condition the large sized zooplankton shows highest fish biomass with smaller phytoplankton, in mesotrophic condition with medium sized phytoplankton and in eutrophic condition with larger phytoplankton. From the optimization it is clear that any zooplankton ranging from small to large sizes can not adapt the largest sized phytoplankton and in this situation the fish is eliminated from the system due to lack of adequate amount of zooplankton biomass. From all these observations it may be concluded that the adaptation of larger zooplankton varies with different sized phytoplankton in terms of maximum fish biomass according to different nutrient conditions. These changes of association of zooplankton with small to large sized phytoplankton are directly related with the increase of nutrients in the system.

Finally it is necessary to mention that some basic problems of parameter combinations remain unsolved. Combination of proper parameter is more important than the exact representation of different size dependencies. Evidently it is unable to reproduce a priori in the model all the competitive mechanisms, although the optimization approach can help to provide insight into some of the principles. The recent theoretical concern over "structural" problems in aquatic ecosystem has centered on size structure (Sheldon et al., 1972; Kerr, 1974; Thomann, 1978; Yurista, 1999).

Steele and Frost (1977) showed in their simulations result that no profound changes in the proportions of different size groups, at least the changes were far less than would be expected in comparison with natural observations. It can be assumed that some "species" of both phytoplankton and zooplankton should disappear almost completely, which is not the case in the simulations. To explain the above things there are at least two possible reasons: (1) the simulation time may have been too short to manifest such profound differences and no asymptotic stability was reached from the arbitrary initial condition used, (2) the competitive mechanisms leading to population reduction at lower frequencies may be different from those considered.

The present investigation has led to the following conclusions

[1] It is necessary to modify model runs by matching the state and driving variables to situations in different water bodies, for direct comparison with data on size distribution frequencies.

[2] The functional relations of individual process rate parameters to phytoplankton and zooplankton body volumes and their combinations will have to be up dated in the line with recent research, and a more up to date version will have to be used. In this regard separate compartments of detritivorous zooplankton and pisciphagous fish can be used.

[3] Only one goal function i.e. exergy is treated and optimized it in the system. Although Jorgensen (1994) found that two other important goal functions such as ascendency and emergy are well correlated with exergy storage, however paralley ascendency and emergy can also be used to confirm the model result and system dynamics.

The second model is dealt with the movement of system from ordered to chaotic state by gradual decrease of zooplankton body size. A biological system evolves towards the edge of chaos, according to a theory proposed by Kauffman (1991, 1992 and 1993). The second model has examined the movement of ordered to chaotic state by changing the value of two selected key parameters (growth rate and half saturation constant of zooplankton) by gradual reduction of zooplankton body size. The model behavior was followed, as the zooplankton growth rate and half saturation constant are changed. Gradual higher specific growth rate and lower half saturation constant of zooplankton mean that, the zooplankton body size gradually decreases from $V_z^{3.26}$ (during equilibrium condition) to $V_z^{2.24}$ (during chaos) following general allometric relationships (Peters, 1983). Ray et al. (2001a, b), Jorgensen et al. (2002) showed that size combinations between phytoplankton and zooplankton are very crucial for self organization of the system. The system adapts by changing size, but it does not adapt the gradual decrease of zooplankton size and moves towards chaos. In reality it may happen if non selective predation by fish or invertebrates is present, zooplankton yields bigger size (Peter, 1983).

During equilibrium state comparatively minimum specific growth rate and maximum half saturation constant of zooplankton are considered and when gradual smaller sized zooplankton are selected the specific growth rate are increased and half saturation constant decreased the system starts oscillation. Therefore, the maximum information occurs when system is occupied by comparatively larger sized zooplankton.

The size combination of zooplankton and phytoplankton is crucial in any aquatic ecosystem and the combinations are adjusted according to the changing conditions of the ecosystem. When the system selects smaller size of zooplankton it is obviously beneficial for this group to grow fast and when this fast growth continues the phytoplankton will be rapidly exhausted and lead the system into violent oscillatory state and ultimately chaos. During oscillatory state zooplankton with higher growth rate grow rapidly. But as their food phytoplankton becomes disappear, the zooplankton decline rapidly due to lack of food. In the next stage the phytoplankton are thereby more available, and the zooplankton again grow faster and so on. Therefore, the system organizes itself by choosing the appropriate growth rate of zooplankton, which itself adjusts to the availability of phytoplankton resources. This appropriate higher growth rate corresponds to the edge of chaos. This growth rate is possible highest which makes the system most adaptable to the amount of available phytoplankton resources for zooplankton. This situation also coincides with the best possible grazing of zooplankton on available phytoplankton – i.e. to get on average as high a concentration of zooplankton and its food (phytoplankton) as possible.

When non selective predation of fish or invertebrates takes place in the ecosystem, gradually larger size zooplankton become dominant in the system as they are able to escape predation pressure in comparison to the smaller size. Jorgensen (1995); Jorgensen et al. (1998) and Ray et al. (2001a, b) suggested many properties which are associated with self-organizing critical systems. These all properties follow the power law, which are the relationships between body size and abundance for species in ecosystem, the frequency with which observed changes exceeds a given change versus the change, the typical frequencies of 'avalanches' and the occurrence of 'avalanches' according to a well examined ecosystem model be used to explain the underlying causality.

Gradually the zooplankton size are considered smaller in the model (by increasing specific growth rate and decreasing half saturation constant following allometric principle) and system moves from ordered to chaotic state. Jorgensen (1995) proposed it as a tentative hypothesis, linking Darwin's theory with Kauffman's hypothesis (Kauffman 1991, 1992 and 1993) that biological systems evolves towards the edge of chaos. The properties of an ecosystem to adapt to changed conditions are rooted in the interplay between self organization and selection. Kauffman (1993) proposed that at the edge of chaos, where the organisms are able to coordinate the most complex behavior, implies that the ecosystem shows hysteresis phenomena as a consequence of drastic changes imposed on the system (Jorgensen 1999).

REFERENCES

Ahrens, A. M. and Peters, R. H., 1991. Plankton community respiration: relationships with size distribution and lake trophy. *Hydrobiolgia*, 224: 77 – 87.

Ahrens, A. M. and Peters, R. H., 1991b. Patterns and limitations in limnoplankton size spectra. *Can. J. Fish. Aquat. Sci.* 48: 1967 – 1978.

Anderson, V., Nival, P and Harris, R., 1987. Modelling of a planktonic ecosystem in an enclosed water column. *J. Mar. Biol. Assoc.* U. K. , 67: 407 – 430.

Blueweiss, L., Fox, H., Kudzma, V., Nakashima, D., Peters, R. H. and Sams, S., 1978. Relationships between some body size and some life history parameters. *Oecologia* (Berl.), 37: 257 – 272.

Calder W A III, 1984. *Size, Fuction and Life History*. Harvard University Press, Cambridge.

Chattopadhyay, J., Sarkar, R. R., 2003. Chaos to order: preliminary experiments with a population dynamics models of three trophic levels. *Ecol. Model*

Desortová, B., 1976. Productivity of individual algal species in natural phytoplankton assemblage determined by means of autoradiography. *Arch. Hydrobiol.* Suppl. 49, Algol. Stud., 17: 415 – 449.

Dickie, L. M., Kerr, S. R. and Boudreau, P. R., 1987. Size dependent processes underlying regularities in ecosystem structure. *Ecol. Monogr.*, 57: 233 – 250.

Eppley, R. W., Rogers, J. N. and McCarthy, J. J., 1969. Half-saturation constants for uptake of nitrate and ammonium by marine phytoplankton. *Limnol. Oceanigr.*, 14: 912 – 920.

Eppley, R. W. and Sloan, P. R., 1966. Growth rate of marine phytoplankton: correlation with light – absorption by cell chlorophyll a. *Physiol. Plant,* 19: 17 – 59.

Falco, J. W., Kerr, P. C., Barron, M. B. and Brockway, G. L., 1975. The effect of mass transport on bio-stimulation of algal growth. *Ecol. Model.,* 1: 117 – 131.

Findlay, I. W. O., 1992. Effects of external factors and cell size on the cell division rate of a marine diatom *Coscino discus pavillardi* Forti. *Int. Rev. Ges. Hydrobiol.*, 57: 523 – 533.

Gaedke, U., 1992. The size distribution of plankton biomass in a large lake and its seasonal variability.*Limnol. Oceanogr.*, 37: 1202 – 1220.

Gaedke, U., 1993. Ecosystem analysis based on biomass size distribution: A case study of a plankton community in a large lake. *Limnol. Oceanogr.*, 38: 112 – 127.

Gavis, J., 1976. Munk and Riley revised: nutrient diffusion transport and rates of phytoplankton growth. *J. Mar. Res.*, 34: 161 – 179.

Gillooly, J. F., 2000. Effect of body size and temperature on generation time in zooplankton. *J. Plankton Res.*, 22: 241 – 251.

Gliwicz, Z. and Umana, G., 1994. Cladoceran body size and vulnerability to copepod predation. *Limnol.Oceanogr.*, 39: 419 – 424.

Gould S J, 1966. Allometry and size in ontogeny and phyllogeny. *Bio. Res*. 41, 587 – 640.

Harris, G. P., 1986. *Phytoplankton Ecology. Structure, Function, and Fluctuation. Chapman and Hall.*

Hastings, A., Powell, T., 1991. Chaos in three-species food chain. Ecology 72, 896 – 903.

Hemmingsen A M, 1950. The relation of standard (basal) energy metabolism to total fresh weight of living organisms. *Rep. Steno. Mem. Hos*. 4, 1 – 58.

Hemmingsen A M, 1960. Energy metabolism as related body size and respiration surfaces, and its evolution. *Rep. Steno. Mem. Hos*. 9, 7 – 110.

Hrbáček, J., 1962. Species composition and amount of the zooplankton in relation to fish stock. *Rozpr. Cesk. Akad. Ved., Rada. Mat. Prir. Ved.*, 10: 1 – 116.

Humphrey, G. F., 1975. The photosynthesis: respiration rate of some unicellular marine alga. *J. Mar. Biol. Ecol.*, 18: 111 – 119.

Jørgensen, S. E., 1976. A eutrophication model for a lake. *Ecol. Model.*, 2: 155 – 212.

Jørgensen, S. E., 1986. Structurally dynamic modelling. *Ecol. Model*. 31: 1 – 9.

Jørgensen, S. E., 1992c. Development of models able to account for changes in species composition. *Ecol. Model*. 62: 195 – 208.

Jørgensen, S. E., 1994. Review and comparison of goal functions in system ecology. *Vie Milieu* 44: 11 – 20.

Jørgensen, S. E., 1995. The growth rate of zooplankton at the edge of chaos: Ecological Models. *J. Theor. Biol*. 175, 13 – 21.

Jørgensen, S. E. and DeBernardi, R., 1997. The application of a model with dynamic structure to simulate the effect of mass fish mortality on zooplankton structure in Lago di Annone. *Hydrobiologia*, 356: 87 – 96.

Jørgensen, S. E., Mejer, H., Nielsen, S. N., 1998. Ecosystem as self-organizing critical systems. *Ecol. Model*. 111, 261 – 268.

Jørgensen, S. E., 1999. Biomanipulation and ecological modeling. In: Tundisi, J. G., Straškraba, M. (Eds.), *Theoretical Reservoir Ecology and its Applications*. Backhuys Publishers, Leiden, The Netherlands, pp. 547 – 564.

Jørgensen, S. E., Patten, C. and Straškraba, M., 2000. Ecosystem emerging: 4. growth. *Ecol. Model.*, 126: 249 – 284.

Jørgensen, S. E., Ray, S., Berec, L., Straškraba, M., 2002. Improved calibration of an eutrophication model by use of the size variation due to succession. *Ecol. Model.* 153, 269 -277.

Kauffman, S. A., 1991. Anti-chaos and adaptation. *Sci. Am.* August, 78 – 84.

Kauffman, S. A., 1992. The sciences of complexity and "origins of order". In: Mitten, J. E., Baskin, A. G., (Eds.), *Principles of Organization in Organisms*, Proceedings of the Workshop on Principles of Oranization in Organisms. Santa Fe, pp. 303.

Kauffman, S. A., 1993. *Origins of Order: Self Organization and Selection in Evolution.* Oxford University Press, Oxford.

Kerr, S. R., 1974. Structural analysis of aquatic communities. *Proc. Ist. Int. Cong. Ecol.*, 69 – 84.

Kiefer, D. A. and Austin, R. W., 1974. The effect of varying phytoplankton concentration upon submarine light transmission in the Gulf of California. *Limnol. Oceanogr.*, 19: 55 – 64.

Kirk, J. T. O., 1975. A theoretical analysis of the contribution of algal cells to the attenuation of light within natural waters. I and II. *New Phytol.*, 75: 11 – 36.

Kirk, J. T. O., 1976. A theoretical analysis of the contribution of algal cells to the attenuation of light within natural waters. III. *New Phytol.,* 77: 341 – 358.

LaBarbera M, 1986. The evolution and ecology of body size. In *Patterns and Process in the History of Life*, ed. by Ranp M and Jablonsky D (Springer – Verlag, Berlin, Heidelberg,) pp. 69 – 98.

Laws, E. A., 1975. The importance of respiration in controlling the size distribution of marine phytoplankton . *Ecology*, 56: 419 – 426.

Lebedeva, L. P., 1971. A model of the latitudinal distribution of the numbers of species of phytoplankton in the sea. *J. Cons. Int. Explor. Mer.*, 34: 341 – 350.

Liu, S. H., 1976. Oxygen diffusion in a spherical cell with nonlinear oxygen – uptake kinetics. *J. Theor. Biol.*, 60: 449 – 457.

McMahon T A, 1973. Size and shape in biology. *Science* 179 1201 – 1204.

Mandal, S., Ray, S., Roy, S., Jørgensen, S. E., 2006, Order to chaos and vice versa in an aquatic ecosystem. *Ecol. Model.*, 197, 498-504.

Mandal, S., Ray, S., Roy, S., Jørgensen, S. E., 2007, Investigation of thermodynamic properties in an ecological model developing from order to chaotic states. *Ecol. Model.,* 204, 40-46.

May, R. M., 1987. Chaos and the dynamics of biological populations. *Proc. R. Soc. Lond.* 413, 27 – 44.

Moloney, C. L., Bergh, M. O., Field, J. G. and Newell, R. C., 1986. The effect of sedimentation and microbial nitrogen regeneration in a plankton community: a simulation investigation. *J. Plankton Res.*, 8: 427 – 445.

Moloney, C. L. and Field, J. G., 1991. Modelling Carbon and nitrogen flows in a microbial plankton community. In: P. C. Reid, P. H. Burkhill and C. M. Turley (Eds*.), Protozoa and their Role in Marine Processes*. NATO ASI Series. Springer – Verlag, New York, pp. 443 – 474.

Mostajir, B., Simengando, T., Demers, S., Belzile, C., Roy, S., Gosselin, M., Chanut, J. P., Demora, S., Fauchot, J., Vidussi, F. and Leveassseur, M., 1999. Ecological implication of changes in cell size and photosynthetic capacity of marine Prymnesiophyceae induced by ultraviolet – β – radiation .*Marine Ecol. Prog. Ser.*, 187: 89 – 100.

Nikulina, B. N. and Gutelmacher, B. L., 1974. *Photosynthetic activity of individual species of algae in Lake Krivoe.* Ekologia Moskva, 4: 101 – 104 (in Russian).

Nyholm, N., 1975. Kinetic studies of phytoplankton limited algal growth. Ph.D. Thesis, Dept. *Appl. Biochem*., Technical University, Copenhagen.

Nyholm, N, 1978. A simulation model for phytoplankton growth and nutrient cycling in eutrophic, shallow lakes. *Col. Modelling*, 4: 279 – 310.

Pasciak, W. J. and Gavis, J., 1974. Transport limitation of nutrient uptake in phytoplankton. *Limnol. Oceanogr.*, 19: 881 – 888.

Peters, R. H., 1983. *The Ecological Implications of Body Size*. Cambridge University Press, Cambridge.

Platt, T., 1985, Structure of the marine ecosystem: its allometric basis. In: R. E. Ulanowicz and T. Platt (Eds.), Ecosystem Theory for Biological Oceanography. *Can. Bull. Fish. Aquat. Sci.*, 213: 55 – 64.

Platt, T., Mann, K. H. and Ulanowicz, R. E., 1981. *Mathematical Models in Biological Oceanography.* The Unesco Press, Paris.

Radtke, E. and Straškraba, M., 1980. Self- optimization in phytoplankton model. *Ecol. Model.*, 9: 247 - 268.

Ray, S., Berec, L., Straškraba, M., Jørgensen, S. E., 2001a. Optimization of exergy and the implications of body sizes of phytoplankton and zooplankton in an aquatic ecosystem model. *Ecol. Model.* 140, 219 – 234.

Ray, S., Berec, L., Straškraba, M., Ulanowicz, R. E., 2001b. Evaluation of system performance through optimizing ascendency in an aquatic ecosystem model. *J. Biol. System* 9, 1 – 22

Schmidt-Nielsen K, 1975. Scaling in biology: the consequences of size. *J. Exp. Zool.* 194 287 – 308.

Schmidt-Nielsen K, 1984. *Scaling: Why is Animal Size So Important?* (Cambridge University Press, Cambridge).

Sheldon, R. W., Prakash, A. and Sutchiffe, W. H., Jr., 1972. The size distribution of particles in the ocean. *Limnol. Oceanogr.*, 17: 327 – 340.

Smith, W. O., Jr., 1977. The respiration of photosynthetic carbon in eutrophic areas of the ocean. *J. Mar. Res.*, 35: 557 – 565.

Smith, W. O., Jr., Nelson, D. M. and Mathot, S., 1999. Phytoplankton growth rate in Ross Sea determined by independent methods: temproral variations. *J. Plankton Res.* 21: 1519 – 1536.

Steel, J. A., 1973. Reservoir algal productivity. In: A. James (Ed.), *The Use of Mathematical Models in Water Pollution Control.* University of Newcastle – upon – Tyne Symp., pp. 107 – 135.

Steele, J. H. and Frost, B. W., 1977. The structure of plankton communities. *Phill. Trans. R. Soc. London, Ser. B*, 280: 485 – 534.

Straškraba, M., 1976. *Development of an analytical phytoplankton model with parameters empirically related to dominant controlling variables.* Abh. Akad. Wiss. DDR. Jahrgang, 1974: 33 – 65.

Straškraba, M., 1979. Natural control mechanisms in models of aquatic ecosystem. *Ecol. Model.*, 6: 305 – 321.

Straškraba, M and Gnauck, A. H., 1985. *Freshwater Ecosystem, Modelling and Simulation.* Elsevier, Amsterdam. 309 pp.

Taguchi, S., 1976. Relationship between photosynthesis and cell size of marine diatoms. *J. Phycol.* , 12: 185 – 189.

Tang, E. P. Y., 1995. The allometry of algal growth rates. *J. Plankton Res.*, 17: 1325 – 1335.

Thomann, R. V., 1978. Size – dependent model of hazardous substances in an aquatic food chain. Environ. *Protect. Agency. Ecol. Res. Ser.*, EPA 600 / 3 – 78 – 036: 1 – 39.

Tilzer, M., 1972. Dynamik and produktivitat von phytoplankton and pelagischen bakterien in einem Hochgebirgesee (Vorderer Finsttertaler Sec. Osterreich*). Arch. Hydrobiol. Suppl.,* 40: 201 – 273.

Williams, R. B., 1965, Division rates of salt marsh diatoms in relation to salinity and cell size. *Ecology*, 45:877 – 880.

Wetzel, R. G., 1975. *Limnology*. Saunders, Philadelphia.

Yurista, P. M., 1999. A model for temperature correction of size specific respiration in *Bythotrephes ecderstroemi* and *Daphnia middendoffiana*, *J. Plankton Res.,* 21: 721 – 734.

INDEX

A

B

C

D

E

I

J

K

L

M

N

O

P

Q

R

S

T

U

V

W

X

Y